ST N
Facu
Ruse

D0777054

Instructional Tech.

Developing Auto-Instructional Materials

Instructional Development 2

Developing Auto-Instructional Materials

From Programmed Texts to CAL and Interactive Video

A J Romiszowski

Kogan Page, London/Nichols Publishing, New York

Copyright © A J Romiszowski 1986
All rights reserved

First published in Great Britain in 1986 by
Kogan Page Ltd, 120 Pentonville Road, London N1 9JN

British Library Cataloguing in Publication Data
Romiszowski, A.J.
   Developing auto-instructional materials: from
   programmed texts to CAL and interactive video. —
   (Instructional development; 2)
   1. Individualized instruction   2. Teaching—
   Aids and devices
   I. Title   II. Series
   371.3'94   LB1031

   ISBN 0-85038-911-9

Published in the United States of America by
Nichols Publishing Company, Post Office Box 96, New York, NY 10024

Library of Congress Cataloging in Publication Data
Romiszowski, A.J.
   Developing auto-instructional materials.

   Bibliography: p.
   Includes indexes.
   1. Programmed instruction—Authorship.   I. Title.
LB1028.6.R65   1985   371.3'944   85-3106
ISBN 0-89397-208-8

Printed and bound in Great Britain by
Anchor Brendon Ltd, Tiptree, Essex

Contents

Preface to the Instructional Development Series

This two-volume work on the development of instruction is planned as a companion to an earlier book – *Designing Instructional Systems*.

This earlier book dealt with the decision-making process involved in overall course planning and curriculum design – the initial macro-design stages of a project. The present work continues on to the micro-design stages of lesson and instructional materials development. The work is divided into two volumes. Volume 1, *Producing Instructional Systems*, deals with lesson planning for individualized instruction in the conventional classroom environment, as well as the planning of small group-learning situations, simulations and games. Volume 2, *Developing Auto-Instructional Materials*, deals with the development of many different types of materials, including programmed instruction, structural communication, various styles of structured writing, audio and audiovisual instruction and the many types of computer-based materials now being introduced in both education and training.

Taken together, these two volumes give extensive coverage of practical techniques for the development of instruction.

It is quite useful to draw a distinction between instructional *design* and instructional *development*, although some authors seem to use the two terms synonymously. It is true that in some cases it is difficult to separate 'design' (what happens on the drawing board) and 'development' (what happens in the workshop). The two processes are interrelated, forming an iterative cycle of design, development, re-design, etc. However, development requires the existence of a practical try-out situation and prototype products or services ready to be tried out. This is not a requirement of the initial design stages of a project, which may be a largely theoretical exercise, based on the experiences gained in other projects or on principles gleaned from a study of the literature. We differentiate between design and development in the context of instruction and it is this that distinguishes the present work.

Designing Instructional Systems laid the foundations for a systematic approach to the planning of instruction in both educational and training contexts, leading the reader through two levels of decision making:

1. Is an instructional system really necessary?
2. What should be its overall structure?

These are the political and strategic levels.

Instructional Development takes the reader on to practical application of the plans prepared at these earlier levels. Naturally, this involves further, more detailed, design, as well as the actual production of prototype lessons and materials. These prototypes, when tested, may lead to further detailed re-design or even to changes in the initial overall plans. However, the emphasis is on production and testing and the further design decisions are very detailed, topic-specific or 'tactical'. These decisions may also be conveniently classified into two levels:

3. The decisions involved in the detailed planning of lessons and exercises.
4. The decisions involved in the preparation of instructional materials.

Very often these two levels of decision-making are performed by one and the same person, as in the case of a teacher who prepares his own materials to be used as part of the lessons that he gives. It is also very common, however, for the materials production tasks to be the responsibility of specialists, the lesson planner merely selecting appropriate existing materials or orienting the specialist materials producer.

One reason, therefore, for dividing this work into two volumes is to aim more precisely at these two groups of readers.

Another reason for the division is the sheer size of the task. To deal thoroughly with these two levels of instructional development in one book either would have produced an unwieldy tome, or size and space restrictions would have prevented the inclusion of many examples (an essential feature of a practical treatment of the techniques discussed).

However, bearing in mind the possibility of two distinct groups of readers – lesson planners and material producers – every effort has been made to make the two volumes self-contained. Essential basic concepts are defined in both volumes, so the reader may accompany the theoretical argument without necessarily having to refer to both volumes.

The same independence is achieved, in relation to the earlier *Designing Instructional Systems*, by the summary, in early chapters, of the principal conceptual schemata developed earlier and now used as the basis for practical decision making.

However, there is a continuity and coherence within the books, best illustrated by the two 'world maps' of instructional design and development presented on the following two pages. The first map summarizes the principal content of the earlier book, *Designing Instructional Systems*.

The second map summarizes the principal aspects of instructional development covered in the present work. Volume 1 is devoted to Level 3, and Volume 2 deals with Level 4. This map shows clearly how these more detailed levels of decision making are dependent on earlier decisions taken at the initial design levels. The reader who has read *Designing Instructional Systems* will discover a strong thread of continuity running throughout the series. Some of the present chapters, which summarize the basic theoretical approach adopted, may be found to be redundant in this case. Clear indications are given, in overviews, of those chapters which may be safely omitted if earlier books in the series have recently been read. Readers may prefer, however, to refresh their memories on the concepts and schemata developed earlier which now come to play an important part in the instructional development process. This may be particularly important if some time has elapsed since the earlier book was read.

It is hoped that the structure of the content in the present work will enable the two volumes to be used conveniently as both initial reading or later reference material.

Alexander Romiszowski, April 1984

Stages in the systems approach

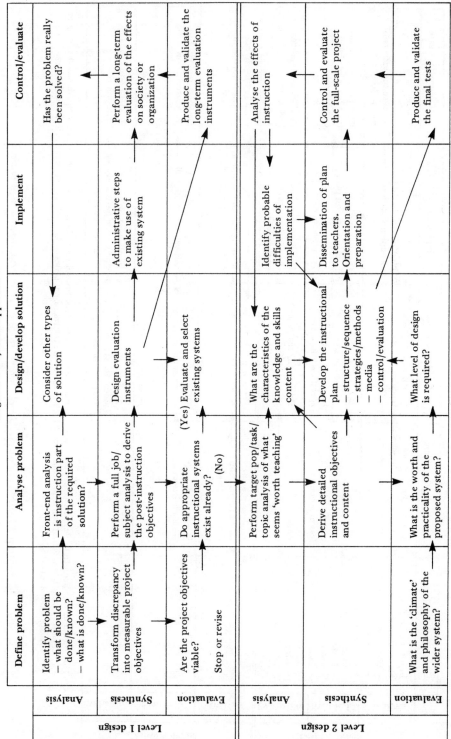

Map 1 *World map of the 'initial design' stage of an instructional design project*

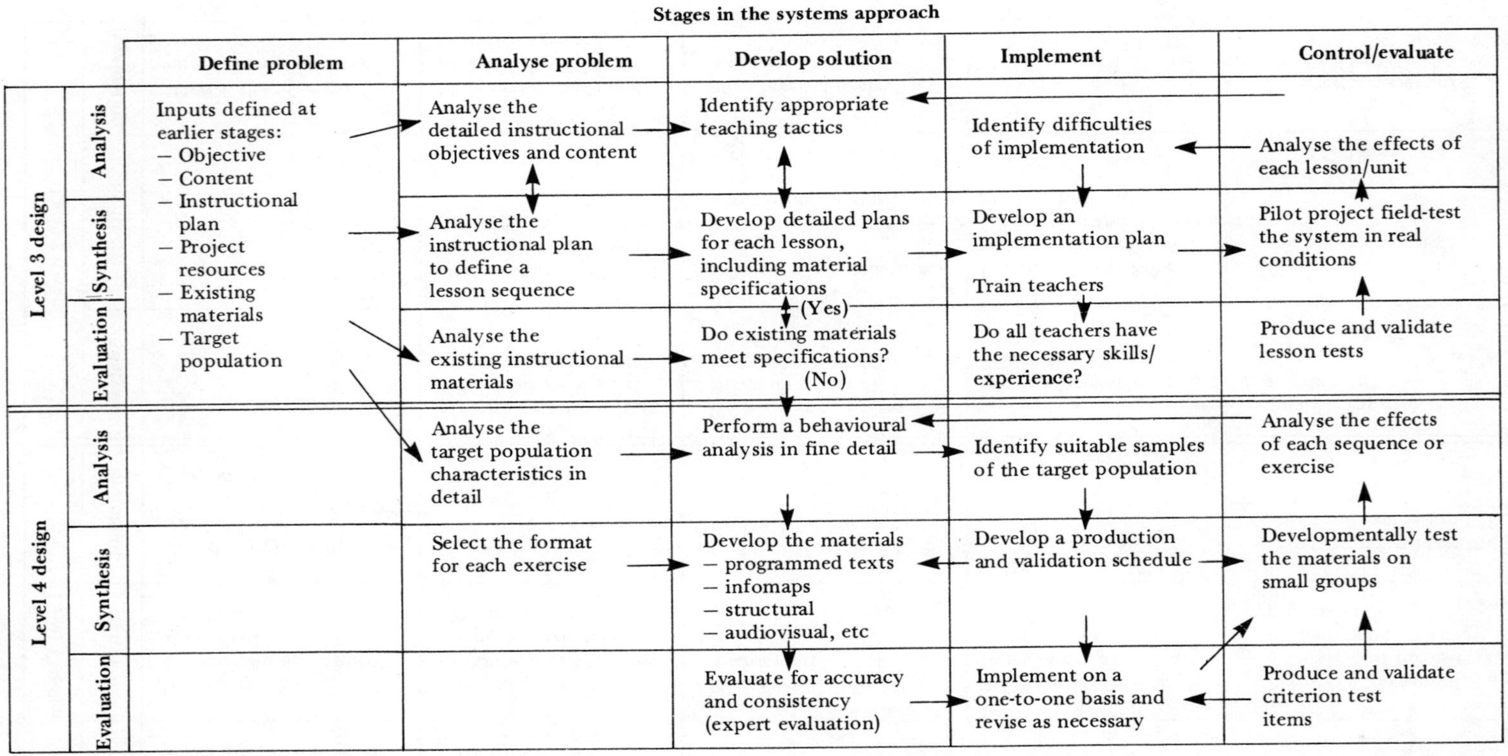

Stages in the systems approach

	Define problem	Analyse problem	Develop solution	Implement	Control/evaluate
Level 3 design — Analysis / Synthesis / Evaluation	Inputs defined at earlier stages: — Objective — Content — Instructional plan — Project resources — Existing materials — Target population	Analyse the detailed instructional objectives and content Analyse the instructional plan to define a lesson sequence Analyse the existing instructional materials	Identify appropriate teaching tactics Develop detailed plans for each lesson, including material specifications Do existing materials meet specifications? (Yes) (No)	Identify difficulties of implementation Develop an implementation plan Train teachers Do all teachers have the necessary skills/experience?	Analyse the effects of each lesson/unit Pilot project field-test the system in real conditions Produce and validate lesson tests
Level 4 design — Analysis / Synthesis / Evaluation		Analyse the target population characteristics in detail Select the format for each exercise	Perform a behavioural analysis in fine detail Develop the materials — programmed texts — infomaps — structural — audiovisual, etc Evaluate for accuracy and consistency (expert evaluation)	Identify suitable samples of the target population Develop a production and validation schedule Implement on a one-to-one basis and revise as necessary	Analyse the effects of each sequence or exercise Developmentally test the materials on small groups Produce and validate criterion test items

The 'problem' has been fully defined at Levels 1 and 2, as has the overall form of the solution. Level 3 is concerned with the detailed tactics of each lesson — what learners and teachers should do at each stage of the instructional process. Level 4 is concerned with the development of special instructional materials.

Map 2 *World map of the 'development' stage of an instructional design project*

Introduction to Developing Auto-Instructional Materials

The intended readership of this volume is composed of those teachers or instructors who intend to develop any form of self-instructional material or package, those who specialize in the development and production of teaching materials and media (whether print-based, computer-based or audiovisual) and the instructional designer or student of instructional design who wishes to integrate a coherent approach to materials design and development into a more general, broadly based approach to curriculum and lesson planning.

We limit ourselves to the discussion of auto-instructional materials, partly because the whole field of media and materials is so vast that justice cannot be done to it in a book this size, partly because we are dealing with the design and development of instructional systems and not with the skills of production of artwork for visuals, photography, audio recording or studio work. Teacher-delivered instruction was dealt with in the previous volume, *Producing Instructional Systems* and so now we address ourselves to instructional systems in which the bulk of the teaching function (message delivery and/or control of learning and/or corrective feedback and guidance) is automated.

Auto-instruction generally implies a high degree of individualization. Therefore the first part of *Producing Instructional Systems* is especially relevant as a starting point for the study of the present volume. In case this is not available, or was read some time before, Chapter 1 has been included in order to summarize our views on the individualization of the instructional process. We analyse the critical aspects of this concept and examine the role of mediated auto-instruction.

The development of mediated instructional materials should be undertaken in the broader context of curricular and lesson planning. There are exceptions to this rule, when instructional designers are called upon to develop a self-contained package on a given topic, not directly related to any specific course, curriculum or training need. This is often the case, for example, when CAL courseware is developed by a programming 'software house'. However, the dangers of ignoring the reality in which the package will be used are all too apparent – for example, a package on French grammar would not be well received in a curricular context which tended to avoid the direct and formal teaching of grammatical rules; a package on set theory as an introduction to arithmetic would not have much acceptance in schools that have abandoned, or never used, the 'modern maths' approach to curriculum in mathematics. Therefore, the materials developer must be at least aware of the overall 'macro' processes of curriculum design, as well as the more 'micro' considerations of lesson planning. These were fully treated in earlier books: curriculum design (Levels 1 and 2 of the four-level ID model) in *Designing Instructional Systems* and lesson/activity planning (Level 3 of our model) in *Producing Instructional Systems*. We realize that not all readers would feel the same immediate need to study these more general levels in detail. We have therefore included two introductory chapters – Chapters 2 and 3 – which summarize the overall treatment of the ID process that was developed in the earlier books. Chapter 2 presents the theoretical model of learning which will be of direct use to the materials developer as well. Chapter 3 presents an overview of the practical procedures of curricular and lesson planning, to act as a 'back-drop' for the present work. These two chapters are structured in the form of 'maps', to facilitate rapid reference and selective reading. Each 'map' is a synthesis of one or more

A guide to the reading of *Developing Auto-Instructional Materials*

	Part 1 *Theory base:* The general content for materials development	Part 2 *Print-based materials:* Instructional programming and its evolution	Part 3 *Computer-based materials:* The impact of the computer	Part 4 Techniques for audiovisual material development	Part 5 The evaluation of instructional materials
Analysis Understanding the problem	1. Individualized instruction: the philosophical context	4. The fourth level of analysis: task-centred and topic-centred approaches 5. A study of change in programmed instruction: analysis of the topic-centred and task-centred programming techniques	10. CAL, CAI and CMI: definitions, scope and potential	14. Audiovisual media: informational, motivational or instructional? 15. Audiovisual instructional packages	18. The validation and evaluation of instructional materials
Synthesis Designing a solution	2. A model of learning: the theoretical context 3. Four levels of instructional design: the practical context	6. Mathetics revitalized: tactics for producing expositive S-I materials 7. Structural communication 8. Structured writing or 'mapping' 9. Experience programming: case studies and role-plays	11. Initial design and development 12. The range of CAI designs 13. Coding the courseware for the computer	16. Developing audiovisual instructional materials	19. Developmental and field testing: organization, execution and interpretation
Evaluation Looking at the proposed solution — and rethinking				17. Interactive video and the new media: an evaluation	20. Evaluation: final words

chapters of the earlier volumes. References are given in each 'map', to enable the interested reader to locate more extensive explanations.

The bulk of the book is divided into four further parts. Parts 2, 3 and 4 deal with the three categories of auto-instructional materials development which were used as an organizing framework in *Producing Instructional Systems*:

* print-based auto-instructional packages;
* computer-based auto-instructional packages;
* audiovisual packages (including any type of audiovisual media).

Finally, Part 5 deals with the question of validation and evaluation of instructional materials during the process of development.

The treatment in Part 2 is organized in a historical sequence, commencing with techniques for print materials that were developed in the early 1960s, proceeding to later refinements and more sophisticated approaches. Part 3 then follows with a discussion of computer-assisted learning, tracing early techniques and their similarity to those already used in the print-based domain, then continuing to later developments which attempt to use to a greater extent the special opportunities presented by the electronic computer.

Part 4 follows a similar path, commencing with an analysis of audiovisual packages in education and the techniques used some decades ago for the production of truly instructional audiovisuals, and then proceeding to the latest developments of interactive, multi-media presentations and the new opportunities they present to the instructional developer.

Part 5 closes our discussion in a general way: now that we have established that, in the field of instruction, there is an overall set of basic principles that may be applied to materials development (whatever the media of delivery), it is natural to present a general media-independent model for developmental and field testing of instructional materials.

The chapters in most of the five parts of the book are sequenced to move from the *analysis* of basic concepts and principles, to the *synthesis* of schemata, strategies, techniques and special instruments that may assist the materials developer in his or her task. Finally, these draft materials are themselves subjected to *evaluation*, through the study of specific examples of their application to typical students. In order to assist the reader to be selective in reading the chapters that are of most immediate relevance, the overall grid structure of the book is presented in the table that follows. Given a reasonable level of prior knowledge of instructional design and development, the reader should be able to use the book selectively, either by reading a specific row or column of the matrix. Those interested in analysis and in basic concepts may read across the top row chapters of all three parts. Those primarily interested in tools and techniques may opt for the second row. Those who are primarily concerned with evaluation may choose the third row, although this one is not all that independent of the previous two rows. The five parts may also be read independently if the reader is already familiar with the earlier books *Designing Instructional Systems* and *Producing Instructional Systems*.

Alexander Romiszowski, February 1985

PART 1.
The Context for Materials Development

Overview

This first part is in reality a form of extended introduction to the main topic of the book. For this reason, the three chapters included here have already been described in the overall 'Introduction to Developing Auto-Instructional Materials'. It remains at this point only to comment on some of the more detailed aspects of content and structure, and to give some guidance on appropriate reading strategies.

This book is the third in a series devoted to instructional design and development. As such, it is based on the application of a general philosophy and specific concepts that were first expounded in the earlier books. The reader who is familiar with both *Designing Instructional Systems* and *Producing Instructional Systems* may find little that is new in this introductory part – its main function is to give an overview of the general philosophical and conceptual context employed by the author, to those who are *not* familiar with the previous books. The chapters are short, condensed, and make use of 'maps' in order to facilitate rapid reference to key ideas. Also, there are many detailed cross-references to specific chapters in the earlier books of this series (and also to other sources), to enable the 'new' reader to follow up in more detail any specific aspect of interest. The reader with practical objectives in mind should, however, find all the necessary background concepts defined in the three chapters that follow, without the need for extensive reading of other sources. The reader with more academic objectives may, on the other hand, use these chapters as a starting point to more extensive study of the literature, following up the references quoted on a map-by-map basis. The reader who is already familiar with the author's earlier books may be tempted to 'skip' this introductory part, at first, referring back, if necessary, whilst reading the later chapters. There are, however, one or two 'new' points introduced in each of the three chapters included here.

Chapter 1, which is the only one not 'mapped', discusses the concept of individualization as the 'frequency with which an instructional system adapts itself to the needs or preferences of the individual learner'. This concept was discussed at length earlier, in Chapters 1, 2 and 3 of *Producing Instructional Systems*. However, the present chapter goes somewhat further in discussing the role of the computer as an instrument for individualization and explains why the computer is given such prominence in later chapters.

Chapter 2 presents the 'theory-base' on which the models, later applied, rest. It is in two parts. The first half presents a series of maps, each one devoted to a specific theoretical model, or position, which was influential in forming the author's own approach. The second half presents a series of maps that summarize, in a highly condensed manner, the author's model which is here used as the basis for the development of instruction. This model was developed by the author in earlier books and copious references are included to chapters where more detail may be found. However, there are some new aspects included in this chapter, which will not be found in the earlier references. One new aspect is a greater emphasis on the 'inner self' – on personality factors, deeply held beliefs, ethics or memories, all of which influence the individual learner's approach to a specific learning task and, therefore, should be taken into account by the instructional designer. Another new aspect is the attempt to use the structured writing, or 'mapping' technique, firstly as an example of one methodology of materials writing, and secondly to justify and explain its own basis in sound theoretical principles of learning and communication. Both these 'new' aspects are touched on only briefly here, as this is not intended to be a theoretical text. They are to be developed further in later works. In the present context, however, they are mentioned in order to give an updated view of the author's conceptual schemata at the time of writing.

Chapter 3 presents, in summary, the author's practical model for instructional design and development. This is a 'four-level' model, ranging from the initial design of a project to develop an instructional system (Level 1), through curriculum design (Level 2) and detailed lesson planning/development (Level 3), to the development and production of specific exercises and instructional materials (Level 4). Levels 1 to 3 were the subjects of the two earlier books in this series and are summarized here in the form of highly condensed 'maps' and copious cross-references. Level 4 is, of course, the subject of this present book. Readers may, at this point, notice that the 'levels of individualization' model, described in Chapter 1, maps perfectly on to the 'levels of instructional design' model described in Chapter 3. We are thus brought to the point of identifying a cardinal principle – the design of self-instructional materials is in essence the search for appropriate ways to automate the adaptation of the teaching-learning process to the needs or preferences of the individual learner, with a frequency of adaptation that is appropriate to the learner, the task to be learnt and the structure of the information to be communicated.

1. Individualized Instruction: The Philosophical Context

1.1. Approaches to individualization

This chapter is devoted to identifying the main characteristics of individualization, investigating the many practical schemes or plans for the individualization of instruction, and suggesting a method for describing in a standardized way exactly how a particular system is in fact individualized.

Individualized learning of instruction? We also discuss the importance of mediated instruction in most systems of individualization.

It has often been said that all learning is individualized. After all, the learner learns – nobody can do it for him.

Teachers can *help* learners to learn, and this they can do in a variety of ways. Some of these ways take the individual learner into consideration more than do others. In this sense they are more individualized.

Teachers may help learners to learn in ways other than by instruction. They may exercise guidance or counselling functions, or remove obstacles to efficient learning (such as cramped or noisy study conditions), or provide extra resources (by lending books), and so on. However, our concern here will be restricted to the processes of communicating information to the learner, stimulating relevant learning activities, evaluating the result of those activities and taking remedial action if necessary. This we have termed *instruction*. We are therefore concerned with the individualization of instruction.

The development of interest in individualization Of course some instruction has always been individualized – right back to the one-family unit of the Stone Age, through the age of Socrates and Aristotle, to the Oxbridge type of university system. It is only in the more recent history of mankind that the major part of instruction has ceased to be individualized (or small-group) and has become large group-centred. Many educators felt that this 'mass' education was in many ways inferior. Thus efforts were made to devise systems that stimulate the (economically unfeasible) one learner – one tutor relationship.

The development of individualized instructional programmes began in the later decades of the nineteenth century (Harris, 1960). Although the main current of educational practice has continued, fixed in its course, an increasing number of programmes that make schooling more adaptable to differences among students have been proposed and developed (De Haan and Doll, 1964; Shane, 1962). Arguments for breaking down uniformity of instruction gained support with the appearance of instruments for measuring human abilities shortly after the turn of the century. It has become clear that students differ not only in intelligence but in creativity (Wallach and Kogan, 1965) and in various elements of intellect (Guilford, 1967). It also became clear that great differences between individuals in competence and performance exist (Gilbert, 1969; 1978), and that inequalities in intellect, physical ability, and social behaviour, marked in childhood, 'increase as students move through the grades' (Thomas and Thomas, 1965).

The diversity of approaches to the individualization of instruction Some of the early proposals for changing the traditional system were referred to as individualized programmes, others exhibit the hallmarks of individualization even though they do not bear the title, and to these may be added still others, not necessarily associated with the school but obviously individualized: tutorials, correspondence courses, and informal programmes of independent study.

Together such programmes constitute a diverse family. They are based on different interpretations of individualization. They are inspired by different philosophies and theories and influenced by different technologies.

The self-instruction trend

The boom of programmed instruction in the early 1960s was one interpretation: individualization as self-paced individual study of prescribed material (usually common to all the students in a group). Briefly, the self-pacing element caught the imagination of educators and linear programmes were seen to be the road to individualization.

However, people soon remembered that there were other factors involved in learning than the pace of reading. Branching programmes (Crowder, 1963) appeared on the scene, soon to be followed by even more complex adaptive teaching machines such as the SAKI keyboard instructor (Pask, 1960) and the Edison Responsive Environment or talking typewriter – a multi-media teaching system which adapts to the individual student's learning pattern in complex ways (Moore, 1962).

This trend has continued with the development of computer-based learning systems, commencing as rather sophisticated branching teaching machines (eg PLATO I), moving to programmes of drill-and-practice in routine skills (Suppes *et al*, 1968) and finally computer-based instructional systems which simulate real-life problems in business technology or society. We are on the verge of developing viable systems of conversational programming which would enable the simulation of the one-to-one tutorial situation in all its essential aspects – the general-purpose adaptive teaching machine.

The trend to student-directed learning systems

Yet another approach to individualization has been the growth in independent study: project-based work, Quest programmes, student-directed learning, resource-based learning and so on. Such independent study may be carried out by the individual alone, or it may be designed for small-group work, but it is still termed individualization, as it breaks down the larger group and ensures that each member of the small group has a task to perform.

The use of small-group learning systems

Yet other types of group situations have been developed with the individual in mind. In sensitivity training, T-groups are used so that individuals can learn to react appropriately to other individuals. In more specific training situations, role-playing and other simulation techniques are employed. Games are now commonly employed as instructional techniques, not only in elementary schools, but right up through the school system, into university, business and the professions. A well-structured instructional game ensures that each individual engages in learning-directed activities, whilst involved in a group situation.

1.2 Sorting out the approaches

With so many varieties of approaches to the individualization of instruction it is necessary to develop some way of classifying and describing them.

Group-based and individual-based strategies

One may adopt a variety of parameters. Maurice Gibbons, in his book *Individualized Instruction: A Descriptive Analysis* (1971), adopts a classification of strategies for individualization into strategies which are employed in the group situation and strategies which impinge on the individual. He then subdivides each into active, responsive or permissive in function of who directs the activity (who makes the decisions).

His final subdivision is concerned with the teacher-student interaction (in group situations, whether this is with the whole group or with sub-groups, and in individual learning situations, whether the teaching is direct from a human teacher, or is indirect, ie mediated). We present part of this classification in Figure 4.13 later on in this book.

This 'family tree' classification is somewhat rigid. Many instructional systems are partly individual study, partly group study. Indeed MacPherson (1972) suggests that any well-designed system probably would be a mixture. Some decisions regarding options in the course will be taken by the learner; others must be taken by the teacher, or even the system itself. Some activities will involve teacher-student contact; others will be mediated by packaged presentations (both self-instructional and group).

The level of decision-making as regards individual-ization

Another approach is to consider the level within a course at which individualization takes place. Is it:

1. At the course level: do students simply exercise their option to take or not to take a given course? Do they do this, incidentally, on the basis of a systematic consideration of the course objectives or simply by their individual whims? (Level 1 of individualization).
2. At the course unit level: course options are planned in the light of the overall objectives of the course, the inter-dependencies of the units and the resources available. Neglecting the constraint of resources, reasonable course options are few in most courses with tightly defined, specific objectives, and many when objectives are ill-defined. (Level 2 of individualization).
3. At the lesson level: lessons usually form a sequence, one building on the other. If this is so, is it reasonable at this level to talk of learner-selected objectives? Do we mean simply that learning rate, sequence and perhaps to some extent the methods and media, are individualized? Or do we mean that the learner can select some objectives (eg enrichment), over and above the common core of essential objectives? (Level 3 of individualization).
4. At the individual, detailed objective level: the options would appear much the same as at the lesson level (ie learning rate, sequence, methods/media) but on a more micro scale and therefore individual choice is exercised more often (Level 4 of individualization).

One may consider these levels of individualization as another way of expressing the frequency with which decisions are made – in terms of four broad frequency bands.

Four aspects of individual-ization

We have identified four relevant questions that one may ask of any instructional scheme that purports to be individualized.

1. What is to be individualized?
2. When (with what frequency) will the course adapt to the individual?
3. Who decides?
4. How does the system adapt to the individual?

What?

1.2.1 What may be individualized?

The majority, though not all, of individualization schemes allow the learner to work on his own, at his own pace, for at least part of his study time. However, many other characteristics of a course can be, and in some schemes are, individualized. Some of the more obvious and more important characteristics which may be individualized include:

Pace of study. Students may be constrained to learn at a predetermined pace (as when listening to a lecture or viewing a TV programme) or they may be allowed to work at varying paces (as in programmed instruction, independent study, small group discussions).

Materials or media. Students may be allowed to choose (or be assigned to on the basis of past performance alternative versions of a lesson in different media, or alternative lessons leading towards the same objectives.

Methods of study. Students may receive alternative lessons differing in the instructional strategy adopted (eg expositive or discovery) and/or in the detailed tactics of instruction (eg choice of examples, number of problems to be worked, amount of hints given to the learner, sequence of topics, etc).

Content of study. Students may receive alternative lesson content, either as a means of tailoring a course to the individual's own objectives or as a means of selecting material familiar or interesting to the individual to be the vehicle for the attainment of broader educational objectives.

Objectives of study. Course objectives may be varied – either for liberal reasons, or in order to adapt courses to the different aptitudes of individuals or the different needs of the organization.

When?

1.2.2. When does individualization take place?

Individualization is a relative thing – a matter of degree.

I have elsewhere suggested (Romiszowski, 1976; 1981) a scale of frequency bands depending on whether individualization decisions are taken:

1. for a course;
2. for each unit within a course;
3. for each objective within a unit;
4. for each learning step taken to achieve the objective.

Who?

1.2.3. Who decides?

The decisions to individualize may be taken by:

1. *The student himself*, when he chooses a particular course option, or a particular textbook.
2. *The teacher*, who may prescribe individual objectives or media or extra content.
3. *The system itself* which may have built into it a diagnostic device which automatically adapts the presentation to the individual student. This is the case with the branching style of programmed learning (Crowder, 1963) and with many computer-based learning systems.
4. *Any combination of student, teacher and system*. This is the most common situation, involving a joint decision between student and teacher.

How?

1.2.4. How does the system adapt to the individual?

Two factors are identified as of importance:

1. The style and type of instructional materials and media employed in the system, and the way in which the learners select and have access to the materials.
2. The role of the tutor or instructor (if any) in the system, both as a medium of instruction and as a medium for management and control.

It is apparent that there are innumerable different ways of going about the How? question. In reality, a reply to this question generally requires a detailed description of the system in question. However, the What?, When? and Who? questions may be used to compare the general structure and philosophy of different schemes.

1.3 The mediation of instruction

We shall now consider the first of the two factors identified in the previous paragraphs: the use of mediated instruction as a means of individualization.

In his excellent book *Resources for Learning*, Taylor (1971) says:

> We can conceive of a number of different systems of learning being used in schools. Foremost among them is the teacher-based system familiar to us all in the classroom. Other systems already tried here or abroad are: book-based, book-and-boy-based, assignment-based, radio-and-television-based computer-based. At first sight, it seems a daunting complexity, but we may find the appearence of conflict between these systems misleading...
>
> We can start by noting that whatever system is used in a school the teacher is assumed to be present. The most extreme exponents of programmed learning threatened to reduce the teacher to a cipher, but with regard to schools this was *1984* stuff and scarcely conceivable. Teachers are plainly of critical importance in caring for and about children; the inspiration, encouragement, control, guidance they provide matters profoundly. Further, no one but the teacher on the spot can perceive and supply the particular needs of a particular child at a particular moment. Neither the teacher's pastoral nor his tutorial functions can be replaced. When we talk about alternative systems of learning, then, we are asking only where, principally, the *burden of instruction* should rest.

The teacher and resource-based learning

Taylor then goes on to make the point that the essential difference between teacher-based and media-based instruction is that the accent moves from the teacher teaching to the learner learning. The teacher in the 'traditional' teaching situation (whatever that is) supplements his presentation with visual aids, refers the learners to textbooks and sets reading assignments and so on. However, he remains the principal medium of instruction, the principal learning resource at the learners' disposal – perhaps that is what we mean by 'traditional'. The systems which will interest us in this book, however, are principally the 'non-traditional' systems where the bulk of basic instruction is taken over by resources other than the teacher. Such systems, as Taylor points out,

> are made from the same cloth frayed out into threads or, in the case of the computer, overlaid with the dazzle of electronic millinery. For brevity we can group all these systems together and, since they rely on collections of learning situations and materials, call them 'resource based' or, alternatively, 'package based'.

1.4 The control of instruction

Prescriptive and student-controlled systems

We now turn to the second How? factor mentioned earlier. Many of the various viewpoints on teaching may be classified as being either highly 'prescriptive' or highly 'democratic'. The prescriptive approach supports individualization on the basis of a comparison between the individual student's profile and some ideal model. Such a comparative diagnosis leads to an individual prescription of learning activities for the student.

The democratic, or student-centred, approach supports individualization for the student's own sake – to adapt the course to his needs, to give him responsibility and to develop him as an individual.

Cybernetically controlled systems

There is a certain amount of partisan warfare between these two points of view, on philosophical grounds. Currently, an alternative to both of these opposed views is being constructed by the cyberneticians. Work such as that of Pask suggests that machine-based systems can be constructed which can learn from the learner, can adapt to the learner's strategy for learning and can redesign the presentations (on a conversational-tutorial pattern) in ways superior to those achieved by human tutors. This was obviously the case with machines such as the SAKI keyboard trainer for complex high-speed sensori-motor skills. Recent work in the field of computer-assisted learning (CAL) is beginning to achieve similar results in the conceptual and problem-solving area. Many teachers administer courses which are as prescriptive and as rote-learning-oriented, as the most outdated of linear programmed instruction. They do this because they have learnt in that way and because they have not the skills necessary to enter into an adaptive, conversational tutorial mode of instruction with their students, even if class numbers allowed them to attempt it. Others adopt a student-directed approach without themselves becoming a useful resource: they fail to supply information, guidance or feedback.

It is here that I see the greatest justification for the study and development of computer-based, mediated, teacher-support systems (perhaps even teacher-replacement systems). We have no alternative. The traditional approach to teaching has been singularly unsuccessful, more so as mass educational opportunities expanded.

The future for media-based cybernetically controlled systems

It is not the function of this chapter to investigate or comment on the quality of teachers in any great depth. Suffice it to observe that quality and quantity are inextricably linked (by the laws of supply and demand) to the value that society appears to put on education. As long as priorities (as expressed by salary levels) remain as they are, educational and training systems will continue to be short of highly qualified teachers, particularly in certain key areas such as mathematics.

It is for these reasons, of both quantity and quality, that I see a bright future for the continued development of mediated individualized instruction.

Present economic difficulties and the resulting cutback in expenditure on education and training have perhaps slowed down the process somewhat. But in the long run, these same economic difficulties will lead to the search for more cost-effective means of instruction. In large-scale applications this will generally imply an increase in the use of carefully prepared mediated instruction, both in the mass-media mode of distance education and in the individualized mode of learning packages.

The aspect of cybernetic control – adaptive, conversational interaction between instructional system and learner – will figure prominently in these systems. This is, indeed, the most important aspect of CAL.

We shall return to this and the other aspect of CAL in later chapters that deal specifically with the production of computer-based learning materials.

1.5 A schema of systems for individualizing instruction

We identified three characteristic positions regarding the control of an instructional system.

1. The *prescriptive* approach, which attempts to measure students' individual differences and match instructional strategies to them according to a predetermined algorithm.
2. The *student-directed*, open, or free-learning approach, which attempts to let the student have maximum control over the choice of learning strategies and media (even sometimes content and objectives).
3. The *cybernetic* approach, which attempts to set up an interactive system, adaptive to the student's needs in an on-line manner, based on what the system has learned concerning the student's needs, learning styles, difficulties, etc.

A matrix for the classification of media-based systems

The chart appended here (Figure 1.1) is an attempt to construct a schema for the classification of systems of instruction. The classification, in three columns, according to the primary media used, is combined with the above-mentioned classification (according to the chief philosophy of control adopted) to form a 3x3 matrix. This chart was first presented in *Producing Instructional Systems* (Romiszowski, 1984). The examples listed were all fully discussed there.

NOTES:

1. The rectangles, circles and ovals are not drawn to any scale, they simply indicate approximately where into the suggested classification the systems described fit.
2. The items in the rectangles refer to techniques or materials which will be discussed more fully in later chapters in this book.
3. The items in dotted circles or ovals refer to well-known systems which have been widely implemented and which were described in Chapter 3 of *Producing Instructional Systems*, the first volume of this series on instructional development.

This comparison is necessarily somewhat crude, as many of the systems are capable of being prescriptive or student-directed to greater or lesser degrees (often depending on the personalities of the teachers and learners involved). No quantitative implications should be read into the chart. However, it does help to illustrate certain general trends.

Some observable trends

The examples listed cover the whole field of media/control combinations fairly comprehensively. There appears to have been an early trend, so far as print-based systems are concerned, for a concentration of prescriptive projects in the USA, and student-directed projects in Europe. The big growth projects in the USA have been IPI at the elementary level and the Keller Plan (PSI) in higher education. Both of these systems are directly descended from the programmed instruction movement of the late 1950s and early 1960s, maintaining all of its prescriptive characteristics. In Europe, on the other hand, programmed instruction-based systems, such as IMU in Sweden, or the Kent Mathematics project in the UK, have from the beginning attempted to build in a strong element of student-directed

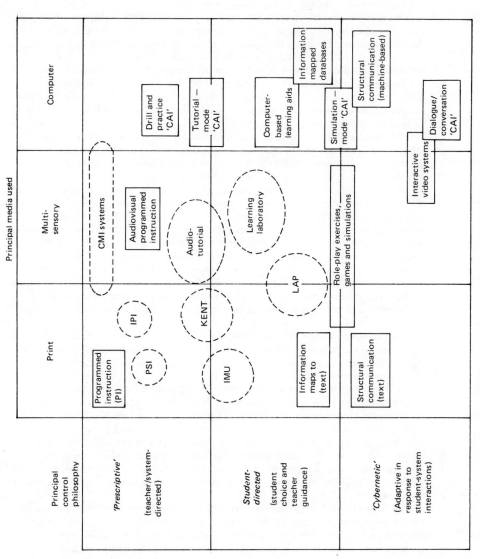

Figure 1.1 *Schema for classifying media-based instructional systems*

learning. The Kent project, from its beginnings, which go as far back as IPI, has organized its learning materials on a matrix structure, specifically to facilitate student choice of pathways.

In the multi-sensory area, once again, the Audio-Tutorial System (Postlethwait *et al*, 1972), which has become very popular in many North American universities, is highly prescriptive. The more student-directed system, such as mathematics laboratories, although much publicized, are reported to be on the wane in the USA but are still popular and growing in the UK.

The area of greatest promise for the future would appear to be the use of the computer and the adoption of adaptive cybernetic strategies of programming. In the conversational or dialogue mode, the computer is programmed to learn from the learner as well as to teach him – the computer learns to interpret the individual learning strategies of each separate learner, his/her preferences for teaching style, media and so on. Thus the need for a theory of individual differences is reduced. One no longer needs to develop general rules for matching learner characteristics to methods and media. These rules are developed on-line for each individual learner.

The simulation mode of CAI is also particularly promising in that it can compress a great amount of 'real-lifelike' experience into a short amount of learning time and can present the learner with opportunities to gain this experience in the safe setting of individualized study on a computer terminal.

The game aspect is also highly important as a motivating factor, whether used in the more complex, experiential simulation games, or in the simpler context of drill-and-practice exercises. Drill-and-practice, though much maligned, has its place, and will continue always to have its place, for the learning of routine skills, standard procedures, facts (eg language vocabularies) and so on. The use of the computer as a base for such exercises offers a careful, error-free, tuition service of infinite patience and the injection of a gaming aspect can do much to relieve the dullness of such learning. Although curriculum changes may reduce the sheer bulk of the memorization content of school learning, they will never eliminate it completely. The more of it that can be transferred to interesting computer-based games, the better.

Without a doubt, the programmed-instruction tutorial mode of CAI (that is the most used in current applications) in fact offers the least innovation. As our chart shows, this mode as currently practised is not much more than the branching programmed instruction of the 1960s or the multi-media audio-tutorial of the 1970s. However, the presentation of the material can be somewhat more slick and sophisticated and indeed, as computer prices topple and availability increases, it may become both cheaper and more convenient to use microcomputer-controlled interactive video than to use any other alternative audiovisual device (or perhaps even multiple copies of texts). There is a parallel trend in the audiovisual field, where, once a user has invested in a videocassette system (perhaps for other principal reasons) it is now both more convenient and much less expensive to produce multiple copies of already existing synchronized slide-tape programmes in videocassettes.

We may very well see a similar trend in the near future as regards the use of computer-assisted learning. Rather than having multi-media systems of a more conventional nature, composed perhaps of several texts, a few audiovisual packages and a simulation game (together with a set of formative evaluation instruments and a study guide), we shall encounter all of these packaged in one single medium – the computer-controlled videodisc. One of the great attractions of the laser videodisc is its ability to store any digitally encodable information – moving and still pictures, pages of text, graphics, and the computer code necessary to control the presentation and use of the material. The costs (at the time of writing) are rather high, but they have already tumbled considerably and, to judge by past history, will continue to tumble. As the cost decreases, it will no doubt become economically attractive to place all one's instructional materials in a single delivery medium, just as it has already become the rule, for reasons of both cost and convenience, to transfer synchronized slide-sound packages to videocassettes when multiple copies are to be used in several training centres.

The interactive videodisc also offers new and interesting possibilities which earlier media (both non-computer-aided audiovisuals and non-audiovisual CAI) did not have. Whether we shall develop instructional programming techniques and skills which will enable us to exploit these possibilities in a pedagogically rational manner, remains to be seen. Up to now, much of CAI and the few currently available interactive videodiscs do not begin to exploit the possibilities of the new media. We shall be devoting the whole of Chapter 17 to considering just how the link-up of computer and videodisc may best be used in the educational and training context.

There is no doubt in our minds that the new developments described in Chapter 17 will lead to an integrated approach to the packaging of multi-media instruction. Whether we shall see a similar integration of approaches to instructional design and development, is another matter. This book is largely devoted to the fostering of such an integrated approach.

1.6 Towards an integrated methodology of ID

The trends observed in the analysis of mediated systems of individualized instruction add up to give a picture of fragmentation.

Current multiplicity of media design approaches

There seem to be traditions of print-based materials development that have little in common with the traditions of audiovisual materials development. More recently, the CAL brigade has started to forge its own 'traditions', without always taking into account what has been learned in the other media. Often, it seems, instructional developers lose sight of the basic fact that, whatever the medium of communication in which they are working, the end they hope to reach is essentially the same – effective learning by the learner.

The baby and bathwater syndrome

We seem to forget that learners have not changed in their physiological or psychological make-up just because we have new media at our disposal. We commit, with regularity, the wasteful error of abandoning partly successful systems and methodologies just because new ones come along. This well-known phenomenon in education was noticed as far back as the 1920s by John Dewey – he called it the 'baby and bathwater syndrome'; throwing away the good with the bad and starting afresh. Then we spend time and energy designing new systems and methodologies, often reinventing the wheel so to speak. New computer-based interactive video systems are developed which are little different, in terms of the pedagogical structure of the material, from programmed instruction or resource-based learning systems that were in use some 10, 20, even 30 years ago. The media have changed, but the message and the way of presenting it have not. Is this a lack of imagination or creativity on the part of the instructional developers? Or is it rather that the characteristics of the learning tasks and of the learners have not changed appreciably over the years, so that good solutions of yesteryear continue to be good solutions today?

Back to basic principles!

We tend to favour the latter explanation and see that one contribution of the present book should be to attempt to show the communality, in terms of basic instructional design principles, that runs through all approaches to the systematic development of instructional materials. We hope to counteract the baby and bathwater syndrome and save some of the energy spent on reinventing wheels, by illustrating the way in which the same basic design principles may be applied in any medium of instruction.

We do not mean to imply, however, that there has been no progress in recent years as regards the science and technology of instruction. On the contrary, there have been great developments in such areas as simulations, games, experiential learning in general and small group learning techniques in particular. Our technology has become the richer and more powerful for these developments. Also the new media and, in particular, the computer, have made possible a whole range of instructional tactics which would be difficult or impossible to implement with the more traditional media. We do mean, however, that the basic principles that guide selection among these new strategies and tactics have been known for a long time, yet often are not applied in practice. We hope that this book, by drawing together in one volume the print-based, computer-based and audiovisual domains of materials development, may serve to illustrate these basic common principles and develop an integrated approach to the design of self-instructional materials in any media.

2. A Model of Learning: The Theoretical Context

2.1 Introduction

The previous chapter has set the scene for self-instructional materials development from the philosophical viewpoint. We have attempted to answer the question 'why develop individualized systems of instruction and why do self-instructional materials play such an important role in the effective individualization of instruction?'.

In this chapter, we proceed to the theoretical bases that will guide our materials development effort. Detailed lesson planning depends on three factors:

1. Knowing the subject matter and/or the tasks to be taught.
2. Knowing the characteristics of the target population.
3. Having some systematic methodology for matching the procedures, or tactics, to be adopted in teaching to the specific characteristics of the subject matter, the learning task and the target population.

The function of this chapter and the next is to present in summary the bases for these matching procedures.

There is no space here to deal fully with all aspects of learning theory which may be of relevance to materials development. As necessary, we shall refer to useful psychological principles in the body of later chapters. In this chapter, however, we present the basic model of categories of learning objectives which we shall use as a decision-making tool throughout the rest of the book. This model is of our own manufacture and we have found it useful both as a personal decision-making tool and as a way of communicating our methodology to others. It is, nevertheless, only a model. There is nothing sacrosanct about it. And, like all models, it is the result of the synthesis of ideas that have to a large extent been picked up from others. In order to give some background to our model we present, in the first part of this chapter, short summaries of the principal theoretical positions and the principal models of other authors that have been influential in forming our own viewpoints.

The reviews of other viewpoints and models are necessarily highly condensed and the reader is directed to other, more extensive, accounts by means of references placed at the end of each description.

The second half of the chapter presents our own model, once more in rather condensed form, but cross-referenced to the chapters in earlier volumes of this series that present a full account.

In order to facilitate the use of this book as a handy review and reference text, we have arranged the presentation in the form of single-topic 'maps'. The map titles should indicate clearly to the reader whether a particular map is of interest at a particular time. The techniques of structured writing which have been adopted for the reference chapters of this book are described further in Chapter 8. We hope that readers will find this form of presentation helpful.

Map 2.1 *Overview of the chapter*

Introduction	The chief purpose of this chapter is to present, in a condensed and easy-reference format, the theoretical model of learning that is used as a basis for the development of later chapters. The technical terminology used later on has quite specific meanings, which should be clearly understood by the reader. In order to facilitate the mastery of this terminology, without the need to refer to other sources, we have presented the essential aspects of the model in a series of 'maps', that may be read in random order for reference purposes. They may also be read sequentially, however, to get an overall idea of the model and its origins. The first half of the chapter is devoted to maps which summarize some of the earlier models which have influenced the author's own model. The second half of the chapter presents a series of maps that outline the author's model.
The learning theories and views that underpin the model	Many views or theories exist which have tried to explain the processes of learning. Some of these are scientifically based theories, whilst others are rather philosophical viewpoints of how education 'should be'. We have singled out the most prominent theories/views as the backdrop against which models of the learning process and of different categories of learning have been developed. The next map (Map 2.2) presents an overview of five views that have been important in forming modern ideas on learning and instruction. These five views are the behaviourist, cognitive, developmental, humanist and, most recently, the cybernetic positions.
The emphasis given to certain aspects of the instructional process	Each of these views has tended to emphasize certain aspects of the teaching-learning process more than others. For example, the humanists have laid great stress on the freedom of the individual in respect of what he/she will learn, when and (sometimes) how. The behaviourists are concerned with stimulus and response (input and output), but particularly with the shaping or conditioning of specific learning outcomes (outputs). The cognitive psychologists (Gestalt, field and developmental) are, on the contrary, more concerned with the internal thinking and memorizing processes, than with the specific outcomes. The cyberneticians prefer to treat the learner as a system in an information-rich environment and analyse the learning process in terms of information transfer and transformation.
The process of development of the model	The present model is, of course, influenced by the earlier work of many authors. The first part of this chapter will summarize the contributions of those most influential on the formation of the present author's views. As an overview, we present here a rough timechart which shows how, in the author's view, the work of various earlier writers has contributed to the development of a more comprehensive and more coherent view of learning and the learning process. Those items that are highlighted on the timechart by being put into rectangles are developed further in later maps.

Map 2.1 *(continued)*

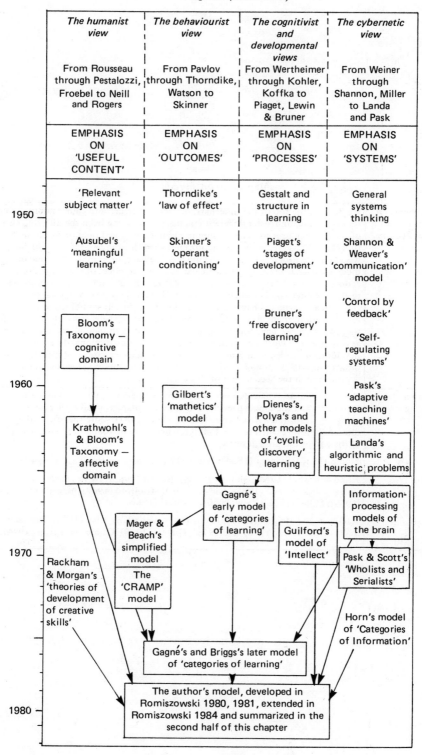

The humanist view	The behaviourist view	The cognitivist and developmental views	The cybernetic view
From Rousseau through Pestalozzi, Froebel to Neill and Rogers	From Pavlov through Thorndike, Watson to Skinner	From Wertheimer through Kohler, Koffka to Piaget, Lewin & Bruner	From Weiner through Shannon, Miller to Landa and Pask
EMPHASIS ON 'USEFUL CONTENT'	EMPHASIS ON 'OUTCOMES'	EMPHASIS ON 'PROCESSES'	EMPHASIS ON 'SYSTEMS'

1950

'Relevant subject matter' — Thorndike's 'law of effect' — Gestalt and structure in learning — General systems thinking

Ausubel's 'meaningful learning' — Skinner's 'operant conditioning' — Piaget's 'stages of development' — Shannon & Weaver's 'communication' model

'Control by feedback'

Bloom's Taxonomy — cognitive domain — Bruner's 'free discovery' learning — 'Self-regulating systems'

1960

Gilbert's 'mathetics' model — Pask's 'adaptive teaching machines'

Krathwohl's & Bloom's Taxonomy — affective domain — Dienes's, Polya's and other models of 'cyclic discovery' learning — Landa's algorithmic and heuristic problems

Gagné's early model of 'categories of learning' — Information-processing models of the brain

Mager & Beach's simplified model — Guilford's model of 'Intellect'

1970

Rackham & Morgan's 'theories of development of creative skills' — The 'CRAMP' model — Pask & Scott's 'Wholists and Serialists'

Horn's model of 'Categories of Information'

Gagné's and Briggs's later model of 'categories of learning'

The author's model, developed in Romiszowski 1980, 1981, extended in Romiszowski 1984 and summarized in the second half of this chapter

1980

Map 2.1 *(continued)*

Comments on the form of the timechart	We have, at the top of the timechart, divided our figure into four columns, which represent four somewhat different emphases that spring from the differing views of the learning process. These columns are not continued right down through the timechart, however, as the distinctions between theoretical origins gradually get weaker as we progress through the models. We have, indeed, placed some of the models rather arbitrarily in a particular column, to reflect our own opinion of the roots of the model (not necessarily in total agreement with the authors of the respective models). Our intent is to illustrate how, in our view, the various theoretical viewpoints have, in practice, been eclectically combined in the more recent work of many writers. The work of Robert Gagné is especially notable for this. However, we have found Gagné's models not as complete, nor as coherent as is required to work right across the spectrum of instructional design and development projects. In order to attack better a wider variety of instructional design projects, we found it useful to develop a new model. This new model is presented in the second half of this chapter.
Further reading	The timechart summarizes a very extensive bibliography. We developed the arguments and the model presented in the next chapter, in an earlier book, *Designing Instructional Systems* (Romiszowski, 1981). More specific readings, quoted there, are listed below. Full references may be found in the bibliography at the end of the present book.
to 1950s	Bloom (1956); Skinner (1954); Piaget (1957); Shannon and Weaver (1949).
1960s	Krathwohl *et al* (1964); Skinner (1968); Bruner (1966); Dienes (1960 and 1973); Polya (1963); Gagné (1965); Gilbert (1961); Guilford (1967); Mager and Beach (1969); Miller (1967); Pask (1960); Ausubel (1968).
1970s	Rackham and Morgan (1977); Landa (1974 and 1976); Pask (1976); Pask and Scott (1972); Horn (1969 and 1973); Gilbert (1978); Gagné and Briggs (1974).

Map 2.2 *Some theoretical positions regarding the learning process (an overview)*

Introduction	Many different theories of learning have been proposed by psychologists over the years. To some extent they have reflected the philosophies and scientific culture of their time. One theory has replaced another in popularity. Often, bitter rivalry has existed between theoreticians each one claiming to have the 'complete' or 'best' explanation of how learning takes place. During the last generation or so, however, there has been a tendency towards a more or less peaceful coexistence between a number of theories and towards the formulation of mixed, or eclectic theories, which use different explanations for different forms of learning. We present here an overview of the currently most popular and influential theoretical positions.
The behaviourist view of learning	The behaviourist view of learning gained popularity in the early part of the century, as a more scientifically testable position than the earlier views of learning (such as 'mental discipline', 'developing the mind-matter', 'exercising the faculties', 'natural unfoldment', etc). Behaviourists define learning as a change in behaviour, or more precisely, a change in the learner's capacity to behave in novel ways, not brought about by mere maturation or growth. It follows that the change in behaviour is brought about by the influence of the environment. Environmental stimuli bring about changes in capacity to respond. Complex learning as explained in the gradual building up of more and more complex patterns of stimulus-response connections. Early protagonists include Thorndike, Pavlov, Watson and Skinner.
Cognitivist view of learning	The cognitivist view of learning also gained popularity, principally in Europe, in the early part of the century. The early cognitive psychologists were principally interested in perception of the form, or 'shape', of objects, patterns or procedures. Hence the German word for form — Gestalt — was associated with these viewpoints. Later workers expanded the cognitive view, coining the concepts of 'cognitive field', 'cognitive schema', 'cognitive structure', to explain how higher order and abstract knowledge is acquired, and indeed often created by the learner. Cognitivists see learning as much more than the building up of complex patterns of stimulus-response associations. Whereas the behaviourists, in general, do not find it necessary to hypothesize about the nature of 'thinking', cognitivists consider that most learning may only be adequately explained in terms of a model of thinking, or cognitive processes. The early Gestalt psychologists postulated a series of 'laws of perceptual organization' to explain how learners identify certain stimulus situations covertly (even when only partial data is available) and 'laws of insightful problem-solving' to explain how learners come to exhibit new behaviour which they had not had the opportunity to observe or practise previously (and therefore could not have gained by a process of stimulus-response conditioning). Later cognitive psychologists postulate more complex models of the internal processes of learning, involving the gradual formation of conceptual schemata, by the assimilation of new knowledge and the accommodation of currently-held knowledge-structures to coexist in harmony with the newly acquired knowledge. Early protagonists included Wertheimer, Kohler, Koffka and Lewis.
The developmental view of learning	The developmental view of learning is an extension of the cognitivist view, which adds the extra consideration of how the learner's mind and learning capacities develop with maturation. The young child, in particular, is seen as passing through a series of stages of development, characterized by the type of learning activities that become possible. Piaget, for example, refers to the 'sensorimotor', 'concrete operations' and 'formal operations' stages of cognitive activity (each stage capable of further subdivision), identifying typical age-levels for each stage. Other developmental psychologists apply a similar concept of 'stages of development' to the development of any complex idea in the mind. Skemp (with adolescents) and Polya (with adult learners) conceptualize the acquisition of new mathematical concepts and principles as necessarily passing through a concrete-operations stage before being learnt as a formal set of abstract and generalizable mental operations.

Map 2.2 *(continued)*

The humanist view of learning	The humanist view of learning is in sharp distinction from the others so far mentioned, in that it is more a philosophy than a distinct explanation. Humanists are more concerned with what education and learning should be like, because of their view of what the human race is and what human societies should be like. As regards explanations of the learning process, humanists are happy to espouse all three viewpoints previously mentioned as 'partial explanations'. They also, on occasion, may employ the more ancient explanation of 'faculty' or 'mental discipline' development, when arguing the humanist case for the inclusion of certain specific bodies of socially significant subject matter in the curriculum. On other occasions, they may employ the nineteenth-century viewpoints of 'natural unfoldment' (popularized in the eighteenth century by Rousseau, then by Pestalozzi and Froebel) in the modern guise of 'self-actualization' (Carl Rogers). As a strategy of education, the humanist view is firmly based on the learner — the human being — as the centre of interest in the teaching-learning process. As an explanatory theory of learning humanism draws eclectically on all other theories.
The cybernetic view of learning	The cybernetic view of learning is perhaps the most recent distinct position to be formulated and popularized, its growth of popularity coinciding with the development of information theory, information-processing applications, new information technologies, systems engineering and the new multi-disciplinary 'science of sciences' — cybernetics. A whole range of new concepts (accompanied by a new technical jargon) has been applied to the explanation of information-transfer (and transformation) processes, human learning being one special case. A variety of information-processing models of the learning process have been postulated. These often incorporate new concepts borrowed from the information sciences (long- and short-term memory, 'bits' of information, etc) in combination with established concepts (such as 'stimulus', 'response', 'cognitive schema', 'accommodation'). The cybernetic view, in common with the humanist's, combines useful concepts from a variety of earlier theories. However, unlike the humanist's eclectic explanation of learning processes, the cybernetic view is integrative, in the sense that new concepts exist which explain quite clearly how it is that behaviourist and cognitive viewpoints may both be correct in specific circumstances. The very concepts of system and sub-system and their application in system analysis, explain how a given learning phenomenon may be treated, at one level of resolution, as a 'black box' receiving information inputs and emitting performance outputs — a view much in line with the simplest of behaviourist models of the learner (eg Skinner's model). At a finer level of resolution, however, the same phenomenon may be viewed as a highly complex system of interacting sub-systems and processes, every bit as complex as the most ambitious cognitive or developmental viewpoints. The important cybernetic principle of 'requisite variety' leads one to accept for granted that, as the learner and his/her environment are both complex and varied, the learning processes are also complex and varied. No one explanation of learning can be expected to hold good in all cases. On the contrary, it is theoretically necessary that different types of learning (information processing) should occur by means of different processes and it is possible that different learners would learn the same information by different processes. The learning processes that occur must be equivalent, in complexity and variety, to the variety that exists among subject matter items, learners and the overall learning environment. Due to this natural complexity and variety, any general theory of learning (if and when our understanding of the process permits such a formulation) will, of necessity, be complex. In the meantime, however, we may proceed to formulate models which come as close as our current understanding allows. Models are always a simplification of the real complexity. However, we judge them by their usefulness — are they an adequate, or best-available, tool for the task in hand?
Comment	In this chapter, we present a model which has proved to be useful to the author and several of his colleagues/collaborators. This presentation is condensed, but references are quoted to more extensive treatments of each aspect presented.
Further reading	Space precludes the more detailed analysis of learning theories here. For more complete accounts, see: Bigge (1982); Bower and Hilgard (1981); Miller, G A (1967); Read, D A and Simon, S B (1975).

Map 2.3 *Some models of categories of learning:*
Bloom's and Krathwohl's taxonomies of educational objectives

Introduction	Many authors have proposed models, or schemes, for the classification of different types of learning. Usually, these models present definitions of learning outcomes, or objectives. Occasionally they also list learning process (eg 'learning to learn') or the content or inputs to be learned (eg 'verbal information').
A basic classification into 'domains' of learning	The most traditional approach to the classification of learning uses broad categories such as knowledge, skills, attitudes, values, etc. Perhaps the best-known model, popularized by the writings of Benjamin Bloom and his collaborators is the three domains of educational objectives.

Domain	Short definition
Cognitive domain	Learning of knowledge; its application, thinking, etc.
Psychomotor domain	Learning practical tasks, that require perception, decision and action.
Affective domain	Learning of feelings, preferences, value systems, etc.

Each of these domains may be further subdivided into specific categories of learning outcomes, or objectives.

Sub-classifications in the cognitive domain		

Category name	Brief description
Knowledge	The remembering of previously learned material. The lowest level of learning outcomes in the cognitive domain.
Comprehension	Ability to grasp the meaning of material. Interpreting, paraphrasing, explaining.
Application	Ability to use learned material in new and concrete situations. Applies, demonstrates.
Analysis	Ability to break down material into its component parts, so that its organizational structure may be understood.
Synthesis	Ability to put ideas together to form a new whole. Proposes, integrates, designs.
Evaluation	The ability to judge the value of material for a given purpose.

Sub-classifications in the affective domain		

Category name	Brief description
Receiving	The student's willingness to attend to particular phenomena or stimuli.
Responding	Active participation. The learner not only attends but also reacts favourably.
Valuing	Demonstration of a preference, belief, or commitment with respect to certain behaviours/phenomena.
Organization	Bringing together different values into a consistent and coherent value-system.
Characterization	The total adoption of a value-system, so that it becomes a personal life-style.

The psychomotor domain	Bloom and Krathwohl did not get round to sub-classifying the domain of psychomotor objectives. Several other authors have tried to do this, but there is no generally accepted taxonomy.
Reading guide	For detailed accounts of this approach, see Bloom, B S *et al* (1956); Krathwohl, D R *et al* (1964); Romiszowski, A J (1981) (Chapters 3 and 11); Simpson, E J (1967).

Map 2.4 *Gilbert's (and Skinner's) model of behaviour categories*

A classification of basic patterns of behaviour	A behaviourally oriented approach to the classification of the outcomes of learning would focus on characteristic patterns of stimulus-response connections that may be formed during the learning process. Thomas Gilbert, for example, presented a model of three basic patterns of behaviour, suggesting that complex behaviours are formed by the combination of the three basic patterns.		
	Behaviour pattern	*Diagram*	*Examples*
	1. *Behaviour chain* A step-by-step procedure, in which the outcome of one step acts as the stimulus for the next response.	$S_1 \to R_1 . S_2 \to R_2 . S_3$ $R_3 . S_4 \to R_4$ (etc)	Tying a shoelace. Dividing two numbers. Operating a photocopy machine.
	2. *Multiple discriminations* A simple decision-making procedure, involving a choice between two or more responses, depending on the stimulus.	$S_1 - R_1$ $S_2 - R_2$ $S_3 - R_3$	Reading a symbolic message like Morse Code, or indeed our alphabet (discriminating between b and d).
	3. *Generalization* A simple classification procedure, in which several possible stimulus situations are seen to have something in common.	S_1 S_2 $S_3 - R$ S_4 S_5	Classifying red objects correctly. Identifying the extroverts in a group. Identifying all situations in which it is unsafe to park a car.
Some combinations	1. What is commonly referred to as 'having a concept' involves the performance of a combination of generalization and discrimination behaviour — classifying correctly any example or instance of the concept and not including any non-examples, however similar they may be. Example: Having a 'well-formed concept' of honesty involves the identification of honest and dishonest practices in a given case — all possible honest reactions should be correctly identified and no dishonest reactions admitted into one's classification.		
	2. Most complex procedures are combinations of chains of activity and discriminations at decision points. Sometimes, the decisions are unique, matter-of-fact choices between predetermined alternatives (plain multiple discriminations). At other times the decisions involve the application of a concept (generalizations and discriminations). Examples: The first case is illustrated by the procedure of following a set of instructions like a travel itinerary or a recipe. The second one is illustrated by procedures for mathematical problem-solving, etc.		
Reading guide	For detailed accounts of this approach, see: Gilbert, T F (1961); Mechner, F (1965); Romiszowski, A J (1981) (Chapter 10).		

Map 2.5 *Models based on the concepts of cognitive development*

Piaget's model of learning categories in terms of stages of development	Piaget developed a classification of stages of cognitive development which a normally developing child passes through at more-or-less predictable ages. 1. *Sensorimotor stage* (0-1½ years): Emphasis on coordination of action; gradual connecting of means to reach goals. 2. *Pre-operational (representational) stage* (2-6 years): Internalization of actions (representation in thought); use of symbols in play, language, and mental imagery; use of elementary functions of one-way relationships. 3. *Concrete-operational stage* (6-11 years): First use of reversible operations signified by conversation; thought structures connected to the concrete — concerned with relations among objects (seriation) and classes of objects (class inclusion). 4. *Formal-operational stage* (beginning around 11 or 12 and expanding during adulthood): Ability to deal with the possible; understanding of relations between relations; ability to verbalize rules used in solving problems.
Models of learning categories in terms of stages of cognitive development — practical suggestions	The age-bands specified in this model indicate that below certain ages, it is unlikely that a typical child would be able to perform certain types of intellectual activity. However, they do not imply that a child, who has passed a certain age, always thinks in the abstract manner characteristic of formal operations. Often, he/she will revert to concrete examples in order to clarify new ideas. Thus, the model of stages of cognitive development may be used for two purposes: 1. to plan when a certain type of learning should be incorporated into a child's educational programme, and 2. to plan a sequence of learning activities, which moves from concrete experience of a problem situation, through solving the specific problem in a practical way (concrete operation) to the identification of general principles involved in the solution of this and similar problems (formal operation). This developmental approach to teaching appears to be equally appropriate for young children and adults and has formed the basis of models or theories of instruction developed by such workers as Dienes, Polya, Bruner and Skemp. Much store is laid, in these models, on the learning of important new principles by means of 'discovery' or 'experiential' methods. The models also have a cyclic character, as illustrated by the table below, which compares the suggestions of Dienes (for the teaching of children) and Polya (for the teaching of adults).

Map 2.5 *(continued)*

Examples	Dienes	Polya
	The preliminary or play stage corresponds to rather undirected, seemingly purposeless activity usually described as play. In order to make play possible, freedom to experiment is necessary.	A first, exploratory phase which is close to action and perception and moves on an intuitive, heuristic level.
	The second stage is more directed and purposeful. At this stage a certain degree of structured activity is desirable.	A second, formalizing phase ascends to a more conceptual level, introducing terminology, definitions and proofs.
	The next stage really has two aspects: one is having a look at what has been done and seeing how it is really put together (logical analysis); the other is making use of what we have done (practice). In either case this stage completes the cycle, the concept is now safely anchored with the rest of experience and can be used as a new toy with which to play new games.	The phase of assimilation comes last: there should be an attempt to perceive the 'inner grounds' of things; the material learned should be mentally digested, absorbed into the system of knowledge, into the whole mental outlook of the learner. This phase paves the way to applications on one hand, to higher generalizations on the other.
	Source: Servais and Varga (1971)	
Further reading	For a more detailed account of Piaget's model, see: Gallagher, J M and Reid, J K (1981). For accounts of practical models that apply cognitive/developmental approaches to the design of instruction, see: Bruner, J S (1966); Dienes, Z P (1973); Polya, G (1963).	

Map 2.6 *Gagné's early (1965) model of the categories of learning*

A hierarchical neo-behaviourist model	As one progresses to higher-order learning, it becomes increasingly difficult to describe the outcomes of learning completely in terms of patterns of S-R connections. The cognitivists, indeed, would say that it is impossible. A number of eclectic, part-behavioural and part-cognitive, models were developed in the 1960s. The best known of these was Gagné's hierarchical model of eight principal categories of learning outcomes. This model combined the behaviourists' model of basic S-R patterns with some higher-order categories, drawn from cognitive psychology. Starting at the simplest level, with 'signal learning' (as is the case of Pavlov's dogs learning that a bell was the signal for food and thus salivating) and single stimulus-response 'association learning' (as is the case of observing that the pulling of a lever delivers food), more complex behaviour, such as the above-mentioned chains, discriminations and generalizations (concepts) are built up. Up to this point the model is strictly behaviourist, combining eclectically the type of learning identified and described by Pavlov, Thorndike, Skinner, and their later disciples. From this point on, the model adopts a somewhat cognitive approach: certain concepts are strung together, by means of a 'language' to form meaningful rules or principles capable of explaining observed phenomena or predicting the outcomes of certain events. Further up the hierarchy, certain principles may be applied to solve certain types of problems. This activity leads to the discovery of more general, higher-order principles.
Diagram	How the eight learning categories described by Gagné relate to each other higher categories require prior mastery of relevant lower ones) 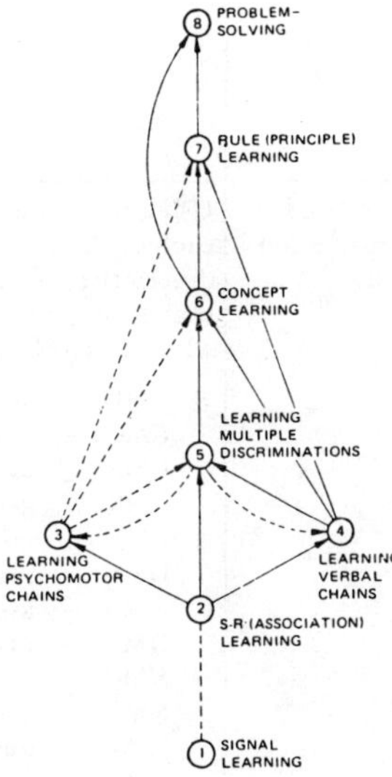
Further reading	For a detailed account of this model see: Briggs, L J (1970); Gagné, R M (1965); Romiszowski, A J (1981) (Chapter 9).

Map 2.7 *Guilford's model of the structure of the human intellect*

A complex three-dimensional model of the human intellect	At about the same time that R M Gagné and others were extending earlier behaviourist models to include categories that could comfortably accommodate higher-order conceptual learning and problem-solving skills, other psychologists were developing models that had little in common with the behaviourist tradition. One such was J P Guilford's three-dimensional model of the 'structure of intellect'. The three dimensions are: 1. Operations — intellectual activities, or general skills that are brought to bear on a task. 2. Products — the outcome of the intellectual activities or what types of new knowledge are formed. 3. Content — the 'language' in which the task situation presents itself. The sub-categories of each of these dimensions and the resultant cubical model of 120 different 'structure of intellect factors' are shown below.
Diagram	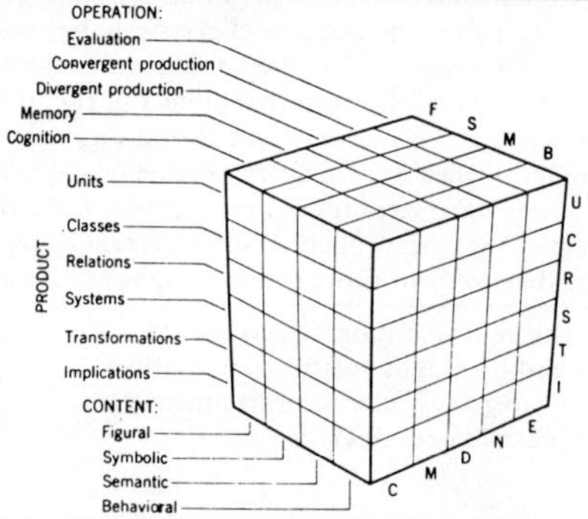
Example of interpretation and use	Using the initial letters of the sub-categories of each dimension, one can give each of the 120 factors a unique three-letter label. They can also be given a more descriptive, functional, name. For example, Guilford and his collaborators found that 12 of the factors were particularly important in intellectual problem-solving. They called these (taking the letters in order operation-content-product): CMU — verbal comprehension; CMC — conceptual classification; CMR — awareness of conceptual relations; CMS — general reasoning; CMI — conceptual foresight; DMU — ideational fluency; DMR — associational fluency; DMT — originality and flexibility; NMC — production of categories of meaning; NMR — production of conceptual relations; NMI — deduction; EMI — sensitivity to problems. Thus, problem-solving is a complex activity based on the use of a series of intellectual capabilities, all of which may, presumably, be developed through appropriate training.
Further reading	For a detailed account of this model, see: Guilford, J P (1967).

Map 2.8 *Models based on an 'information-processing' view of learning*

Introduction	We present here a number of models, or perhaps partial models, for the classification of learning processes, which have been contributed by writers who have taken an information-processing view of the learning process. These partial models tend to deal with one or another aspect of learning – retention, problem-solving, learning style, etc.
Memory and retention	Many writers have proposed models of the mechanism of perception, memorization and long-term retention of information by the human learner. Most of these models (eg those of Atkinson, Schiffirn and Schneider) postulate three different memory systems:

<table>
<tr><td>Short-term sensory storage</td><td>Initial registration of perceived information. A large-capacity mechanism, with a very short holding time – a fraction of a second.</td></tr>
<tr><td>Short-term memory</td><td>Stores selected items of perceived information for a few seconds, but has a very small capacity.</td></tr>
<tr><td>Long-term memory</td><td>Items which relate meaningfully to other knowledge, may be stored in a structured form for indefinite periods.</td></tr>
</table>

	These models emphasize the importance of the structuring of information. George Miller suggests that the human short-term memory can handle only about 5 to 9 items, or 'chunks' of information at a time (the 'magic number 7 ± 2'). However, the chunk may be large or small in terms of information content. The structuring of the perceived information into meaningful chunks therefore increases the total amount of information that may be held in short-term memory. Further structuring must occur in the learner's mind, in order to relate the new information to previous learning, then permitting its long-term retention.
Problem-solving processes	Other writers have concentrated on the classification of the different types of procedures or mental operations that are performed by the learner in processing a body of information. Landa, for example, suggests a classification of thought processes into:

<table>
<tr><td>Algorithmic</td><td>The execution of a fixed sequence of specific operations will always lead to a correct solution of a given type of problem.</td></tr>
<tr><td>Heuristic</td><td>Although they may use the same basic principles as a starting point, different people will plan different procedures for each problem and may solve some, but not necessarily all problems of a given type.</td></tr>
</table>

	Landa showed that some categories of problems (eg planning grammatically correct sentences) were almost entirely soluble by algorithmic procedures, while other categories (eg the resolution of geometry problems) always required heuristic procedures.
Learning style	Yet other writers have focused on the general strategies of learning that lead to success. Gordon Pask, for example, has identified two characteristic types of learning strategy.

<table>
<tr><td>Serialist</td><td>The tendency to follow a linear path of step-by-step deduction, or development of a topic.</td></tr>
<tr><td>Wholist</td><td>The tendency to 'jump ahead' in an attempt to get the 'whole picture' of the topic, returning to study details only as and when it proves necessary for comprehension.</td></tr>
</table>

Map 2.8 *(continued)*

	Pask found that individual learners tended to have strong preference for one or other style and that their learning success was much superior when the information to be learnt was presented in a form that facilitated the preferred learning style.
Commentary	These three sample partial models all stress the importance of the organization of the information to be learnt — an aspect which has tended to be neglected by many of the other models mentioned here.
Further reading	For a more detailed account of these models, see: Atkinson, R C and Schiffirn, R M (1968); Landa, L N (1976); Pask, G and Scott, B C E (1972); Schiffirn, R M and Schneider, W (1977).

Map 2.9 *The 'Mager and Beach' and 'CRAMP' models*

Introduction	For various reasons, the more complex models discussed so far have very often been considered impractical by instructional designers. Nevertheless, the same designers have felt a need for some scheme of classification of learning outcomes, that could be used as a tool for decision-making in respect of teaching strategies and testing. Two such simple 'practitioners' models' developed specifically for vocational and training applications are the 'Mager and Beach' and 'CRAMP' models.

1. Mager and Beach	Category	Description
	Discrimination	Knowing when to do something.
	Recall	Knowing what to do and why to do it.
	Problem-solving	How to decide what should be done.
	Manipulation	How to do what should be done.
	Speech	How to say or explain something.

2. CRAMP	Category		Description
	C	Comprehension	The understanding of a body of subject matter.
	R	Reflex skills	The acquisition of skilled movements.
	A	Attitude	The changing of feelings or reaction to people or situations.
	M	Memorization	The learning and recall of specific facts and definitions.
	P	Procedures	The learning of what to do in a given situation.

Commentary	These two models, both developed by highly experienced training design teams (one in the USA and the other one in the UK), have some similarities, but are most notable for their differences. They do not cover exactly the same set of possible varieties of learning. Yet they were developed as special-purpose design tools for the same educational sector — vocational training.
Further reading	For a more detailed account of these models, see: Industrial Training Research Unit, University of Cambridge, UK (1975); Mager, R F and Beach, K H (1967). For a critique of these models, see Romiszowski, A J (1981) (Chapter 10).

Map 2.10 *Gagné's (and Briggs') later (1974) model of learning categories*

The eclectic model of Gagné and Briggs — description	About a decade after his development of the eight-category hierarchy of learning, Robert Gagné, in conjunction with Leslie Briggs, presented a much-modified and extended model of learning outcomes. This model proposes five principal kinds of learned 'capabilities': Intellectual Skills; Cognitive Strategies; Verbal Information; Motor Skills; Attitudes. The earlier hierarchical model is now considered as a subdivision of the main category 'intellectual skills', although this too has undergone some major revisions. The table represented below (from Gagné and Briggs, 1974), presents examples of each of the categories and sub-categories of the new model.

Capability	Verb	Example
Intellectual Skill Discrimination	Discriminates	Discriminates, by matching, the French sounds of 'u' and 'ou'.
Concrete concept	Identifies	Identifies, by naming, the root, leaf, and stem of representative plants.
Defined concept	Classifies	Classifies, by using a definition, the concept 'family'.
Rule	Demonstrates	Demonstrates, by solving verbally stated examples, the addition of positive and negative numbers.
Higher-order Rule (Problem solving)	Generates	Generates, by synthesizing applicable rules, a paragraph describing a person's actions in a situation of fear.
Cognitive Strategy	Originates	Originates a solution to the reduction of air pollution, by applying model of gaseous diffusion.
Information	States	States orally the major issues in the presidential campaign of 1932.
Motor Skill	Executes	Executes backing a car into driveway.
Attitude	Chooses	Chooses playing golf as a leisure activity.

Commentary	We observe that the new model attempts to be even more eclectic than the earlier one, embodying the (previously largely missing) affective and psychomotor domains of learning, attempting to distinguish between the acquisition of facts (verbal information) and concepts. Also, a distinction is made between concrete and defined concepts. Finally, an attempt is made to distinguish between relatively simple problem-solving that involves the combination of two or more known principles in order to generate a new higher-order principle, and the more complex, creative cognitive strategies (which other writers, for example Guilford, would consider as a form of problem-solving, indeed the most valuable form which involves not only the creative solution of a novel problem, but also its formulation). Whereas this model attempts to be much more complete in its coverage than most others mentioned so far, it falls short of being complete and also contains several apparent inconsistencies.
Further reading	For a detailed presentation of the model, see Gagné, R M and Briggs, L J (1974). For a critical analysis of this model, see Romiszowski, A J (1981) (Chapter 11).

Map 2.11 *A new model of learning categories:*
a simple model of human performance

Introduction	The model presented here conceives of the performer as a system interacting with an environment or problem situation (which may be an abstract problem posed by the performer himself). It is a model which should serve, therefore, for all types of performance — physical, intellectual, emotional, etc.
Description	The performer is conceived as composed of four principal sub-systems: *Receptors*, with which a problem situation or incoming information is perceived and selected for interpretation or action (usually, this involves the senses, especially sight, but the concept also includes internal, reflective-introspection, capabilities that enable a person to 'see' an abstract problem posed by himself or some other person with whom he is conversing); *Memory*, or *store*, in which knowledge gained from prior learning is kept in an organized manner (this is the 'long-term memory' from which the performer recalls necessary information in order to make sense of new incoming information and where he stores new information that appears to be of relevance or of further use); *Processor*, or *brain*, or *intellect*, with which the performer analyses incoming information, structures new relationships between the new and prior information, sets up hypotheses, tests and evaluates them and so on — in short, where the thinking goes on; *Effectors*, with which the performer realizes the actions that communicate the results of the memory-recall and information-processing that has gone on. In practical performance tasks the effectors are the limbs, in intellectual tasks they are the voice and the writing or typing activities that communicate one's thoughts to others. However, once again, we extend our concept to include the mechanisms of introspection and self-analysis that communicate to the performer the results of his own intellectual activity — speaking to oneself, if you like.
Diagram	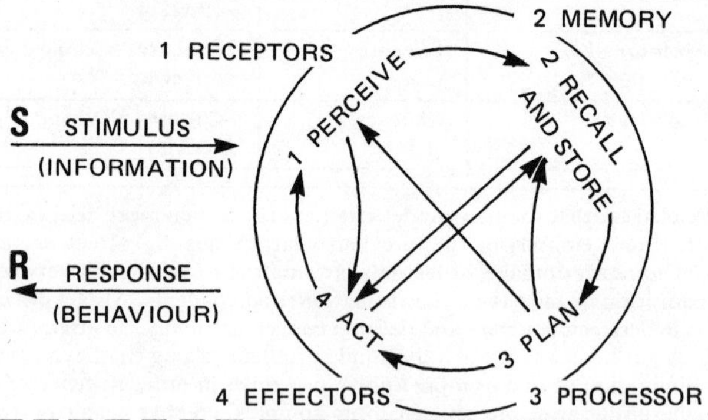

Map 2.11 *(continued)*

Operation	The diagram shows a number of arrows and loops, which describe some of the most common modes of operation/performance.	
	S-1-4-R	The performer perceives a stimulus situation which triggers off an automatic, reflex action that does not involve any conscious use of recall or decision-making capabilities. *Example:* Well-learnt, proficient, typewriting performance.
	S-1-2-4-R	Once the stimulus situation is perceived, a previously learnt, standard procedure (algorithm) is recalled and is applied, step by step — no creative planning or novel decision making is involved. *Example:* Performing long division.
	S-1-2-3-4-R	Once the stimulus situation is perceived and no ready-made algorithm is recalled from store, the performer seeks other stored knowledge (general principles), with which to plan a strategy of action that will be appropriate and then executes the plan. *Example:* Geometry problem solving.
	S-1-2-3-2 S-1-2-3-4-2 3-4-1-2 3-1-2	All these loops are to do with the structuring and storing of new knowledge. There is no externally observable response but the internal process (of planning new relationships between items of information or of executing newly planned procedures) is perceived as of future use and therefore stored in long-term memory. This process may happen in many ways, involving newly perceived information and previously learnt knowledge, in combination or individually, occurring accidentally or planned.
	4-1-4 4-R-S-1-4	The process of perceiving the probable, or actual, results of one's actions and taking corrective measures — self-control.

Map 2.12 *Basic abilities which influence performance*

Introduction	The four principal activities involved in human performance — perception of information, storage and recall of information, processing information to plan and take decisions and acting on the decisions — require certain abilities on the part of the performer. The 'expanded' cycle of skilled performance, shown below, identifies 12 important abilities.
Diagram	
Significance	The presence or absence of one or more of these basic abilities may have a very significant influence on the performer's competence in executing a particular task. Not all the abilities are equally important in a given task. It is necessary to analyse each specific case to discover which abilities are important and which are insufficiently developed and therefore inhibit the satisfactory performance of the task. Most of the abilities may, to a certain extent at least, be developed through appropriate training exercises.
Example of a skills analysis using the cycle	The 'expanded' skill cycle of 12 basic abilities may be used to analyse the probable causes of under-performance of a task, by (1) analysing the task to identify the critical abilities and (2) analysing the performer to identify the existing levels of the abilities previously identified as critical. The table below presents an example of stage (1) of this analysis.

1. Attention	Ability to concentrate on arithmetic tasks.
2. Perceptual acuity	Ability to recognize numbers and division signs, whether presented symbolically or verbally.
3. 'Noisy environment' *(discrimination)*	Ability to recognize the need for division when this is implied indirectly (eg 'Joe gets a third of the money').
4. Ability to interpret	Ability to analyse the problem to identify what is divided by what (eg $3 \div 9$ or $9 \div 3$).
5. Relevant knowledge	Ability to recall the procedure (the algorithm) for long division, in all its steps. Also one's 'tables' (facts).
Further **6.** — schemata, or **7.** — analysis, or **8.** — synthesis, or **9.** — evaluation	Not involved, as long division is a reproductive skill. One simply recalls the algorithm and applies it step by step to the data.
10. Initiation	Merely requires the physical abilities of writing figures.
11. Continuation	May require some degree of perseverance in the case of large numbers.
12. Control	Ability to automatically check the 'scale' of the answer.

Further reading and examples	Several further examples of this form of analysis and a more detailed description of the technique may be found in: Romiszowski, A J (1984) (Chapter 9).

Map 2.13 *The 'inner self' that drives the performer*

Introduction	Our model of human performance is like an onion — we peeled one layer off our 'black box' model, to reveal four principal sub-systems (receptors, memory, processor, effectors). We then peeled off another layer, to reveal a short list of 12 basic abilities (or sub-skills) that may influence the level of skilled performance of any task. However, the existence of the required abilities at the requisite level of development does not, in itself, guarantee that skilled performance will occur. The four sub-systems must be 'switched on' and 'tuned in'. If we peel off yet another layer, we come to the 'inner self' — the prime mover that powers up and runs the sub-systems we have discussed so far.
Description	The 'inner self' is a handy term for all those individual factors that drive, motivate, or 'turn on' a person to engage in a given form of activity — personality factors, affect, experiences and beliefs, physiological factors such as left/right brain differences and so on. Other authors may refer to some of these by different terminologies — ego, id, etc. Yet others postulate various kinds of basic drives.
Diagram	
Significance	The complex array of factors which we have conveniently lumped together here is of extreme importance, in that it is the driving force which, in the ultimate analysis, controls whether learning occurs efficiently and tasks are performed to satisfactory standards. Many psychologists have devoted their lives to the study of just some of the factors involved. Many techniques exist which claim to modify or influence one or other of these factors. They include therapy, group dynamics, behaviour modification, direct chemical or other intervention, persuasion, leadership, etc.
The role of instruction	The role of instruction, as a means of influencing the 'inner self', is quite restricted. Certain simulation or role-play exercises may have profound effects on personality. Some textual and mediated communications have profound emotive effects, through the imagery and storytelling skills embodied in them. Some teaching techniques, such as 'Suggestopaedia', make direct efforts to involve the 'inner self' in the instructional process. However, the bulk of instructional systems and mediated instructional materials aim to develop specific knowledge or skills. Due to space limitations and other reasons, we shall restrict ourselves in this book to these aspects, leaving the analysis of the 'inner self' for another occasion.

Map 2.14 *Knowledge, skill and attitude — basic distinctions*

Introduction	It is very common to talk, colloquially, of instruction as the development of knowledge, skills and attitudes. The three-domain taxonomy of Bloom, Krathwohl and others also seems to refer to these three categories. However, earlier discussion has shown that this is not so: the cognitive domain, as defined by Bloom, includes both knowledge-objectives (knowledge, comprehension) and skill-objectives (application, analysis, etc). We shall try to maintain a clear distinction between knowledge and skill in our model. Attitude will be treated in an unusual manner, as a composite of personal skills (self-control) and feelings/ emotions (part of the 'inner self').

Knowledge and skill: definitions

Knowledge	Skill
Information acquired and stored, in an organized manner, in the mind.	The capacity to do something (perform) with a given degree of effectiveness and efficiency.
1. Knowledge of the steps to be followed in order to divide two numbers.	The ability to divide numbers with the necessary degree of speed and precision.
2. Knowledge of the Laws of Ohm and Kirkhoff.	The ability to solve electric circuit problems correctly by applying the laws.
3. Knowledge of basic principles of human relations that apply to supervision.	The ability to resolve conflicts between staff and supervision, tactfully, justly, fast, etc.
4. Knowledge of the general principles of scriptwriting, etc.	The ability to write a script for a film that applies the principles well and is 'great'.

Attitudes: the two meanings

The term 'attitudes', when used colloquially, is often ambiguous, having at least two possible interpretations:

1. How a person 'really feels' in relation to some type of situation, thing or person. In this sense, attitudes are part of the 'inner self' — feelings, affect, personality, beliefs.
2. How a person 'actually behaves' in relation to some type of situation, thing or person. In this sense, attitudes are the observed outcomes of the personal skills of self-control, when applied to the control of feelings, emotions, personality traits.

Examples

Attitudes as feelings/emotions	Attitudes as self-control skills
A teacher of music sets out to develop a love of classical music in his students.	A teacher sets out to develop a more positive attitude towards doing homework on time.
A teacher sets up a simulation game, designed to reduce the levels of racial prejudice in a group of students.	A trainer sets up a role-play exercise, designed to give practice in the avoidance of racial discrimination in selection/recruitment.
A supervisor makes every effort to ensure that new staff are happy in their work.	A supervisor makes every effort to ensure that the sales staff always show an attitude of respect to customers.

Map 2.14 *(continued)*

Observations	The examples of knowledge and skill illustrate that skill usually is dependent on the presence, in the performer's memory, of certain necessary knowledge. However, knowledge may be acquired, without the necessary development of an associated skill in its use. Whereas knowledge is a scalar quantity (one either has, or has not, acquired particular 'bits' of information), skill is a vector quantity in that one may develop a given skill to varying degrees in varying types of problem situations. The examples of the two meanings of 'attitudes' illustrate that, in general, instruction is usually an appropriate approach for the development of self-control skills, but seldom appropriate for the changing of feelings, emotions, beliefs or personality. Also, whereas skill-development is part of the job of every teacher, personality-changing may be outside the competency of some teachers and even morally untenable in many adult training situations.

Map 2.15 *A classification schema for skilled performance*

Introduction	Earlier in the chapter we summarized some models for the classification of different categories of learning. The outcomes of the learning process are specific capabilities to do something new — to perform in specific ways to specific standards. The objectives of instruction are therefore generally specified in performance terms. We may therefore develop a schema for the classification of learning outcomes, or objectives, in terms of categories of performance.
Dimension 1: the 'domains' of performance	Bloom, Krathwohl and others have postulated three domains of learning: cognitive, affective and psychomotor. Earlier discussion has shown that social and interpersonal skills are difficult to accommodate in this schema. We prefer a four-domain classification as follows:

Domain of performance	Description/examples
1. Intellectual, or cognitive performance.	Thinking or reasoning. The skills of 'using one's head'.
2. Physical, or psychomotor performance.	Sports and practical tasks. The skills of 'using one's body'.
3. Reactive performance.	Personal behaviour. The skills of 'controlling oneself'.
4. Interactive or interpersonal performance.	Interacting with others, leading, selling, persuading. The skills of 'controlling others'.

Dimension 2: the 'reproductive/ productive' scale	Our model of human performance, and the various loops embedded in it, suggests that skilled performance varies in terms of the degree of creativity or productive thinking which is involved in the execution. There is a scale on which skills may be placed, which goes from the totally automated, reflex, or 'reproductive' type of performance, to the highly creative or 'productive' type. We may define the two extremes of this scale as follows:

Reproductive skills ⟷ Productive skills	
Executed in a totally automated manner, involving no analytical thinking, planning or evaluation of alternatives. Depend on the application of previously learned algorithms.	Executed in a non-automated manner, involving the analysis and evaluation of alternatives and planning of appropriate strategies of problem-solving. Depend on the application of concepts and general principles.

Map 2.15 *(continued)*

A two-dimensional classification schema for skilled performance	Combining the two dimensions defined above, it is possible to construct a matrix on which any type of skilled performance may be classified.	

	Reproductive skills	**Productive skills**
Cognitive or intellectual skills	Division, multiplication, reading a circuit diagram.	Solving geometry problems. Designing a novel circuit.
Psychomotor or physical skills	Typing, writing by hand, kicking a ball straight, running.	Playing piano artistically. Playing football strategically.
Reactive or personal skills	Personal habits, punctuality, hygiene, truthfulness, etc.	Personal lifestyle – living in accordance with a set of values or beliefs.
Interactive or interpersonal skills	Interpersonal habits, good manners, tone of voice, listening skills, etc.	Leadership, persuasion, selling, supervision, negotiation, interviewing, teaching and similar complex/creative skills.

Observation	The reproductive-productive scale is a continuum. The examples shown above could in fact be spread out by greater or lesser degrees to left or right, to illustrate relative differences in the degree of creative planning involved in their execution. The schema is intended to act as a conceptual model of the variety of skilled performance types, not as a set of 'pigeonholes'. Some complex skills may lie partly in two or more of the cells of the matrix (as some of the examples above suggest).

Map 2.16 *An approach to the classification of knowledge and information*

Introduction	As we have defined knowledge as 'information acquired and stored in an organized manner in memory', it follows that any schema for the classification of knowledge categories should serve equally well for the classification of information categories. A common distinction is made between factual information or knowledge and conceptual information/knowledge. The former 'stands on its own' as it were, whilst the latter must be 'understood'/'exemplified'.

Primary distinction: factual and conceptual knowledge	Factual knowledge/information	Conceptual knowledge/information
	Description: Information that stands on its own, without the need of examples to make it meaningful. Names, events, places, fixed procedures, which are applicable in unique ways or to specific tasks.	Description: Information that is of general applicability in a series of example situations. Concepts, general principles, theories or strategies that can explain, or deal with, various problems of some given category.
Examples	The symbols used in electric circuit diagrams to identify batteries, resistors, etc.	The concepts (the meaning of the terms) 'potential difference', 'resistance', etc.
	The procedures (algorithms) for division, multiplication and calculation of square roots, etc.	The concepts of number, the laws (principles) of association and distribution, etc — which explain why the procedures work.
	The procedures used to find a defect in a given machine or electronic equipment — the set of steps and measurements that are given to a worker to follow, when trouble-shooting.	The concepts of science and the laws (eg Ohm's Law) that explain how/why the machine/equipment works, the principles of trouble-shooting that lead to the design of an efficient search procedure.

Secondary distinction: 'WHAT' or 'HOW'	Another way to classify information is in terms of the kind of question that it answers: WHAT is this?; HOW is this done?; or HOW can this be explained?	
		Factual	**Conceptual**
	WHAT	*Specific facts:* Names of people, things, etc; events that happened, symbols, conventions, etc.	*Specific concepts:* Definitions of terms, ideas, viewpoints, etc; classifications, etc.
	HOW	*Procedures (algorithms):* Set of fixed steps by means of which a given task is to be executed, or a given goal achieved.	*Principles (rules/laws):* Relationships between ideas or viewpoints or observed phenomena, which explain how or why things are, how one should act, etc.
Examples		**Facts**	**Concepts**
	WHAT	Foreign vocabulary (the Latin words for chair, table, war, justice).	The concept of 'chair' — what it is, what it does, the difference between a chair and a stool, etc.
		Descriptive account of Pizarro's overthrow of the Inca empire.	The meanings of 'war', 'justice', 'religion', 'avarice', etc, that enable one to understand the account of Pizarro's conquest.

Map 2.16 *(continued)*

		Procedures	Principles
	HOW	The procedures to be followed in constructing grammatically correct sentences in Latin.	The general rules of composition which enable one to judge the quality of a text, or to write creatively/persuasively, etc.
		The algorithm for the calculation of an employee's income tax deduction.	The theorems of geometry, which are applied to the solution of a range of varied and novel problems.
		The procedures used for hoisting the sails on a yacht, steering, reading the compass, sextant, etc.	The principles of sailing and of navigation, which enable a sailor to get the yacht to a certain desination, with safety, speed, etc.

Map 2.17 *More detailed subdivisions of the knowledge categories*

Introduction	Many writers have classified learning categories in greater detail. Some of these more detailed classifications may be useful to identify specific instructional tactics that should be used for specific objectives in a lesson. We may subdivide our four basic categories of knowledge in many ways. This map presents some subdivisions that have proved to be useful.
Facts: concrete and symbolic (defined)	Factual knowledge may be gained by direct observation of one's environment. Psychologists refer to this kind of learning as the forming of 'concrete associations' — associating observable things, events, etc, with particular places, people, names, etc. A great deal of factual information, however, is gained through some form of 'symbolic communication' between people (this is often mediated by means of print, etc). Gagné and Briggs refer to this category as 'Verbal Information'. However, one should include other symbolic (non-verbal) 'languages' of communication in the category — mathematical symbols, road signs, etc. Communication and learning only occur if the 'language' is known.
Concepts: concrete and defined	Many psychologists find it useful to distinguish between concrete and defined concepts. The first may be learnt directly, through experience/observation, while the second require definitions and therefore imply that some form of 'language' is used in their communication. *Concrete concepts:* the colours 'red', 'blue', etc; ideas such as 'bigger', 'higher', 'near', 'far', 'heavy', 'light', etc. *Defined concepts:* the general ideas of 'colour', 'size', 'position' require the use of simpler (concrete) concepts as examples to illustrate their meaning and some form of communication that establishes the 'example-general idea' relationship; the general idea of the 'physical properties' of an object takes one to an even higher level of abstraction, using previously defined concepts (colour, shape, size, weight, etc) as examples of the more general concept.
Procedures: sequential and decision-making	Except for the most simple of procedures, one can identify two component types of information. 1. Sequences of steps to be performed in a preset order. We might call these linear sequential procedures. The execution of such procedures is seen as a chain of actions — behaviourally oriented psychologists would call them 'chains' (see Gilbert's mathetics). Note that it is possible to distinguish between the knowledge of the steps (or the information that describes them) and the execution of the steps in a competent manner (which is a skill). 2. Decisions between alternative sequences of action. Once more, it is important to distinguish the taking of the decision (a skilled action) from the information which must be known in order to be able to take a decision, at the appropriate moment, between valid alternatives.

Map 2.17 *(continued)*

Principles: of nature and of action	General principles may be relationships between ideas or phenomena in our environment, which can be observed to hold good over a specific range of problem situations. They help us to explain observed phenomena and predict the outcomes of them. The laws of physics and of other sciences are good examples. We shall refer to them here as laws, or 'principles of nature'.
	It is useful to distinguish another type of principles— those which guide our actions, in a general way, when we attempt to make sense of something or plan what to do in a given problem situation. They are 'how to proceed' guidelines, but differ from algorithmic procedures in their lack of a fixed sequence of action or guarantee of a solution if followed. Many writers refer to these types of principle as 'heuristics'. We can also refer to them as 'principles of action'. Whereas principles of nature explain how a given system works, principles of action suggest how we should approach the study of, or interaction with, the system. The way that one should approach the solution of certain types of problem to maximize the chances of successful solution, would be principle of action. Such principles as 'try to separate the known data from the unknown', 'compare the problem with some similar previously solved one and identify the similarities and differences', are examples of heuristic principles of action useful in mathematical and some other types of problem-solving activities.

Map 2.18 *The structuring of knowledge in the mind*

Introduction	Both short- and long-term memory require that information to be learnt be as highly structured as possible. George Miller showed up the incredibly small capacity of short-term memory by a series of experiments which demonstrated that most people can only hold about seven different items of information in short-term memory at one and the same time. However, the size, or information content of these items could be quite small or quite large. A person treating one-digit numbers as separate entities, can deal with about seven of them at one time, but the same person, after applying some form of technique for 'chunking' the digits in groups, can deal just as easily with seven groups ('chunks') of digits. By structuring telephone numbers in groups of three and four digits, one can remember seven complete telephone numbers about as easily as one could previously remember the seven isolated digits of just one telephone number. All manner of tricks and mnemonic devices have been developed to help learners to memorize factual information by 'chunking' it into larger units that are in some way meaningful. In the case of conceptual information, larger units of knowledge are formed by the derivation of more general concepts and principles from simpler ones. Thus, a whole range of basic concepts and relationships that deal with bodies in motion, may be better remembered, once they are interrelated in the form of more general 'principles of conservation of energy'.
Fact-structures ('chunks')	Concrete associations and verbal/symbolic information are generally stored in memory in close relation to other similar or related information. One item brings to mind another. Thus, the memory of one line of a poem serves to aid the recall of the whole poem. A sketchy drawing of a nose-shaped object and a couple of curves are all that is necessary to convey not only that the drawing is representing a face, but calls up the memory of a specific face of someone we know, or who appears on the media. Given a few key facts, we can complete the message, due to the way that prior memories are structured.

Map 2.18 *(continued)*

Concept- structures ('schemata')	Both concrete and defined concepts convey meaning — some idea of a generally applicable nature. Single ideas may be combined into more abstract, more powerful and often more memorable ideas. The structuring of concepts in the mind into new and different relationships is a continuous process, which aids memory and recall. Any new concept, to be retained, must be meaningfully related to the previously stored conceptual knowledge — it must 'hang' comfortably on the existing structure of knowledge, or the existing structure must adapt to accept (rather than ignore/reject) the new knowledge. Piaget refers to these two interrelated processes as 'accommodation' and 'assimilation': new concepts are assimilated to form part of existing conceptual structures, or else the structures change in order to accommodate the new. Thus, conceptual structures, or as they are often called, schemata, undergo continual change and growth. Each person has a unique set of schemata, or ways of organizing the conceptual information in his/her memory.
Principle- structures (theories and strategies)	General principles are also interrelated in the mind, to form more structured, more powerful ways of explaining observed phenomena or taking complex decisions. A set of interrelated 'rules of nature' form a general theory. For example, the theory of conservation of energy embodies a set of interrelated simple principles that may be formulated in a set of equations of motion, work, etc. A set of interrelated 'rules of action' form a complex decision-making, or problem-solving strategy, like for example, the Systems Approach or the Scientific Method of Enquiry. Of course, one also forms specific problem-solving strategies that are mixtures of the two types of rules.
Procedure- structures (structured algorithms)	Simpler procedures (like the procedures for addition) become the steps in more complex, multi-level, procedures. One remembers the complex procedure by 'chunking' it into sub-procedures, and the structuring enables very long procedures to be stored as one step. The importance of this structuring procedure is seen in computer programming. The languages that allow 'structured programming' are more powerful and easier to use creatively. The principal source of the claim that LOGO helps children to learn to think, is that its use prompts the hierarchical structuring of procedures into ever more complex 'chunks' that can then be used as building blocks in the design of more powerful procedures.

Map 2.19 *The structure of knowledge — a summary*

Introduction	We may bring together the various categories of knowledge in the form of a hierarchical tree, in order to illustrate the structure of the classification schema which we are developing. We may also use this diagram to illustrate that in reality one often stores knowledge as mixtures of different categories. This forms interrelations between different types of the knowledge at our disposal, creating complex cognitive structures.
Illustration (A) The 'formal' model of knowledge categories (B) The way we conceive that knowledge is structured in memory (C) Some specific types of structures useful in problem-solving	
Specific procedure definition	A specific procedure has only one application. It does not apply to a range of example situations. Its execution depends on the identification of specific factual information which triggers off the steps of the procedure in the correct sequence.
Example	The procedure to follow to get from A to B in a town.
General procedure definition	A general procedure applies to a range of example situations. It may be a fixed algorithmic procedure but the data on which it operates may change. Such procedures involve the use of some concepts at certain decision points. Some are semi-algorithmic/semi-heuristic.
Example	The procedure of long division.
Theories definition	Theories are complex structures of interrelated ideas, mainly concepts and principles, which attempt to make sense of a whole range of interrelated phenomena.
Examples	The theory of relativity. The theory of behaviourism.
Heuristic strategies	Heuristics are 'rules of action' of a general nature. Heuristic strategies are interrelated sets of such rules, which guide one's approach to a whole range of problem situations.
Example	The Systems Approach to problem solving in complex systems.

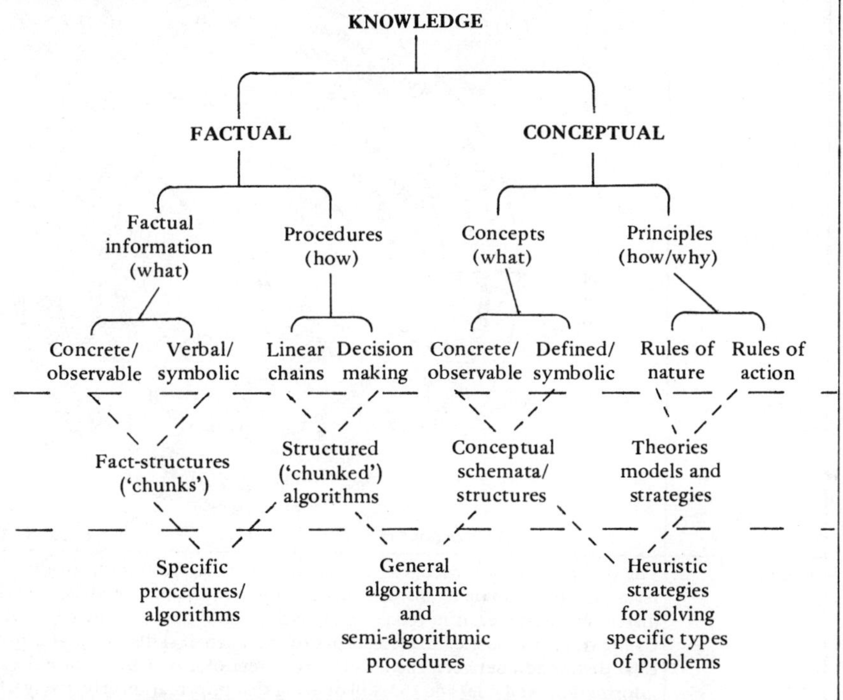

The diagram in the Illustration row shows:

KNOWLEDGE
branching into FACTUAL and CONCEPTUAL.

FACTUAL branches into:
- Factual information (what)
- Procedures (how)

CONCEPTUAL branches into:
- Concepts (what)
- Principles (how/why)

Sub-branches:
- Factual information (what): Concrete/observable, Verbal/symbolic
- Procedures (how): Linear chains, Decision making
- Concepts (what): Concrete/observable, Defined/symbolic
- Principles (how/why): Rules of nature, Rules of action

(B) The way we conceive that knowledge is structured in memory:
- Fact-structures ('chunks')
- Structured ('chunked') algorithms
- Conceptual schemata/structures
- Theories models and strategies

(C) Some specific types of structures useful in problem-solving:
- Specific procedures/algorithms
- General algorithmic and semi-algorithmic procedures
- Heuristic strategies for solving specific types of problems

Map 2.20 *A schema of knowledge and its structure in the mind*

Introduction	To wind up this chapter, we bring together some of the ideas regarding the structure of knowledge, in the form of a circular diagram, of several 'layers'. In the centre, we have the complex and personal schemata that all persons form in order to store knowledge in memory. Although we may not know the structure of an individual person's cognitive schemata, we may postulate, in general, that it is composed of different types of knowledge. These are shown in the outer layer. In between, we indicate the structured manner in which individual items of knowledge are stored.
Diagram	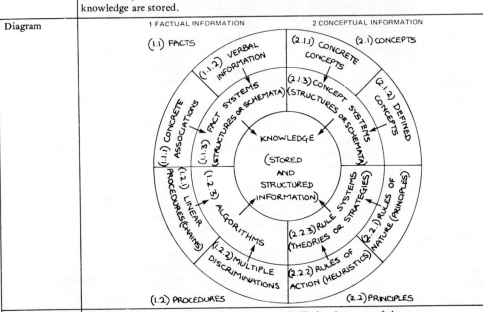

Comments	1.	The outer ring of knowledge-types has great similarity, in terms of the terminology, to some of the categories defined in Gagné's models. One significant difference, however, is in referring to all these as types of knowledge, whereas Gagne refers to most of them as types of intellectual skills. We prefer to keep a clear distinction between the knowledge aspect of, say, a procedure (having the information available) and the skill of using that information effectively/efficiently. One may have the knowledge without the need to develop a skill (as in the case of a supervisor in a factory). One may develop a skill in the execution of a procedure without mastery of the knowledge (as in the case of using a job-aid as a guide). One should be clear as to which aspect (or both) is of importance to the learner.
	2.	The use of the knowledge and skill schemata will be further illustrated in the following chapters. The rationale for this model and its development may be further studied in: Romiszowski, A J (1981) (Chapters 11 and 12); Romiszowski, A J (1984) (Chapter 4).
	3.	The circular nature of the diagram presented here is chosen to emphasize that knowledge is in fact stored in an integrated way, not as separate categories in separate boxes. Nor is it necessarily organized in a hierarchical manner (though it may be sometimes).
	4.	Finally, remember that the schema is a personal conceptual tool of its author. Therefore, it is subject to change. And your schema may be somewhat different. What is important is that it works — is useful as a tool for thinking and decision-making.

3. Four Levels of Instructional Design: The Practical Context

3.1 Introduction

In this chapter we review the principal stages and procedures of the instructional design and development process. As was the case in the previous chapter, our intention is to furnish an overall context for the specific theme of this book. The development of instructional materials is, after all, the final phase of a much more extensive instructional design and development process. Even if the materials developer was not personally responsible for, or even involved in, the initial planning of a training project or new curriculum, it is a good idea to be aware of the techniques and procedures used in such planning.

The later stages of instructional development and lesson planning are of even more relevance, as it is at these stages that someone, somehow, took the decision to use a specific type of self-instructional package in order to achieve certain specific objectives and may even have gone so far as to develop an outline of the package, in the form of a detailed lesson plan. If this 'someone' was not the person who will develop and produce the final materials, it is important that effective communication takes place between the two. Thus, the materials developer must understand the earlier stages of the instructional design and development process.

These earlier stages have been dealt with in some depth in earlier volumes of this series and it is not our intention to repeat all this earlier discussion here. It is expected that the reader already has some idea of the instructional design process as a whole. It may be, however, that the reader has not read the earlier volumes referred to above. The summary presented here should be sufficient to give a general view of the approach that was suggested. In case the reader wishes to refer to more extensive treatments of certain topics, we have given topic-by-topic references to books and chapters.

The material presented here has been organized in the form of single-topic 'maps' in order to facilitate the use of the chapter as a review and reference source.

Map 3.1 *Systems approach*

Systems approach - definition	An overall approach which involves tackling problems in a disciplined manner keeping priorities in mind. The sub-system making up the overall system can be designed, fitted, checked and operated so as to achieve the overall objective efficiently (Rowntree, 1974).
Properties of the systems approach	Inputs, outputs and process are defined in relation to each other. A change in one part will affect all other parts. Each decision is justified in terms of pre-planned objectives. Systems models are used which show how each phase fits into the next and feedback loops facilitate revision and review. Environmental restraints which impinge on the school or teaching centre are considered. Systematic consideration of the suitability of solutions to problems as compared to their alternatives is carried out.
Description	The systems approach is a problem-solving method which helps to: 1. Define the problem as clearly as possible. 2. Analyse the problem and identify alternative solutions. 3. Select from the alternatives and develop the most viable solution mix. 4. Implement and test the solution. 5. Evaluate the effectiveness and worth of the solution.
Comment	The systems approach is not necessarily a step-by-step process. Analysis, synthesis and evaluation are recurring stages repeated throughout the process and not necessarily in the traditional format of beginning, middle and end. Therefore heuristic problem-solving techniques are often better suited than algorithmic procedures. The heuristic process features the creative use of general principles rather than the employment of specific rules. Although it does not lead necessarily to a solution at all times, it does increase the possibility of arriving at a viable solution.
Example	Polya's approach to mathematical problem solving is an example of the systems approach in practice (see Polya (1945) *How To Solve It*). 1. You must first understand the problem. 2. You must find the connection between data and the unknown and obtain a plan of the solution. 3. Carry out the plan. 4. Examine the solution obtained.

Map 3.2 *Instruction*

Introduction	Generally we think of education as 'what goes on in schools' and training as 'what goes on in industry'. However, because education involves some level of training and training involves some form of education (unplanned learning) it is difficult to use these two terms discriminately. Hence, we as instructional designers shall define the concept of instruction a bit more precisely.
Definition	Instruction is a goal-directed teaching process which is more or less pre-planned.

Diagram			Specific objectives exist?	
			Yes	*No*
	Pre-planned study	*Yes*	Instruction	Visits to theatre, museum, study tours
	Resources exist?	*No*	Projects Apprenticeships	Incidental learning

Example One	A grade 11 maths teacher's objectives are to teach the basics of trigonometry using the lesson plans as described in the school board's chosen textbook *Introducing Trigonometry — A Step by Step Approach.*
Example Two	Train a new group of tellers in the bank's policy regarding hold-ups using the book and film titled *What to Do in a Hold-Up.*
Non-example	Formal education is not always as structured as instruction. There is usually some kind of goal; however, the route is not as well defined. Consider the examples listed to the right in the diagram above.
Comment	Whether it be education or training, both require 'instruction' — that is, an attempt to impart certain defined skills or knowledge — at times.
Analogy	Education provides a 'map' to the learner and allows him to wander through the 'field' at his own pace and in his own direction. The 'map user', however, requires *instruction* to learn to read the map.
Further reading	*Designing Instructional Systems* (Chapters 1 to 4).

Map 3.3 *Four levels of instructional design: an overview*

Introduction	We can identify four broad levels of decision-making when designing instruction. These correspond to four levels of system with which the designer has to deal.
Level 1	The course level (overall objectives and structure, philosophy, etc).
Level 2	The lesson level (instruction necessary to achieve each overall objective — a curriculum or 'instructional plan').
Level 3	The instructional event level (the specific acts that should be made in order to achieve effective learning of a specific objective — 'lesson plans').
Level 4	The 'learning step' level in which each instructional event is planned in detail (instructional materials).
Comments	The four levels are characterized by increasing detail. The four levels must be executed systematically, ie applying the systems approach in all its aspects and in particular those aspects to do with creative problem-solving.

The four-level ID process in summary	**Applying our:**	**In the light of:**	**Determines our:**
	1. Philosophies and theories of instruction	☐ Final objectives ☐ Target population ☐ Wider system	Instructional objectives and overall policy
	2. Overall instructional policy	☐ Detailed objectives ☐ Entry skills ☐ Actual resources and constraints	Instructional strategies and *plans* (sets of *methods*, in sequence)
	3. Instructional plan	☐ Content ☐ Enabling objectives ☐ Knowledge and skill taxonomies	Instructional *tactics* (for each step of each lesson)
	4. Instructional tactics decisions	Actual practical experience in applying them to specific learning/teaching problems	Specific instructional *exercises* (in any medium)

Map 3.3 *(continued)*

The procedures of analysis and synthesis	Level of analysis ⟶	Chief outcomes at this level of design ⟶	Instructional decisions commonly made at this level of design
	1. Job analysis. Subject analysis. Front-end analysis.	Final objectives. Tasks to teach. Topics to teach.	Final evaluation. Syllabus/curriculum. Overall sequence (of units in the course). Overall choice of principal methods/media.
	2. Task analysis. Topic analysis.	Intermediate objectives. Prerequisites. Task/topic structure.	Formative evaluation. Diagnostic tests. Course structure. Sequence of lessons. Selection of methods/ media for each unit of course.
	3. Knowledge and skills analysis.	Enabling objectives. Type of learning for each objective.	Detailed lesson plans. Instructional events for each objective. Methods/media matched to each objective type.
	4. Detailed analysis of the learning behaviour/problems typically encountered.	Exercise design for each learning step.	Programmed learning exercises — in any *suitable* media (text, practical, audiovisual, human presenter, computer).

Map 3.4 *The four levels in greater detail*

Level 1: Purpose	In order to identify the worth of a subject/job, Level 1 analysis: — defines the subject/job system — analyses subject/job structure — selects key topics/tasks — decides what needs to be taught.
Products	This analysis results in: — main topics/job description — topic network/task listing — key topics/tasks — course/training objectives.
Level 2: Purpose	Level 2 further analyses by: — examining structure and content — establishing a hierarchy of topics/objectives/tasks — listing what is necessary in the mastery of a topic/task — identifying where problems may arise — assessing target population.
Products	Level 2 analysis results in: — entry prerequisites and standards — possible learning/teaching sequence — general plans (methods, media) — the 'curriculum' — tests — intermediate and enabling objectives.
Analysing worth	Before Level 2 analysis and after, an analysis of worth must be done to determine whether further analysis should go on or indeed if the project should continue. This analysis is done on the following bases: — value — cost — potential development problems — potential implementation problems.
Level 3: Purpose	Objectives from Level 2 are matched with instructional tactics (classification methods, taxonomies of learning, etc).
Models	Gagné's categories and conditions of learning Mager and Beach's performance-type approach The model presented in Chapter 2.
Products	Lesson plans Storyboards Group exercises.
Level 4: Purpose	Level 4 analysis is performed: — only if complexity of problem warrants it — only if such data is needed — only if it can be afforded. Objectives from Level 2 are broken down into behaviour or skill elements. This allows for more precise details of instruction (used most frequently for learning difficulties).
Models	Seymour's skills analysis Gilbert's mathetics approach Landa's mental operations approach.
Products	Instructional materials Scripts.

Map 3.5 *Level 1 instructional design: an overview*

Purpose of Level 1 analysis	*Level 1:* Defines the overall instructional objectives for our system, as well as certain other non-instructional actions that should be taken to ensure success in overcoming the initially defined problems.

Two stages of Level 1 analysis	Stage 1	Is there a problem? What is the problem? Is the problem worth solving?
	Stage 2	What causes the problem? Can instruction contribute to the solution? How exactly?

Stage 1: Define the problem to be solved	'Real' problems are those that generate enough dissatisfaction to justify that something should be done to reduce them. Someone is sufficiently dissatisfied with 'what is' in order to pay the cost of achieving 'what should be'. This cost may be simply the inconvenience caused by a simple change, or it may be a complex of real costs: time, money, and other resources required to develop and implement a solution. The amount one is prepared to pay is a measure of the 'worth' of a successful solution. One may sometimes be able to calculate this worth objectively, from the value of changing the 'what is' into 'what should be'. In industry and commerce, productivity increase, reduction in waste or overtime or turnover can all be quantified to establish the worth of a successful solution (perhaps training is the solution). In general education, it is somewhat more difficult to establish the worth of an innovation, but this is no excuse for not trying.
Procedure	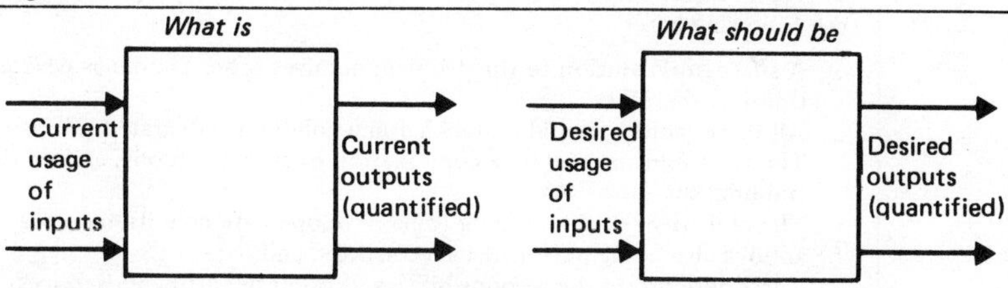 First, identify the system which best defines the problem in input/output terms. Second, define the problem as a *discrepancy* between the *current* state of the system and the *desired* state. Do this in *input/output* terms only. If you find difficulty in doing this you are probably looking at the wrong system.
Tools	Organization analysis Systems analysis Performance appraisal Critical incident analysis Etc, etc.

Map 3.5 *(continued)*

Stage 2: Analysis of causes and solutions	2.1 Is the problem caused by a human performance deficiency?
	  — If NO: Analyse the other causes (material, procedural, etc)
	  — If YES: Proceed to 2.2.
	
	2.2 Is the human performance deficiency caused by lack of knowledge or skill?
	  — If NO: Analyse the other causes (supervision, motivation, feedback, consequences, etc)
	  — If YES: Proceed to the analysis of the objectives and content of instruction.

Tools	Job analysis/Occupational analysis/Job synthesis Performance problem analysis (front-end analysis) Clientele (target population) analysis Future (development) needs analysis Human resources (supply and demand) analysis Subject matter (content) analysis.

Products of Level 1 instructional design	*A project proposal*
	  — Objectives (long term)
	  — Overall resources allocation
	  — Overall implementation plan
	  — Overall evaluation plan.
	
	An 'interdisciplinary contract'
	This is an agreement by all concerned to implement the proposal, made on the following bases:
	
	  — A successful solution to the problem involves other factors as well as training, therefore
	  — All these factors should be tackled in a coherent, integrated way, so that
	  — The final solution will be a combination of several actions, among them training, so
	  — The objectives of the training to be developed are conditional upon other actions also being performed successfully, and
	  — These actions are the responsibilities of other departments/services/individuals in the organization, so
	  — We should establish, jointly, objectives for the various components of the solution and work as a team to develop it, so
	  — We shall distribute responsibility for achieving the objectives as follows, and
	  — We shall have regular review meetings to monitor and evaluate progress.

Further reading	*Designing Instructional Systems*, Chapters 3, 4, 5, 6, 7, 8. *Human Competence: The Engineering of Worthy Performance* (Gilbert, 1978). *Analysing Performance Problems* (Mager and Pipe, 1970).

Map 3.6 *The process of Level 1 design in practice*

Example:
Level 1 —
the process
applied in
the training
context

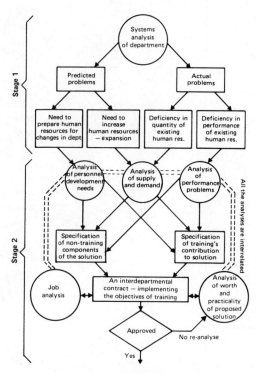

To Level 2: Design of the solution

Comparison
of Level 1
decisions
in education
and training
(stage 2 only
is compared)

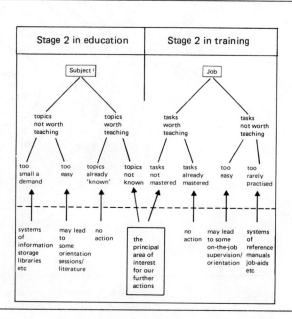

Map 3.7 *Some schemata for Level 1 analysis*

Example 1.
A schema for
front-end
analysis of
performance
problems

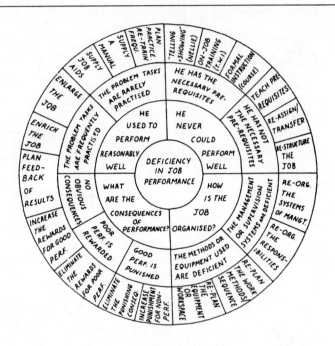

Example 2.
A schema for the
analysis of the
worth and
practicality of
a proposed
project

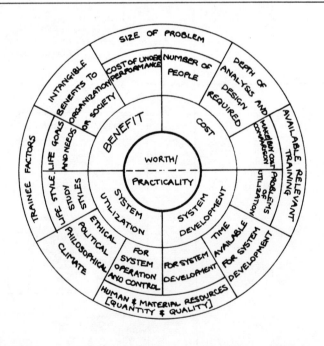

Further reading

Example 1: *Designing Instructional Systems*, Chapter 6.
Example 2: *Designing Instructional Systems*, Chapter 8.

Map 3.8 *Level 2 instructional design: an overview*

Purposes of Level 2 1D	Thus the purposes of a deeper, Level 2 analysis are to have a closer look at the tasks/topics deemed to be worth teaching at Level 1, in order to: ☐ Examine their structure and content ☐ Identify what is involved in their mastery ☐ Identify which parts give the learners most problems ☐ Generate ways of helping them to overcome these problems.
Products of Level 2 ID	The chief products of Level 2 analysis are: ☐ The final test instruments and exercises ☐ The entry or prerequisite tests and standards ☐ The learning/teaching sequence (or alternatives) ☐ Curriculum plans (with some idea of the methods/media) ☐ Lesson tests (for on-line evaluation during the course).
The tools (comment)	Task analysis Topic analysis Target population analysis Knowledge and skill analysis — this last analysis sets out to identify appropriate methods, media, exercises, evaluation procedures, learning sequences, etc, for each specific objective. Various models or taxonomies exist which may help one to perform this analysis. In this book, we are using the model developed in *Designing Instructional Systems* and presented in summary in the previous chapter.

Example 1
A method of task analysis

No.	Steps in performing the task	Type of performance	Learning difficulty
1	Note the plug location relative to the cylinder	Recall	Easy
2	Remove all spark plugs	Manipulation	Easy
3	Identify the type of plugs	Discrimination	✓
4	Decide whether to adjust or replace plugs	Problem-solving	Moderately difficult
5	Clean plugs, if necessary	Manipulation	✓
6	Adjust plugs, if appropriate	Manipulation	Moderately difficult
7	Replace spark plugs in engine	Manipulation	✓
8	Connect ignition wires to appropriate plugs	Recall, Manipulation	Moderately difficult
9	Check for performance	Discrimination	Very difficult

Example 2
A method of topic analysis

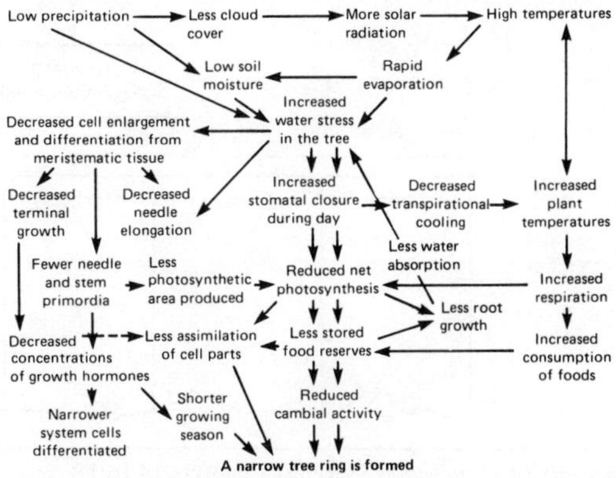

Further reading	*Designing Instructional Systems*, Chapters 9 to 19.

Map 3.9 *The process of Level 2 instructional design*

Stage 1	The preparation of a formal, but not too detailed project proposal, which specifies the overall objectives, specific intermediate objectives and their sequence or interrelationship, any overall method or media decisions which are forced on the project by general constraints or previous Level 1 decisions/policy.
Stage 2	The preparation of an overall instructional plan, or curriculum document, defining: — objectives and their sequence/sequence alternatives; — content (knowledge and skills to be mastered); — methods/techniques to be used for each objective; — specific media/materials required for each objective; — estimates of resources, time, etc.
Flow chart	

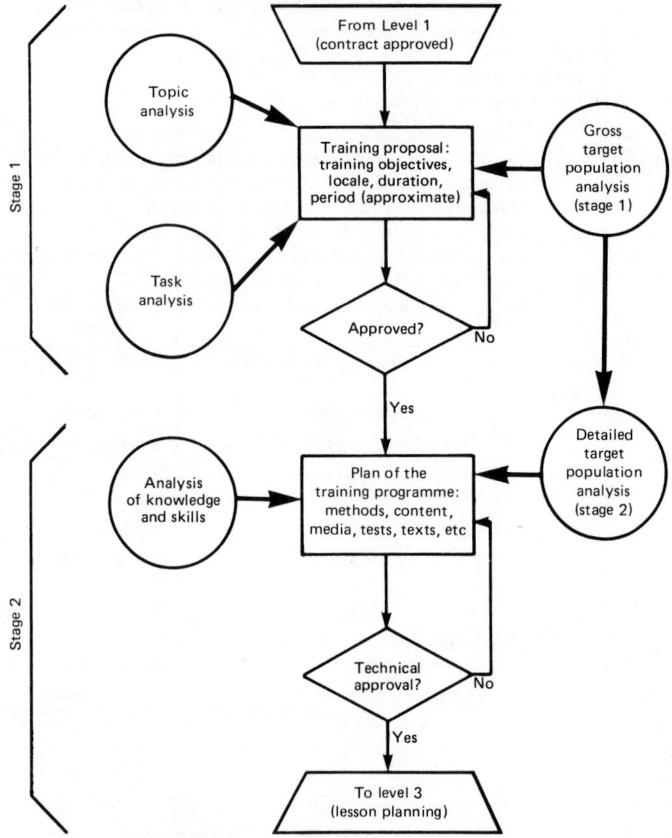

Worksheet	Objectives	Tests	Content	Methods	Media	Time
	1. . . . . . 2. . . . . . 3. . . . . . etc.					

Further reading	*Designing Instructional Systems*, Chapters 14 to 18. *Producing Instructional Systems*, Chapters 5 and 6.

Map 3.10 *Level 2: target population analysis at stage 1*

Introduction	Throughout the process of instructional design, one should keep in mind the final consumer of the course — the student or trainee. It is never too early in the process to begin to analyse his characteristics, needs, habits, levels of knowledge or skill, etc. Rather than a definite stage on its own, this analysis continues throughout. It is a good idea to remember, right at the outset that our 'client' (who is commissioning the course and paying the bill for instructional design) is not necessarily one and the same as the 'consumer' and he may well have different viewpoints and motivations. This has been whimsically termed the 'dog-food syndrome'. While satisfying the client, do not forget the consumer.
Definition	The term 'target population' will be used to denote the 'consumer' of the course we are designing — the students or trainees that we expect (we are aiming) to cater for.
Purpose	The chief purposes of an analysis of the target population are: (a) to determine the 'entry level' to the course (what can we assume concerning the initial levels of relevant knowledge or skills) (b) to determine the structure of the course (what existing attitudes, habits, study skills or relevant experiences do we expect and how will these influence instructional design). At Level 1, the information sought is fairly general (which of the tasks/topics identified as worth teaching *actually need to be taught*). In addition, in the educational context, one needs to check that the topics *we* think are worth teaching, are considered to be *worth learning* by the *target population*.
Methods	In general, one would investigate, therefore: — what the target population *have done* in the past (experience); — what the target population *can do now* (that is relevant); — what the target population *tend to do* (attitudes and habits). The table below suggests some of the investigations that might be carried out at Level 1.

Information required	Suggested methods of collection	
	Educational context	Training context
What they have done		
(a) Previous learning	Analysis of curricula and results of any past courses	Restrict analysis to prior job-related learning
(b) Previous relevant performance	Interview teachers familiar with the target population	Teachers and/or supervisors who know the TP
What they can do now		
(a) Knowledge/skill related to the planned course	Interviews of a sample of the TP (best leave formal diagnostic tests to a later stage)	Observation of on-the-job performance of a sample of the TP plus interviews perhaps
(b) General skills, learning skills, etc		
What they tend to do		
(a) Interest in course and in study generally	Try to discover the TP's objectives. Perform a consumer survey	Study the job's market, turnover rates, motivation, rewards, etc
(b) Existing learning styles and habits	Informal interviews with teachers who know the TP and with samples of the TP	

Map 3.11 *Level 2: target population analysis at stage 2*

Introduction	At Level 1 one sought fairly general information about the target population (see Map). At Level 2 the analyst seeks more detailed data. The data required can best be organized as: job/subject-specific data and general data.
Job/subject-specific data	This is information of a more detailed nature on the existing knowledge/skills level of the probable learners. If the course design team has little or no experience of teaching the topics/tasks to the type of learners expected, there will probably be a need for the testing of a sample of the target population. This can be affected by using the final and intermediate test items already isolated as diagnostic tests.
Diagnostic tests	One argument for producing the tests early in the course design process is that they are then available for target population analysis. Usually, the tests should be followed by informal interviews with the subjects tested, in order to examine in depth any difficulties which have been noted. These tests may sometimes be actually attempting the task, but usually a more careful testing/observation procedure is required to identify the precise sources of learning difficulty.
Entry test	One other result of this stage is the ability to specify the knowledge and skill level that will be assumed as the minimum prerequisite for entry to the course. This can be best presented straight away as a prerequisite or entry test.
General data	This is information that is not directly related to the *content* of the proposed course, but to other aspects of importance, such as course/lesson structure, sequence, methods and media.
Course structure	What amount of new information can the target population digest at one go? How many times must a new item of information be repeated in order to be learned? How much practice time/repetition is needed for a specific new skill to develop?
Sequence	Does the target population prefer a fragmented 'parts' presentation of the topic, or is an integrated 'whole' approach more acceptable? When a topic must be learned in its parts, will the learners prefer a *synthetic* (from the parts to the whole) or an *analytic* (from the whole to the parts) sequence? Would the learners prefer to master the theory fully before applying it in practice, or would they prefer to pick up the theory through practical exercises?
Methods	Do the learners prefer to study in group or individually? Do they tend to study in an active, participative manner or do they tend to be passive recipients? Do they need constant supervision and control, or do they work efficiently unsupervised?
Media	Are the target population in general 'visual' or 'aural' learners? Do they prefer to study verbal information through the spoken or the written word? Are they familiar with the various symbolic languages (eg graphs, flow charts, non verbal communication) that the study of the topic/task requires? Are they happy with inanimate media, or do they need the presence of an instructor?
The general context	Finally, one should investigate the general context in which the learning will take place and from which the learners will come. What related experience have they had which would influence the choice of meaningful examples/analogies? What are the social customs/personal habits/working habits of the target population and how would these influence the course design?

Map 3.12 *Some schemata for Level 2 analysis*

Example 1 Schemata for selecting strategies for the knowledge content	

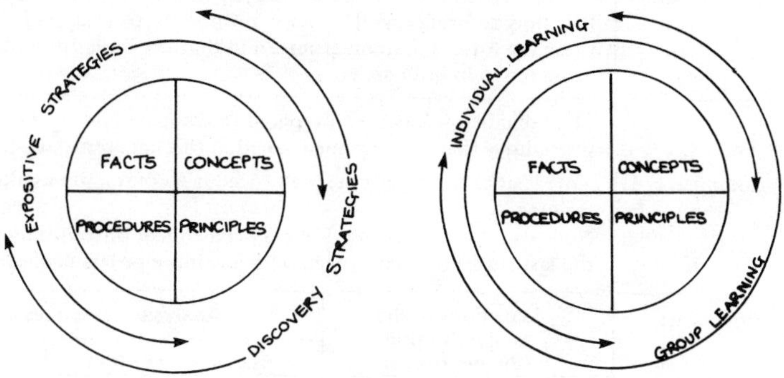

Example 2 General principles for the planning of skill development

			Reproductive skills	Productive skills
	Step 1	Imparting the knowledge content	Expositive or discovery methods (dependent on the type of knowledge)	Discovery methods (principle learning is always involved)
	Step 2	Imparting the practical application	Expositive methods (demonstration and prompted practice) *Note:* Imparting the knowledge and skills content may be combined	Expositive methods (demonstration and prompted practice)
	Step 3	Developing proficiency	Supervised practice of whole task and/or special exercises Continuing feedback of results	Discovery methods (guided problem-solving) Continuing feedback of results

Example 3 Schema of factors that influence media selection decisions	

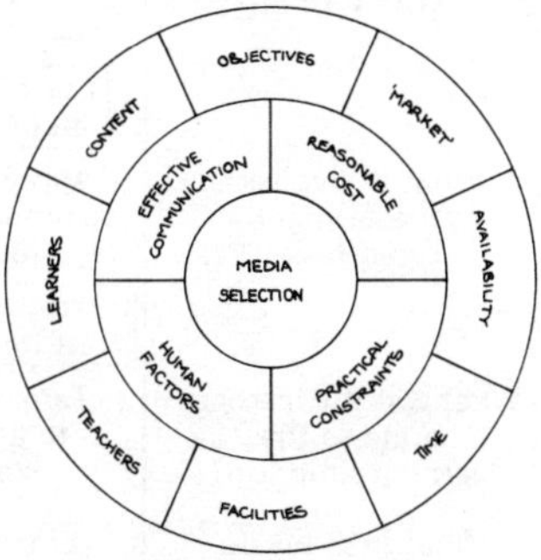

Further reading	*Designing Instructional Systems*, Chapters 15, 16 and 17. *Producing Instructional Systems*, Chapter 5.

Map 3.13 *Levels 1 and 2 of instructional design — a summary*

Introduction	We have studied two levels of analysis in two contexts (education and training). It is time to bring together our concepts and techniques in order to show that there is a lot of common ground and that the basic 'systems approach' that we are using serves in both areas.
Comments 1. 2.	The table below has been prepared to compare and to contrast the practical procedures that we have mentioned in this chapter to date. Note that there are alternative procedures commonly used in education and training. Note also that the distinction between education and training gradually blurs, the last stages of each procedure being more or less equivalent.

Flow chart	Questions in the analyst's mind (the heuristics)	Analysis techniques and procedures	
		Education context	**Training context**
	AT LEVEL 1	Subject	Job
	What is the system of interest? What is its boundary?	definition of the subject area/aims, definition of a notional job	definition of the job's duties and responsibilities and objectives
		Chief topics or Master performer	Job description
	What are its components? What is its structure?	analysis of structure of subject	analysis of the job's structure
		Topic network or model or Topic matrix or map	Task listing or mapping
	What is worth teaching/learning?	Selection of 'key' or useful topics from expert's view	Selection of key, difficult and frequently practised tasks
	What needs to be taught?	Assess usefulness from learner's view. Target population analysis (existing knowledge/skill)	Specify job aids or practice as required. Target population analysis (existing related performance)
		Syllabus (course content) or Course objectives	Training objectives
	AT LEVEL 2		
	How is each task/topic done or structured?	Analyse topic into information elements Analyse objectives into sub-objectives	Analyse task into steps or operations
		Teaching points or Objectives hierarchy or Task structure or list of steps	
	What is involved in mastering the topic/task?	Apply a taxonomy (eg Bloom) Look for prerequisite skills or knowledge essential for satisfactory performance	
		Intermediate test items Intermediate/enabling objectives and test items	
	What will the learners bring to the learning situation?	Target population analysis of an in-depth nature, to assess suitability of sequence, pace, specific examples, methods, media, and entry levels	
	What shall we do about it all?	Entry test + Learning sequence + Lesson plans + Lesson tests + Final test	
		(The overall instructional plan)	
	TO LEVEL 3	Continue detailed lesson planning	

Map **3.14** *Level 3 instructional design — an overview*

| Introduction | The flow chart presented here outlines the main activities carried out at the more detailed, lesson-planning, stages of instructional development. In our model, we call this the Level 3 of the ID process. Note that at this level, the distinction between the 'education' and 'training' contexts disappears. We are, in both cases, developing detailed instructional sequences for pre-defined specific objectives. |

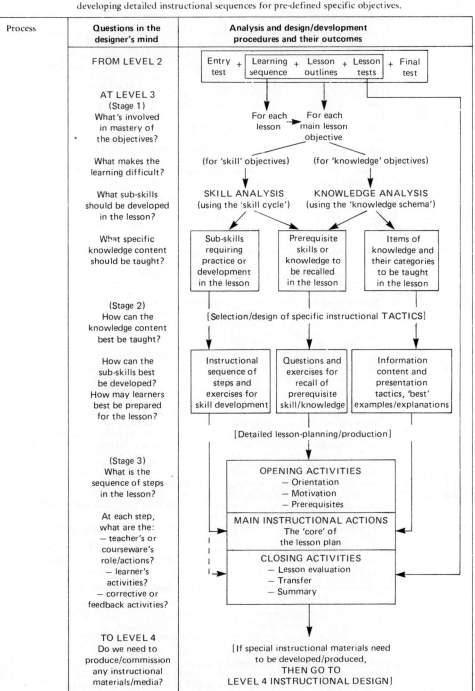

Process	Questions in the designer's mind	Analysis and design/development procedures and their outcomes

FROM LEVEL 2

Entry test + Learning sequence + Lesson outlines + Lesson tests + Final test

AT LEVEL 3
(Stage 1)
What's involved in mastery of the objectives?

For each lesson → For each main lesson objective

What makes the learning difficult?

(for 'skill' objectives)   (for 'knowledge' objectives)

What sub-skills should be developed in the lesson?

SKILL ANALYSIS
(using the 'skill cycle')   KNOWLEDGE ANALYSIS
(using the 'knowledge schema')

What specific knowledge content should be taught?

Sub-skills requiring practice or development in the lesson | Prerequisite skills or knowledge to be recalled in the lesson | Items of knowledge and their categories to be taught in the lesson

(Stage 2)
How can the knowledge content best be taught?

[Selection/design of specific instructional TACTICS]

How can the sub-skills best be developed?
How may learners best be prepared for the lesson?

Instructional sequence of steps and exercises for skill development | Questions and exercises for recall of prerequisite skill/knowledge | Information content and presentation tactics, 'best' examples/explanations

[Detailed lesson-planning/production]

(Stage 3)
What is the sequence of steps in the lesson?

OPENING ACTIVITIES
— Orientation
— Motivation
— Prerequisites

At each step, what are the:
— teacher's or courseware's role/actions?
— learner's activities?
— corrective or feedback activities?

MAIN INSTRUCTIONAL ACTIONS
The 'core' of the lesson plan

CLOSING ACTIVITIES
— Lesson evaluation
— Transfer
— Summary

TO LEVEL 4
Do we need to produce/commission any instructional materials/media?

[If special instructional materials need to be developed/produced,
THEN GO TO
LEVEL 4 INSTRUCTIONAL DESIGN]

Map 3.15 *Level 3 ID — background concepts and principles*

Introduction	The flow chart shown in Map 3.14 suggests that there are three stages to the instructional design/development process at Level 3. These stages answer the questions:
	1. what exactly should a given lesson achieve?
	2. how can this best be achieved and implemented?
	3. what are the details of each step of our lesson plan?
	It also shows that stages 1 and 2 involve parallel sets of activities devoted to the analysis of the skills and the knowledge content of the lesson. The results of these analyses are then brought together in stage 3, as the detailed lesson plan. In this map we review the rationale and methodology for this detailed approach to lesson planning.
Why distinguish between skill and knowledge?	Skill is 'the competent performance of a task/activity'. As such, it requires to be developed through appropriate practice exercises and corrective feedback, repetition, etc. Knowledge is the possession, in the mind, of INFORMATION. This information, to be useful, must be organized in such a way as to be readily recalled when needed. The lesson planner must consider the nature and structure of the knowledge to be formed in the mind of the learner, to best equip him/her to perform skilfully. Thus, although skill and knowledge are generally both involved in most tasks, the separate analysis of these two components helps to focus the lesson planner's attention on the two aspects:
	— informational/instructional MESSAGE design, and
	— appropriate learner activities/exercises.
How is skill structured? (a)	Performance (physical, intellectual, etc) is seen as a CYCLE of internal activities, involving:
	☐ PERCEPTION of incoming information/problem situations, etc;
	☐ RECALL of relevant knowledge, necessary in order to
	☐ PLAN an appropriate course of action, or directly
	☐ ACT in an appropriate manner. (See Map 2.11.)
(b)	Each of these four types of activities depends on the presence of various basic abilities which, for specific tasks, may need to be developed to greater or lesser degree. Many of these abilities depend on specific types of perceptual, memory (recall), planning (thinking), or performing (dexterity, eloquence, etc) skills. (See Map 2.12).
(c)	Some (especially the recall abilities) also depend on the presence, in the mind, of specific knowledge stored in a specifically useful manner. Others (especially the action or performance abilities) may depend on personality or attitudinal factors specific to each learner, which may or may not be capable of modification through the process of 'instruction'. (See Maps 2.13 and 2.14.)
(d)	The development of a skilled level of performance may require the special development of a series of sub-skills which are part of the necessary 'basic abilities'. Each deficient 'sub-skill' may need special exercises to be incorporated in the lesson(s). Different overall forms of exercise are used for productive and reproductive skills. (See Map 2.15 for definitions of these terms.)
How is knowledge structured? (a)	'Knowledge' may be classified into different categories in terms of what it is used for. Overall, one can talk in terms of FACTUAL knowledge, which is simply recalled and directly applied in practice (this forms the bulk of the knowledge requirements for most reproductive skills) and CONCEPTUAL knowledge, which is used in the planning of a given course of action/explanation/problem solution (this form of knowledge is essential to the execution of creative, or productive skills). (See Map 2.16.)
(b)	These two basic knowledge types may be further classified into FACTS, PROCEDURES, CONCEPTS and PRINCIPLES (see Map 2.16) and each of these four categories yet further sub-classified (see Maps 2.17, 2.18, 2.19 and 2.20).
How is knowledge learned?	There are two basic processes by which a person acquires new knowledge:
	— RECEPTION of pre-planned messages (verbal or other), which are perceived, understood and stored in the memory, associated with other prior knowledge, and
	— DISCOVERY of a relationship, concept or mere fact, by the analysis and interpretation of experienced or observed phenomena, events, etc.
. . . and taught?	When we set out intentionally to teach a given body of knowledge, we apply expositive instructional strategies (which promote reception learning) or, alternatively, experiential instructional strategies (which promote discovery learning). (See *Producing Instructional Systems*, Chapter 5, Section 5.7.1.)

Map 3.15 *(continued)*

Expositive and experiential strategies compared	Basic expositive strategy	Basic experiential strategy
	1. Present information. Definition, explanation, demonstration, etc, of the principles/procedures.	1. Organize experiences. Present opportunities to act, play, interact and observe the consequences of one's actions.
	2. Test reception, recall and understanding. Repeat or rephrase message if it proves necessary.	2. Test for understanding of cause-effect relationships. Present further opportunities to act/interact if necessary.
	3. Organize practice. Learners apply the new information to a range of examples. Test for correct application. Vary the number and difficulty of examples, till objectives achieved.	3. Promote the formation of general concepts/principles. In 'debriefing', test for the learner's ability to generalize from the specific experiences. Repeat 1, 2 and 3 if necessary, until the learners, themselves, reach the desired objectives.
	4. Develop proficiency. Present opportunity to apply the newly-learned knowledge to a variety of real situations.	4. Develop proficiency. Present opportunity to apply the newly-learned knowledge to a variety of real problems and discuss the results with others.

(For deeper discussion see *Producing Instructional Systems*, Chapter 5, Section 5.7.2.)

Selecting strategies for teaching of knowledge	Expositive strategies are generally preferred for the teaching of facts. Experiential strategies are preferred for the teaching of principles. Either of the two basic strategies may be preferred for the teaching of concepts and procedures depending on the sub-categories involved, as well as some other content-specific factors. There are also reasons for occasionally breaking these general rules-of-thumb. (See *Producing Instructional Systems*, Chapter 5, Section 5.7.2 for a deeper analysis. Also *Designing Instructional Systems*, Chapter 1.)
Planning the tactics for the teaching of knowledge	The specific tactics selected for the implementation of a given strategy are of two types: ☐ *Specific to the objectives category* — tactics especially effective for the teaching of a particular sub-category of knowledge (eg defined concepts, principles of nature, concrete concepts, algorithms, etc). ☐ *Specific to the lesson content and the students* — use of particular examples, analogies, problems, etc, known to be particularly effective in teaching the specific lesson content to specific types of students. (We shall be dealing more fully with the selection and planning of specific instructional tactics for the teaching of knowledge in Map 3.17.)

Map 3.15 *(continued)*

How are skills learned?	Skills are learned through practice. However, just repetitive practice is not always a sufficient condition. When the overall task is mainly (or even only in part) composed of productive skills, then a 'threshold' of learning is reached, beyond which merely extra practice does not 'make perfect'. What is needed in these cases is practice in the intellectual, planning, sub-skills which are used for creative (productive) decision-making. In some (both productive and reproductive) skill learning, 'learning plateaux' may develop — performance ceases to improve with extra practice for a time and then gives another 'spurt' of improvement. This usually occurs when one or more of the sub-skills associated with the 'essential basic abilities' is deficient (see the 'skill cycle' — Map 2.12) or when some personality or attitude factor, for a time, inhibits progress (see the 'inner self' — Map 2.13).
. . . and taught?	An overall strategy for the teaching of skilled activities must therefore take into consideration whether the skill is basically productive or reproductive. In the latter case, 'practice makes perfect', but in the former, one must provide opportunities for self-analysis, reflection, and exchange of ideas with other equally, or more, skilled practitioners. Hence the importance of briefing and, even more so, debriefing sessions in instructional designs that seek to develop productive skills. A second consideration is that most skills (except the simplest and most reflex type) involve the recall and use of certain specific knowledge. Unless this knowledge has all been learned previously in different contexts and is now simply to be applied in a new situation, it is usually best to combine the instruction of the new knowledge with its skilled application, in the same lesson. A third consideration is that generally, skilled activity which is considered worth teaching is of some use to the learner, so the sooner that the learning can be put to practical use, the better. A fourth consideration is that, generally, the skilled (ie competent) performance of a task is usually enjoyable (or rather 'reinforcing'), so the sooner that some results are seen by the learner, the higher the level of interest and motivation to continue learning.

Selecting strategies for teaching of skills	For all these reasons, an overall approach to the teaching of skills can be suggested, as shown in the schema reproduced as Example 2 in Map 3.12 of this chapter. This is based on three basic steps:
	    1. imparting the knowledge essential to 'get started';
	    2. imparting the basic procedures of practical application;
	    3. developing proficiency and competence.
	The steps are different for productive and reproductive skills.

Planning the tactics for teaching of skills	Any overall strategy must, necessarily, be tempered by the special learning requirements of the specific skill to be taught and the specific type of learner to be taught. As with the tactics for knowledge, skills may require:
	    ☐ *Tactics specific to the objectives' sub-category:* special exercises or sequences designed to develop specific sub-skills identified as 'critical' to competent performance (for this we use the 'skill cycle');
	    ☐ *Tactics specific to the content and the students:* any special exercises, drills, demonstrations, etc, that may have proved to be especially effective in practice, in overcoming learners' inhibitions, or eliminating 'learning plateaux'.
	(We shall be dealing more fully with the selection and planning of specific instructional tactics for the teaching of skills, in Map 3.16.)

Map 3.16 *Identifying critical sub-skills: use of the skill cycle*

Introduction	The basic idea of the 'skill cycle' and its 'expanded' form of 12 'basic abilities' (which may be more or less 'critical' to competent performance of a given skill), was discussed in the previous chapter (Maps 2.11 to 2.13). The four tables which follow, explain and give examples of the analysis procedures which may be used to identify critical 'basic abilities' and, hence, the sub-skills and the knowledge-structures that must be taken into consideration in the detailed design of instruction. The examples and discussion presented in these tables serve to llustrate most of the 12 'basic ability' categories listed in the 'expanded skill cycle'. Examples quoted later illustrate the process in practice.
(1) Perception	Inability to perceive the stimulus may be caused by a low level of perceptual acuity, as for example: ☐ (Cognitive skill) – inability to 'see' a problem. ☐ (Psychomotor skill) – inability to discriminate colour, tone, size, shape, etc to the degree necessary. ☐ (Reactive skill) – inability to notice the signs and events occurring around one (a lack of attention). ☐ (Interactive skills) – inability to notice the responses (including non-verbal responses) of other people. In all these cases the performer shows a low level of acuity in perceiving the necessary stimulus information even when that information is quite clearly presented. However, sometimes the information is not clearly presented. There are distractions. All manner of other irrelevant information is being picked up as well (engineers call this 'noise'). Our performer may be adept at perceiving the stimulus when it is presented on its own, but may have trouble in picking it out in a 'noisy environment'. For example: ☐ (Cognitive) – ability to notice punctuation error in a given single sentence, but inability to notice it when the sentence is part of a larger paragraph. ☐ (Psychomotor) – the car driver who can identify danger signals in light traffic conditions but fails to identify them in heavy traffic conditions. ☐ (Reactive) – the music lover who perceives opportunities to listen to good music when directly invited, but fails to notice them in the general life of his community. ☐ (Interactive) – the manager who can identify signs of employee insecurity in a specific interview situation, but fails to identify the same signs in a more general casual conversation.
(2) Prerequisites	Inability to recall prerequisites may be caused by a lack of these prerequisites. The performer simply does not know what to do in a particular situation. The relevant procedure has not been learned (or has been forgotten). The relevant principles that would enable him to invent or develop an appropriate procedure are not available from his memory store. Alternatively, the performer may fail to recall the relevant knowledge, although it is in store, due to a failure on his part to interpret the perceived stimulus information in the correct way. The new information is compared with the stored experience (knowledge structures or schemata) and is misclassified. Thus the wrong procedure is recalled and applied: ☐ (Cognitive skill) – a given Portuguese noun (the stimulus information) is misclassified as to gender; this leads to the recall of the wrong form of the adjective to be coupled to it. ☐ (Psychomotor skill) – a given road sign is misinterpreted by the motorist, leading to the recall of an incorrect strategy. ☐ (Reactive skill) – a student's examination errors are misinterpreted by the teacher as the result of laziness, leading to the development of an unduly negative attitude towards the student in question. This negative feeling (reaction) may later influence actions (interactions). ☐ (Interactive skill) – the salesman who misinterprets a potential customer's reactions and as a result applies an inappropriate selling strategy. Thus we have two aspects involved in recalling prerequisite knowledge schemata, procedures or principles: 1. Ability to interpret the stimulus information in order to identify what knowledge is required. 2. Having that knowledge in memory store in a usable form.

Map 3.16 *(continued)*

(3) Planning	Inability to plan may also have two main causes. One is planning one's immediate actions. This involves considering the alternatives open to us and deciding among these. The causes of failure in the planning of an action, may be due to inability to generate the list of possible alternative courses of action, or to inability to make the best choice. The first implies inability to use the relevant principles in order to 'invent' alternative procedures (assuming of course that the relevant principles are in store). The second implies inability to evaluate the alternatives by thinking through the implications of each one. For example, the structural engineer may (or may not) come up with, say, four alternative solutions for the construction of a given bridge and he then may (or may not) select the most cost-effective solution. The manager faced with an industrial relations problem may (or may not) consider all the alternative courses of action open to him (eg sackings, warnings, suspensions, ignoring the problem, etc) and he may (or may not) evaluate correctly the hazards of each one (strikes, loss of productivity, etc).
(4) Performance	Inability to perform can also spring from two types of deficiencies — inability to initiate the necessary action or inability to 'see it through'. Assuming, once again, that the performer has perceived the stimuli, interpreted them correctly, recalled relevant knowledge, considered all the alternatives and decided which is the 'best' one, he may yet fail to perform. Examples: ☐ (Cognitive) — having 'seen' the problem (in, say, maths), the student works out the stages of the 'best' solution, but he does not actually work out the solution (due to lack of motivation, time, relevance, etc); alternatively, he begins to work it out, but gets bogged down in the detail of calculation and gives up (due to lack of persistence or mental stamina, etc). ☐ (Psychomotor) — the industrial operative responds as is expected, but his productivity/quality of work is below standard (due to deficiencies in strength, stamina, dexterity). ☐ (Reactive) — practical difficulties are encountered in attempting to live by one's values, leading to compromises. ☐ (Interactive) — a supervisor fails to initiate necessary disciplinary action (due to lack of moral courage) or fails to see it through correctly (due to lack of tact).
Example: (a) Use of the 'skill cycle' to identify deficient and critical sub-skills	A brief description of the application of this form of analysis to the task of 'long division', has been presented in the previous chapter (see Map 2.12). This example shows that the analysis involves two stages: 1. analysis of the task itself, to identify the abilities which are 'critical' to competent performance, and 2. analysis of typical learners (a sample of the target population), to identify which of these critical abilities are likely to be under-developed, define the deficient sub-skills and select appropriate types of special exercises to be incorporated in the lesson(s). In the case quoted in Map 2.12, one might find, for example, that the expected target population have no real problems as regards attention and perceptual acuity — they are able to concentrate adequately on arithmetic tasks and recognize the symbols and numbers used in the mathematical presentation of division problems. However, they are found to have a low level of ability of recognizing a division problem when presented verbally, especially when the presentation is 'noisy' (contains other irrelevant information, that should be ignored). Therefore, in relation to perceiving the problem, special exercises should be incorporated to develop this deficient sub-skill.
(b) Analysis of the type of sub-skill to select objectives-specific tactics	Now we come to the 'objectives-specific' aspect of our analysis. The tactics most successful for developing this sub-skill are those which have been tried and tested by generations of mathematics teachers who have faced up to this specific learning difficulty. We can see, from our general analysis, that the learning problem involves discrimination training — distinguishing between problems (expressed in words) which are and which are not 'division'.
(c) Use of teaching experience to select content-specific tactics	However, which specific 'word problems' and how many such problems, we should include in our lesson, is best decided on the basis of previous teaching experience. If the instructional developer does not have this experience, then experienced teachers should be consulted. Another approach is 'creative invention' followed (as in all cases) by careful 'developmental testing' on a sample.
Further reading	A deeper discussion of the use of the 'expanded skill cycle' for analysis of the 'critical' abilities, may be found in *Producing Instructional Systems*, Chapter 9, a general account in Section 9.2 and further examples in Section 9.3.

Map 3.17 *Planning the lesson tactics for the knowledge content*

Introduction	The selection of appropriate tactics for specific lessons is a process of creative design (invention/adaptation) of:		
	— explanations, definitions, demonstrations; — illustrations, examples, analogies; — learning activities, drills, games, exercises; — evaluation tests, exercises, questions, observations; — feedback, correction, comment, debriefing, etc.		
	All these should be appropriate to the objectives and also to the content and target population. As regards the objectives, a deep analysis of the categories and sub-categories of knowledge to be included, may be used to define lesson tactics and sequence. As regards the content and target population aspects, it is necessary to have subject matter expertise and teaching experience to draw on, as well as a good measure of creative inventiveness. In this map, we shall deal with the general aspect — planning the tactics for specific categories of knowledge.		

Planning the tactics for factual knowledge	If the knowledge content of the sub-objective is . . .	and if the strategy selected is:	
		expositive, then . . .	experiential, then . . .
	FACTS — *concrete* (Specific objects, people, places that have some significance often associated with their names, parts of a machine, names of college students, street names, <u>etc.</u>)	(RECOMMENDED) 1. Present the real object/pictures in a form, sequence and frequency conducive to rote memorization.	(NOT RECOMMENDED) 1. Organize a 'stimulus-rich' environment, which may allow for the frequent observation of the facts.
		2. Test for recall and repeat the presentations as often as required.	2. Test whether the required facts have been observed and memorized. Further enrich the stimulus environment if not successful at first.
		(There is no stage 3 as there are no examples or general principles to which learning should be transferred.)	
		FINALLY, proficiency may be developed by frequent use of the facts later on, as part of other learning exercises.	
	symbolic (Verbal information, which states matters of fact, such as descriptions of events; logic and mathematical equations, when viewed as just statements of fact; telephone numbers; codes of motor parts; Morse code; map symbols; road signs; etc.)	PREREQUISITE: adequate knowledge of the symbolic 'language' used for presenting the information.	
		(RECOMMENDED) 1. Present a clear description of the information, using the symbols in correct manner.	(NOT FEASIBLE) As, by definition, verbal, or symbolic, information uses some form of man-made 'language' of symbols (words, maths notation, visual conventions), it follows that one never 'discovers' such facts, but always 'receives' them by 'reading' some form of man-made message.
		2. Test for recall and understanding of the message. Repeat/rephrase the message, if necessary.	
		3. or 4. Use the new factual information frequently, to prevent forgetting.	

Map 3.17 *(continued)*

	(RECOMMENDED)	(NOT RECOMMENDED UNLESS THE DEVELOPMENT OF MEMORIZATION SKILLS IS A PRIME OBJECTIVE)
systems of facts Factual information is often complex, made up of many interrelated items of information — a long verbal description of a historical event contains many related facts; Morse code contains over 30 similar but different symbols to be learned; maps are made up of many different symbols which can be organized into sets according to what they describe — roads, frontiers, water, elevation, landmarks, etc. Research shows that the structuring of such fact-systems in 'chunks' greatly assists initial learning and long-term memorization. Giving meaning to the structure or to individual bits of information by charts, mnemonics, colour codes, etc, further facilitates learning.	1. Present a map/table/diagram/chart/list of the information, organized so as to form easily visualized and memorized units or 'chunks'. If possible, develop some mnemonics/other mediators to give some 'meaning' to the structuring/chunking used in the presentation.	1. Present the items of information in unstructured form. Ask the learners to 'play' with the information and 'discover' or invent some useful way(s) of structuring or 'chunking' it into easier-to-learn form. Learners draw their own learning aids.
	2. Test for recall. Check the validity and effectiveness of the learning/memory aids which have been used. Revise if necessary.	2. Test for recall. Check whether the learners have made progress in learning the fact-structure they have developed. Correct if necessary.
	3. Learners practise recall and use of the information, with the learning aid gradually being withdrawn/internalized.	3. and 4. Continue as in the left-hand column. (Note that in reality the learners have not actually 'discovered' the facts or information which they have learned. They received these in the form of an unstructured message. They 'discovered' a way to organize it for easy learning and memorization.)
	4. Use the newly learned knowledge frequently, to avoid forgetting.	

Observation 1.	Experiential strategies are generally not recommended for the teaching of factual information, as there is very little, if anything, to be gained in terms of enhanced learning or retention, in return for the extra time and effort that has to be put into the instructional design and the execution of the plan, for experiential strategies.
Observation 2.	The only possible exception to this rule, is if the development of special memorization/rote learning skills is sought as one of the objectives. However, we are at present discussing the teaching and learning of knowledge — it is best not to mix up the two in our presentation, for now.
Further reading	A fuller discussion of the selection of tactics for the teaching of factual content may be found in *Producing Instructional Systems*, Chapter 8, Section 8.2.

Map 3.18 *Tactics for teaching concepts*

Planning the tactics for the teaching of concepts	If the knowledge content of the sub-objective is . . .	and if the strategy selected is:	
		expositive, then . . .	experiential, then . . .
	CONCEPTS *concrete* (Also called 'primary' concepts. Includes classes of real objects or situations. Ex: red, blue, bigger, higher, circular, triangular, etc. The learning of a concept, unlike the learning of a fact, enables the learner to deal correctly with a whole range of cases or instances. These should be classified as examples or non-examples of the concept. The closer/finer the distinction, the more 'refined' is the concept held by the learner.)	(RECOMMENDED FOR ADULT LEARNERS) 1. Present a number of selected examples and non-examples of the concept.	(RECOMMENDED FOR EARLY CHILDHOOD) 1. Organize an environment 'rich' in examples and non-examples of the concept to be learned.
		2. Test for concept formation, by the presentation of new instances for classification by the learner as examples or non-examples of the concept to be learned.	2. Test whether the learner forms an adequate concept by the experience of how others classify the examples. (Note that the learner really is receiving/imitating and not discovering entirely for himself.)
		3. and 4. Organize graded practice, on ever more 'fine' distinctions between examples and non-examples. Refine the 'boundary' of the concept to the required precision.	3. and 4. As in left-hand column. The practice may be less 'organized' in the formal sense but would involve the gradual enriching and refining of the learner's environment.

Map 3.18 *(continued)*

		(Both approaches may be recommended) PREREQUISITE: adequate knowledge of lower-level concepts, used in definition.	
	defined (Also called 'secondary' concepts. Includes concepts which are classes of other simpler concepts (both concrete and defined). The concepts of 'colour', 'size' and 'shape' would classify the concrete concepts cited above as examples (red, etc). The concept 'property of an object' is a yet more abstract concept, having as some of its examples the above-mentioned defined concepts 'colour', 'size' and 'shape'. Note that as we go 'up the ladder of abstraction', we rely ever more on quite complex definitions, which are composed of other concepts. These are prerequisites to the learning of the new, higher order concept.)	1. State the definition or rules that define the concept and its chief 'attributes'. Illustrate with one or more examples and any commonly confused non-examples.	1. Present a series of examples and non-examples of the concept, identified as such. Ask learner to 'discover' the key attributes and classifying rules that should be used to define the concept.
		2. Test for comprehension, by asking the learner to classify further new instances. Provide corrective feedback and further examples or non-examples.	2. Test that learner has formulated an adequate set of attributes/rules. If not, cite some examples which do not 'fit' learner's definition, forcing him to 'rethink'.
		3. Develop further refinement, by asking learner to supply further 'borderline' examples and classify them correctly.	
		4. Develop proficiency by regular use of the newly learned concept in a wide variety of situations.	
		(Note: This sequence is often called 'RUL-EG', meaning 'from rule to examples'.	(Note: This sequence is often called 'EG-RUL', meaning 'from examples to rule'.

Map 3.18 *(continued)*

	systems of concepts (conceptual structures or schemata)	(Both approaches may be recommended) PREREQUISITES: the individual concepts that are to be built into the schema, must be previously understood.	
	It is believed that concepts are stored in the mind as a network of interrelated and interdependent ideas — referred to as the 'schema' or 'structure' of the learner's conceptual knowledge. The formation of useful and powerful conceptual schemata is an important task of any teacher. Some authors suggest that this task may be best performed by the presentation of model schemata (advance organizers, schematic overviews, etc) to the learner. Others argue that the learner should, whenever possible, construct his own schemata and the teacher's task is to assist and control this process. These two views are reflected in the two alternative approaches to the development of conceptual schemata, outlined in the columns alongside.	5. Present analogies, tables, summaries, visual structures, etc, linking two or (usually) more concepts in a way that shows their relationship, similarities, differences, hierarchical dependence, etc.	5. Organize experiential activity aimed to get the learner to form a useful and valid structure of some form, that reflects his mental image of the relationships existing between a set of concepts.
		6. Test learner's understanding of the structure, by asking him to 're-teach' it, or put it into his own words, or apply it to a given problem. Refine, explain or modify the schema presented, if learners have difficulty.	6. Test learner's structure for validity and usefulness as a tool for problem solving/explanation generating. Set up challenges to get the learner to re-think and re-structure the schema if it is inadequate or incomplete
		7. Organize practice in the use of the schema for a range of problem-solving situations. Continue until the schema has become a useful 'tool-to-think-with' for the learner.	7. Promote the formation of an ever more powerful schema, by the linking of the new schema to other related ones, which have been previously learned.
		8. Develop proficiency in the use of the new conceptual schema by regular application to ever more varied problems.	8. Develop proficiency in the use of the new conceptual schema and in its extension or reorganization, through practice and organized self-analysis or reflection.
Observation	It is important to distinguish between the knowledge of a concept (ie the rules or examples that define its boundaries and attributes) and the skill of applying a concept to a practical situation (ie the precision with which actual examples and non-examples are classified). The above-mentioned procedures invariably involve both of these. Also, further skill analysis may be necessary in order to identify critical sub-skills (for concrete concepts, there are often special levels of visual perception skill which need to be developed).		
Further reading	*Producing Instructional Systems*, Chapter 8, Section 8.3.		

Map 3.19 *Tactics for teaching principles*

Planning the tactics for the teaching of principles	If the knowledge content of the sub-objective is . . .	and if the strategy selected is:	
		expositive, then . . .	experiential, then . . .
	PRINCIPLES *of nature*	PREREQUISITIES: simpler, component concepts and principles, used in the definition or explanation of new principles.	
	Rules that govern the behaviour of our environment — physical, geographical, economical, social and so on. Such rules are mainly cause-effect relationships, but may of course have any number of 'however if' special cases. As we saw in the case of defined concepts (which are after all a somewhat different form of rule, but still a rule), two alternative approaches exist which may be called RUL-EG and EG-RUL. These are reflected in the two procedures described.	1. Define the new principle. Show its application with one or more well chosen and varied examples.	1. Organize experience which should enable learners to identify the cause-effect relationships that are in operation.
		2. Test for comprehension of the principle, by asking learners to give or identify yet other examples of the applicability of the principle.	2. Test for comprehension of the cause and effect relationships identified in the experience(s). Can learners define these relationships correctly?
		3. Provide practice in the application of the principle to an ever wider range of cases or examples, promoting 'transfer of learning'. Continue until the desired objective is achieved.	3. During debriefing, promote the correct formation of the new principle. The learner should be able to 're-teach' the principle, either verbally or by some other form of symbolic language.
		4. Develop proficiency through regular use of the newly-learned principle in a variety of problem situations.	4. Develop proficiency both in the application of the principle to new problems and in explaining 'why it works'.

Map 3.19 *(continued)*

		PREREQUISITES: any other principles or concepts to be used in defining or exemplifying the new principle.	
	of action		
	Rules that govern the behaviour of the holder of the principle, in his approach to life, to interaction with others, to specific types of problem-solving tasks, etc. These are often referred to by the name 'heuristics' — ie decision-making rules for probabilistic situations, in the light of incomplete information about a problem.	1. Define and show application. Set a personal example by applying the principle whenever it is relevant.	1. Set up interactive simulation/game or other experiential situations, which may enable learner to identify the principle.
		2. Test for comprehension (as above).	
		3. Arrange simulated practice sessions, that demand the application of the new principle. Repeat as often as necessary.	3. Promote the correct formation of the principle, by the learners, who should 'teach back' the rule(s) to the teacher during debriefing.
		4. Present ample real-life opportunity to practise.	4. Present real-life opportunities to practise and refine the principle(s).

		PREREQUISITE: the component principles and concepts which are to be linked up in the proposed system of principles.	
	systems of principles		
	(General theories or strategies.) A system of interrelated 'rules of nature' forms a 'theory' which is capable of explaining a much wider range of situations, or predicting the results of a much more complex experiment, than any one of the component rules. Similarly, a system of interrelated 'rules of action' or 'heuristics', forms a 'strategy' which is capable of solving a much wider variety of problems than the component rules on their own. We proceed to ever greater levels of generalization, as we build ever more complex strategies and/or theories out of the individual rules that we learn or observe in our environment.	1. Present/describe the theory or strategy in a clear and complete presentation, using all relevant aids to communication: graphs, visuals, tables, flow charts, etc. Present one or more clear examples of its application.	1. Arrange a simulation/ game experience which will involve the learner in using several already known principles in combination. Learners should identify a set of relations between the principles.
		2. Test for comprehension of the presentation, by asking for explanations of given phenomena by means of the theory under study, etc.	2. Test for understanding of the complex relationships identified in the experience. A challenge to formulate and test hypotheses.
		3. Arrange practice in the use of the theory/ strategy to explain phenomena/solve problems in its domain. Extend the variety and difficulty-range of the problems or phenomena to match the variety that exists in reality.	3. Promote, by questions/criticisms, the build-up of a theory/strategy by the learner. Check validity. Challenge any inconsistency/ incompleteness by posing appropriate situations/problems for analysis by the learners.

Map 3.19 *(continued)*

		4. Develop proficiency by regular use of the strategy/theory on real-life problems.	4. Develop proficiency in both using the new theory/strategy in real life and in the further extension of the knowledge structures formed.
Observation		The teaching of *principles* and of *defined concepts*, is seen to be very similar in our presentation here. This is not surprising, as defined concepts (as the term is used here) are a form of 'rule'. The only difference is that the rule defines 'what is a . . .', instead of defining a relationship between two or more concepts ('if . . . then . . .'). However, in both cases, students must learn (or discover, depending on the type of strategy selected) a verbal definition (sometimes a symbolic definition in some other special 'language'). For this reason, the fact that many authors (eg Horn, Pask and many others) consider principles as part of their definition of 'concept-learning', gives us no problems.	
Further reading		*Producing Instructional Systems*, Chapter 8, Section 8.4. Also, *Designing Instructional Systems*.	

Map 3.20 *Tactics for teaching procedures*

Planning the tactics for the teaching of procedures	If the knowledge content of the sub-objective is . . .	and if the strategy selected is:	
		expositive, then . . .	experiential, then . . .
	PROCEDURES *linear step-by-step* ('chains') (This category includes many simple assembly or production tasks carried out in factories. Also, many basic mathematics or arithmetic procedures are composed of a linear set of steps.) *Observation:* We may use the shorthand notation of D — Demonstrate P — Practise T — Test performance for each of the steps of a linear procedure. Thus P2; T3, means 'Practise step 2 and then Test step 3'. We use this notation in this map	(Two basic forms of expositive procedure exist, with some variations possible on each.) For a three-step procedure, as an example, we may have: — — — — — FORWARD CHAINING 1. D1. 2. P1. 3. D2. 4. P1; P2. 5. D3. 6. P1; P2; P3. 7. T1; T2; T3. — — — — — BACKWARD CHAINING 1. D3. 2. P3. 3. D2. 4. P2; T3. 5. D1. 6. P1; T2; T3. 7. T1; T2; T3. (Recommended when rapid learning of a specific job or procedure is all that is required. Backward-chaining is preferred when it is easy to put into practice.)	The forward-chaining strategy/sequence is not suitable for use in the context of an experiential strategy. It is difficult to 'discover' the first stage of an assembly task, for example, if one has no idea of how the whole task is structured. The backward-chaining sequence is, however, very suitable for the 'discovery' approach. Starting with a completed assembly, one asks the learner to disassemble one step and then to reassemble; then two steps from the end, then three steps and and so on until the learner has discovered all the steps, right up to the first, of correct assembly. (Recommended when reflection on the rationale and the transferability of the procedure are to be encouraged — eg most maths/arithmetic procedures may benefit.)

Map 3.20 *(continued)*

	decision-making (between several alternatives) (also called 'discrimination')	(Unlike the case of the linear procedure, decision-making points, or 'discriminations' between several alternatives, should not be split into a sequence of steps, using 'parts' or 'progressive parts' sequencing. This would artificially divide up the task which, in practice, requires all the alternatives to be considered together.)	
	Decision-making, at its simplest, is the selection of one among a series of options open to the decision maker at a given point. If the decisions are algorithmic, then a precise set of conditions (rule) is associated with each alternative choice. The learner must only 'know the conditions' for each option.	1. Present and explain the alternatives open to the decision maker at a given moment. State the rules that lead to every one of the decisions. Illustrate the decision-making process with examples.	1. Set up an experiential situation, which allows learners to try out all the decisions possible and observe the outcomes. From this the learner should be able to derive the 'conditions' which lead to the selection of each alternative option.
		2. Test basic comprehension of procedure by asking learner to apply it to a representative sample of relatively simple examples.	2. Test for basic understanding of the rules that govern each choice of alternative. Do this by observation of experimental activity and questions.
		3. Arrange practice with a graded set of gradually more difficult/complex decision-making tasks that involve the use of the procedure just learned.	3. Promote the formulation of a set of rules to govern decision-making in the types of situation under study. Check that learners form a complete and correct set of rules.
		4. Develop proficiency by further regular practice of the decision-making procedure.	4. Develop proficiency in the application of the student-discovered procedures to an ever wider selection of problems.

Map 3.20 *(continued)*

algorithms
(More complex procedures, made up of combinations of several linear procedures, linked by decision points. The decision points need not be 'binary' when human decision-making is being analysed. Often, such complex procedures incorporate 'loops' that imply the repetition of one or more of the linear procedures, dependent on the outcome of one or more of the decision procedures. The presence of loops or multi-option decision points makes the algorithm look more complicated, when drawn out, but does not unduly complicate the basic instructional tactics that would be employed to teach it.)
	1. Start at the 'outcomes' or end products of the procedure and 'backward chain' through the whole algorithm in stages. For example, if three linear procedures (*e, f, g*) lead to three outcomes (1, 2, 3) as a result of one decision procedure (Z), we can represent the algorithm as follows:
	
	   (Z?) Procedure (*e*) → Outcome 1
	           Procedure (*f*) → Outcome 2
	           Procedure (*g*) → Outcome 3
	
	The instructional sequence would be: FIRST teach the three linear procedures (*e, f, g*) in any order and by any preferred strategy (say experiential backward chaining); SECOND, teach the decision-making procedure (Z) by any preferred strategy (say, expositive presentation).
	
	2. Continue working 'backwards' from any other outcomes. For example, if this is the complete algorithm:
	
	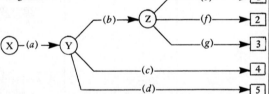
	
	then the THIRD stage will involve the teaching of linear procedures (*b, c* and *d*); the FOURTH stage will be the teaching of decision procedure (Y); and the FIFTH stage will be the teaching of the initial procedure (*a*).
	It is probable that the decision (X), which is 'let's carry out the task', does not need to be taught, as it is no doubt understood from the outset, as the objective of the whole lesson.

Further reading	*Producing Instructional Systems*, Chapter 8, Section 8.5.

Map 3.21 *A procedure for detailed lesson planning*

Introduction	Now the time has come to bring all the tactical decisions together, into a detailed lesson plan. However, there is a particular way of planning a lesson — a particular form of worksheet, which we recommend for the planning of a variety of types of lesson. It may be used for planning conventional teacher-led expositive instruction, group-learning activities, games and simulations, some types of self-instructional print-based materials and some CAL packages. We used this form of planning sheet throughout the first volume of this series (*Producing Instructional Systems*) and we shall be using it again at times in this one.
The lines of communication in a system of 'instruction'	We have, throughout this series of books, considered a 'true' instructional process as one that provides *at least three* types of communication channel between the 'instructor' (whether human or a self-instructional system) and the 'learner'. The following diagram is well known. INFORMATION SYSTEM OF INSTRUCTION (USUALLY THE TEACHER) ⟶ BEHAVIOUR ⟵ SYSTEM OF LEARNING (THE LEARNER OR A GROUP) FEEDBACK AND CONTROL
The three-column lesson-plan	It seems a good idea, when setting out to plan INSTRUCTION, therefore, to provide consciously for the design of each of the three main activities/communication channels that are part of an instructional process. One way to do this is to use a three-column lesson-planning sheet. The columns would be entitled: INSTRUCTOR ACTIVITY / LEARNER ACTIVITY / FEEDBACK/CONTROL Behaviourists (and some other instructional designers) may like to use the alternative column titles: STIMULUS / RESPONSE / REINFORCEMENT Yet another near-equivalent set of titles would be: INFORMATION / BEHAVIOUR / FEEDBACK And in a CAL application, I have used the following: CRT SCREEN INFO. / KEYBOARD/LIGHT PEN / FEEDBACK/CMI This last one has been 'tailored' to a particular system that will deliver the lessons and which can only present new information on the (CRT) monitor screen, can only accept keyboard and light pen responses from the student, but as 'feedback/control' can present corrective feedback on the screen, can make warning noises (beeps) and of course keeps very full records of all student responses and transactions, which are used for summative evaluation, for cumulative-record decisions on what learning feedback to give the student and for courseware formative evaluation.
The three stages of a lesson	In addition to the communication aspects, however, there are sequence aspects to the planning of a lesson. In short, a lesson should commence with some opening activities, then the main instructional activities should be performed, and finally, the lesson should end with some closing activities. There are several possible activities in each of these stages. Not all are obligatory in every lesson, but they should be considered as 'candidates' for inclusion.

Map 3.21 *(continued)*

The lesson planning form: structure and activities	Many writers have suggested typical 'steps', or 'events' (Gagné, 1974) or 'tasks' or 'activities'. We prefer the latter term because the plan describes the activities of the sub-systems involved in the instructional process. The term 'events' conjures up the idea of 'points in time' (as in the PERT project planning/control technique) or also 'objectives achieved' (targets reached). However, the list of activities used in the planning sheet is strongly influenced by Gagné's (1974) suggested list of 'instructional events', as well as the views of many other writers on the subject of lesson planning.		
	INFORMATION	**BEHAVIOUR**	**FEEDBACK**
OPENING ACTIVITIES 1. Attention and motivation	'Now hear this!' 'Have you ever . . . ?' 'A funny thing . . .'	'?????'	
2. Statement of objectives	'This is where we are going . . .' 'This is why it's useful to . . .'	'Aha!'	
3. Recall/ revision of prerequisites	'Do you remember . . .?' 'Can you still do . . . ?'	'@@@@'	'Well, not quite; it's like this . . .'
MAIN INSTRUC- TIONAL ACTIVITIES (for each lesson task or objective in sequence)	(This part of the lesson may loop around several times, depending on the number of specific objectives and the 'parts' into which a given instructional sequence has been divided)		
4. Instructional activity	'Here is the news . . .' 'And this is what you should do . . .'		
5. Learning activity		'Busy, busy, busy'	
6. Feedback activity			'Oops . . . let's try it again . . .'
CLOSING ACTIVITIES 7. Transfer of learning	'And now for something completely different . . .'	'More busy, busy'	'Can you see the relation between this and last week's?'
8. Evaluation of the lesson	'So let's see where we have got to . . .'	'Do the post-test'	Results and comments.
9. Summary and guide to further study	'Here's where we can go now – if we so wish.'		
Observation	The short comments placed here are just a light way of indicating how the lesson planning sheet is completed. The examples which follow in the next map are more indicative of the level of detail that may be registered. However, one may vary the level of detail, the structure and even the exact number of columns in the chart, to suit any special project needs. (We often have a fourth column for timing.)		
Further reading	*Producing Instructional Systems*, Chapter 10.		

Map 3.22 *Lesson planning: an example*

Introduction	The example lesson plan shown here, first appeared in *Producing Instructional Systems*, together with many others. This one has been chosen for reproduction here for a number of reasons:

☐ it is a pretty complete and detailed plan for a fairly well understood topic and uses a well known instructional method (brainstorming);
☐ it loops several times through the steps of the instructional activities, for successive lesson 'parts';
☐ it has multiple and varied objectives (output and process);
☐ the main strategy is experiential rather than expositive;
☐ it involves small group learning activities.

Readers are invited to compare this plan, if possible, with others using the same planning methodology, which are expositive, teacher-led, mediated, computer-based. Such examples can be found in *Producing Instructional Systems* and in this book, in later chapters.

The lesson plan

Lesson title: Solving human relations problems in industry
Target population: Foremen and supervisors from organization XYZ
Number of trainees: 10-15   Time required: 8-10 hours

Lesson objectives:
1. Given typical human relations problems in an industrial setting, the trainees will be able to: (a) analyse the problems, identifying the principal causes; (b) apply known principles of human relations management to the problems in order to suggest appropriate forms of solutions. Both the analysis and the synthesis should be judged as 'correct and practically viable' by a skilled manager of human relations.

2. A subsidiary 'process' objective. Given a problem to be solved by a group of employees, organize the problem-solving session in the form of brainstorming and lead the session through its various stages, keeping to the rules of the technique and achieving, in a reasonable time (one hour) a satisfactory set of possible solutions (satisfactory in terms of the number of different solutions and the probable viability of at least some of them).

Equipment/resources required:
A minimum of six case studies of typical human relations problems of the type commonly encountered in organization XYZ. The cases described should be relatively difficult to solve, but should offer the opportunity to consider a number of different types of solution. Large blackboard or wall space suitable for cards or paper sheets. Writing materials suitable for the group to be able to see the charts and posters comfortably.

Instructor activity	Learner activity	Feedback activity	Time *(minutes)*
Opening events			
1. Attention/ motivation Tell the story of Solomon and the two women claiming the same child, or some other tale that illustrates the use of innovatory solutions to human disputes. Lead the group to analyse the human relations situation in organization XYZ.	Participate in the storytelling and draw parallels between the situations described and the reality of working in the company.	Observe the participation of the trainees in the discussion, and identify signs that indicate reluctance to solve human relations problems, or other motivational deficiencies. Try to correct attitudes before proceeding to the main part of the lesson.	5-15 (depending on the attitudes of the trainees)

Map 3.22 *(continued)*

Instructor activity	Learner activity	Feedback activity	Time *(minutes)*
2. Briefing Define the aims and rules of brainstorming. Explain how the technique will be used in the present exercise.	Suggest a list of advantages that brainstorming may offer in the context of XYZ.	Clear up any doubts that may exist concerning the aims or the methods of the technique.	5
3. Prerequisites By means of a question and answer session, recall the principles of good human relations management (these were taught in earlier lessons).	Reply to the questions, showing adequate mastery of the principles concerned.	Assess and, if necessary, reinforce earlier learning, by means of suitable revision exercises.	Depends on needs. If no revision is needed, 5 will suffice.
Main strategy **4.1 Ideas generation** Present a typical human relations problem as the basis for the first exercise. Ask for suggestions of possible causes. Write down all suggestions, and display them on wall.	Make suggestions of all possible causes of the problem, including unusual and improbable ones.	Control session, quashing any attempts at premature criticism of contributions. Encourage shy trainees to make suggestions.	About 15
4.2 Ideas organization Ask trainees to classify, improve or eliminate items in the list of causes. Identify the most likely causes.	Trainees may now comment and criticize earlier suggestions, as long as they are constructive.	Control any tendency to purely destructive criticism.	5-10
5.1 Generation — phase 2 Now ask for suggestions of possible solutions. Write down and display all the suggestions on the wall or blackboard.	Trainees should make as many and as varied suggestions as they can in the time available.	Control and encourage, as in step 4.1.	10-15
5.2 Organization — phase 2 Ask trainees to classify, improve or eliminate items in the list of possible solutions. Identify the most plausible solutions.	See step 4.2	See step 4.2	5-10
5.3 Ideas evaluation Lead trainees to the evaluation of the alternative plausible solutions and to selection of a solution to be implemented.	Trainees should discuss the pros and cons of each plausible solution previously identified. They should reach a consensus on the most viable one.	Evaluate the process of discussion and critical evaluation performed by the trainees. Evaluate the choice of solution.	15-20
6. More practice Repeat steps 4 and 5, using the other five cases of human relations problems. In each case, appoint a different trainee, or a team of two or three trainees, to organize and coordinate the session.	Each trainee has the opportunity of leading a brainstorming session, as well as ample extra practice in the analysis and solution of human relations problems.	The feedback activities that are listed in steps 4 and 5 above are now executed by the appointed trainees. The instructor observes the session as a whole and the performance of the leaders, in order to redirect them if necessary.	About 1 hour for each of the exercises (total 5 hours)

Map 3.22 *(continued)*

Instructor activity	Learner activity	Feedback activity	Time *(minutes)*
Closing events			
7. Debriefing Start a discussion on the solutions chosen. Question the theoretical bases for each solution. Question the practical viability. Invite analysis of the way in which the solutions were selected. Question the overall usefulness of brainstorming. Question the value of this particular exercise.	Trainees should demonstrate, by their arguments and comments, that they have formed sufficiently powerful and general problem-solving strategies and associated schemata of the principles involved. They should also describe how to implement the proposed solutions in the real-life situation of XYZ.	Evaluate the strategies and schemata of the trainees. Take decisions on whether more practice exercises are required to ensure transfer of the new skills to the job situation. Extend and/or replan the training sessions if necessary.	About 45
8. Transfer of learning Arrange for regular small group discussions, similar to the debriefing session described above, to analyse the real life decisions taken, after training, and to resolve problems occurring on the job.	Trainees should describe to the group how they came to make their decisions. The group analyses the decisions and predicts or comments on the end results.	As in item 7, the instructor attempts to evaluate the structures and strategies that are developing in the trainee's mind, in order to offer useful advice.	No prediction of the time needed

Map 3.23 *Level 4 instructional design: an overview*

Introduction	The flow chart presented here, starts with the 'detailed lesson plan' of Level 4 and illustrates where we go from there, as we descend to the level of development of purpose-designed, objectives-based, individualized, instructional materials. The chart acts as an overview of what is to come in the remaining sections of this book. This is its major function. It does not intend to be an exhaustive or comprehensive model of Level 4, but rather just a brief summary/overview, to get us started.

Process	Questions and activities of design team	Analysis and design/development procedures and their outcomes

Questions and activities of design team:

From Level 3
The detailed lesson plans.
The materials specifications.

At Level 4
ID EXPERT:
What message design principles should we follow?
Learning step size?
What learning tasks?
When to test-out?
What CML records?

SM EXPERT/
TEACHER:
What language level?
What examples known?
What analogies?
What jargon to use?
Common misconceptions?
Common errors?
Good motivators?

MEDIA EXPERT:
What media mix?
What author style?
Use of humour?
Use of visualization?
Use of colour?
Use of movement?
Special effects?

EVALUATOR:
Developmental test time-chart?
Developmental test subjects?
Field-test time-chart?
Field-test subjects?
Formative procedures?
Summative procedures?
Overall production project evaluation?

Analysis and design/development procedures and their outcomes:

	Information	Behaviour	Feedback	Time
1.				
2.				
3.				
4.				

Further analysis of lesson plans and materials specifications

Message design outlines (story-boards)	↔	Exercise design outlines (Roles/cases)	↔	Feedback and control outline plans

Further analysis of target population and subject matter content

— Texts — Scripts — CAL screen plans	↔	— Role-plays — Games — Simu-lations — Case studies	↔	— Observer- Evaluation sheets — Study guides — CMI control algorithms

Analysis of media options, techniques, available resources, visual communication, etc

Prototype materials ready for developmental (and field) testing

(First wave of revision)

Revised prototype materials for field pilot test

(Second wave of revision)

Final (first issue) version of courseware for further pilot testing

Observation	The four 'expert' functions defined in the left-hand column are rarely performed by four different people. At this stage, the ID expert is the evaluator and in some projects may be media or subject matter expert as well. Think of the categories as 'functions', not 'people'.

PART 2.
Instructional Programming and its Evolution

Overview

In this part, we deal with various techniques for the development and writing of largely *textual* auto-instructional materials. The specific techniques selected are a reasonably representative sample of well tried and tested approaches, based on different theoretical principles, but all sharing the aim of 'individualized self-instruction'. They are organized in a more or less chronological sequence, in order to illustrate how the approaches have changed with time and with the successive predominance of one or other theoretical viewpoint on how learning occurs or what is important to teach. The specific examples are also selected in order to illustrate that, despite the differences in approach, there are certain basic ID principles that may be identified in all of them. We shall see in later parts of the book, when dealing with computer-based instruction and audiovisual instruction, that the same general principles appear and that many of the techniques of ID described here, in relation to the preparation of printed instructional texts, may also be effectively employed when working in other media of communication.

We open, in Chapter 4, with an overview of techniques for the more detailed analysis of the intended learning objectives, the learning content and the learner's individual difficulties. This must be carried out in order to ensure that the self-instructional materials to be developed will indeed be effective. We progress, historically, from the *skills analysis* methodology developed by Douglas Seymour and others in the 1940s and 50s (based on the principles of work study and measurement), through the *behavioural analysis* methodology developed by Thomas Gilbert and others in the late 1950s and through the 1960s (based on behavioural psychology principles), to the *mental operations analysis* methodology developed by Lev Landa and others in the late 1960s and through the 1970s (and based on a combination of principles drawn from both behavioural and cognitive psychology, 'knitted together' by information-processing theory and cybernetics). The chapter is liberally illustrated with practical examples of specific analysis techniques, developed by a variety of authors. Two techniques of analysis are singled out for more detailed study, as they are the basis of examples of materials design methodology to be studied later. These are (a) the behavioural analysis methodology, as developed by Gilbert (1961) in his treatise on MATHETICS, and (b) the content analysis methodology developed by Evans, Homme and Glaser (1962) as part of the RUL-EG approach to the writing of self-instructional materials. Although both of these date from the 1960s and are associated with the early days of programmed instruction, we shall see that they are still currently used in the development of printed, computer-based and even audiovisual instructional materials. It is as well to come to understand the strengths and weaknesses of these approaches.

Chapter 5 is devoted to the analysis of programmed instruction as it appeared in the 1960s and developed through the 1970s. Basing our study on the two analysis techniques studied in depth in the previous chapter, we now analyse the materials-development techniques that utilize these analyses as input. We start with the Rul-Eg approach, probably the first and most commonly used formal methodology for the development of programmed instruction. We analyse a typical programme written in this way. Then we study the use of the MATHETICS approach, by analysing a programme on the same topic, but written from a behavioural analysis, rather than a content analysis starting point. We shall see these two approaches recurring in many of the examples in later chapters.

Chapter 6 presents a how-to-do-it account of the MATHETICS approach, as first developed by Gilbert in the 1960s and later extended and refined by the instructional designers of Learning Systems Ltd, in the United Kingdom, during the 1970s. It is probably fair to say that this company was one of the most successful 'programmed instruction' materials and training system design consultancies to be set up in the 1960s (it is still in existence today in the 1980s) and much of its success must be due to the constant development of practical tools and methodologies for effective and efficient instructional design. The mathetics approach became the backbone of the ID process used there, in the mid-1960s, but was much modified and developed over the years to be, to this date, the source of a set of powerful techniques for the design of instruction, be it instructor-led or self-instructional, be it print-based or audiovisual. This chapter, as indeed the previous one, has been written with the collaboration of some of the instructional designers from Learning Systems Ltd, with whom the author had the opportunity to work and exchange ideas over a period of many years.

Chapter 7 breaks away from the behaviourist-oriented approaches described so far, in describing STRUCTURAL COMMUNICATION, probably the earliest attempt at a formal methodology for the development of self-instructional materials based firmly on the principles of learning espoused by cognitive psychologists. This was developed in the 1970s by Anthony Hodgson and his collaborators, at the Institute for Structural Communication in the United Kingdom. The approach showed much promise, not least in that it could be used in both expositive and experiential learning modes, could be under system control or partial learner control, could evaluate the quality and the stucture of the conceptual schemata that were being formed in the mind of the learner and, right from its initial conception in the late 1960s, was envisaged as a methodology for the design of computer-based instructional systems, as well as auto-instructional texts. This chapter was written in collaboration with Anthony Hodgson and includes some of his earlier writings, now difficult to locate, on the concept-structure analysis methodology used. It also includes an example of part of a learning unit written by means of the structural communication methodology. This example was, of course, print-based and illustrates well the advantages that computer-based presentation would bring. It is surprising, therefore, that when the Institute for Structural Communication ceased to exist, due to lack of funding, the methodology seemed to be largely forgotten. Its potential for this day and age of plentiful and cheap computing power is analysed in this and some later chapters.

Chapter 8 discusses another approach, born at the end of the 1960s and developed as a formal technique during the 1970s, which appears to be 'coming into its own' in the 1980s. This was first developed by Robert Horn and his collaborators, under the name of 'Information Mapping', but this name is now a registered trademark of Information Mapping Inc, of Lexington, Massachussets. Several other names have therefore been coined, for example 'Infostruct', and several other companies and individual consultants offer materials design services which are derivative from Horn's original ideas. The methodology described here is true to the original procedures developed by Horn, who indeed collaborated on the writing of this chapter. We therefore refer to them here as 'Structured Writing according to the procedures of the Information Mapping (TM) methodology', or in short as 'mapping'. The presentation of information in the form of structured 'maps' presents many advantages: our own use of 'maps' in several of the chapters of this book serves to illustrate many of the more obvious of these. Horn's discussions in this chapter will help to identify some of the less obvious ones. One advantage, just hinted at here, is the applicability of the technique to the design and development of computer-based information and instructional systems, suitable for initial learning or rapid reference. In later chapters we shall return to this aspect in more detail. Like structural communication, 'mapping' is another technique of great potential for computer-based instructional system design, which is being largely ignored by the 1980's CAI movement. We hope to illustrate this potential as we progress through this book. In this chapter, however, we concentrate on the use of the methodology for the preparation of textual materials.

Finally, Chapter 9 breaks more new ground, in addressing the general approaches and specific techniques used to write the materials often required in experiential learning exercises – case studies, role-play briefs, simulations, etc. These techniques have been developing in parallel with the previously described 'programmed self-instruction' group and have not, in general, been considered as part of 'instructional technology'. Of late, however, the growth of the simulation/gaming movement within the realm of ID, has brought these two camps together. In this chapter we trace the development of techniques for case study and role-play writing since the 1950s and show where these commenced to merge with instructional development techniques. We note that, despite the fact that most case study, role-play and simulation/game exercises are examples of group-learning, they are nevertheless also examples of the individualization of the teaching-learning process and, thus, qualify for our attention in this book.

4. The Fourth Level of Analysis: Task-Centred and Topic-Centred Approaches

4.1 Individualized instruction and materials design

In Chapter 1 we analysed the concept of individualization of instruction and came to the conclusion that individualization is a question of degree, measured by the frequency with which the instructional process adapts in response to the needs of an individual learner. Instructional systems may be individualized at one level of detail, and not at another.

For example, a university degree course may be individualized in the sense that the student may choose from a wide range of credit courses, in order to adapt the content and objectives of the course to his own needs and interests. At a lower, more detailed level, however, there may be very little individualization in some of the credit courses. Once the course is chosen, the student follows the same sequence of units and lessons as all the other students. In other words, the overall instructional plan (Level 2 instructional design) does not offer much in-built individualization.

At a yet lower, more detailed level, however, a particular professor may inject a fair dose of individualization by adopting student-centred teaching methods, using small group discussion, individual project work, and self-study assignments, with plenty of feedback at every step of the lesson. In Chapter 3 we studied some of the techniques that may be used to individualize the planning and execution of specific lessons. We called this Level 3 instructional design.

Level 4 analysis
In this chapter we descend yet one more level, to consider the design of individualized instructional materials. We have called this Level 4 instructional design. It is the most detailed of the four levels, as the production of instructional materials implies the pre-planning of all instructional decisions, right down to the words that should be used in a communication, the structure of the sentences, the number of examples, the specific examples to be used, the number of repetitions and all the other minute decisions that, in a teacher-delivered lesson, are left to the teacher to decide on-line as the lesson proceeds.

Such a deep level of decision making should be based on an equally deep level of analysis, capable of revealing the individual learning difficulties that occur in the study of a given topic and the causes of these difficulties. The analysis should suggest the possible sequences of the steps of instruction, even the sequencing of the sentences in a communication or of the movements in a skilled activity.

Of course, analysis is only a tool that should be used to achieve certain aims – in this case, to design messages that really communicate or exercises that really develop the skills that the learner needs. Sometimes, it is possible to carry out a successful design on the basis of past experience or by adapting existing materials, without the need for a large amount of time to be invested in detailed analysis.

At other times, however, we are faced with a topic which students always find difficult to understand, or a group of objectives that always require a very long period of learning, or create unusually large learning difficulties for some students. In such cases, it seems, past experience and effort was not sufficient to overcome the inherent learning difficulties. Most teachers seem to have difficulty in teaching the topic successfully. No really effective instructional materials are on the market. In these situations, one of the techniques of Level 4 analysis discussed in this chapter may lead to the identification of the root of the difficulties and thus to the design of effective and efficient instructional materials that many teachers will be able to use.

This aspect of 'multiplication' of the use of a well designed set of materials is very important. The investment of time and effort required to carry out a full Level 4 analysis is considerable, and can only be recouped if the resulting instructional materials are used by many students. This implies the publication of the materials and their distribution to many institutions and individuals, either through bookshops or by means of mass distance education.

4.2 Two approaches to materials design and two forms of analysis

We may distinguish two approaches to instructional design:

1. The *task* or *objectives*-oriented approach.
2. The *topic* or *content*-oriented approach.

Task or topic?

In the first approach, the starting point is the definition of the objectives to be achieved by the end of the course that is being designed. All other decisions spring from the definition of the objectives: the design of valid tests, the sequencing of instruction, the selection of content, the methods, the media and the system of implementation and control.

In the second approach, the starting point is the definition of the content to be communicated, studied and used. The selection of the content – the topics to be taught – is one of the first decisions to be taken. Objectives are not always clearly defined and, when they are, they are derived from the content at some point well into the instructional design process.

We might say that if specific objectives are not defined at some point in the process, then the resultant course, or set of materials, does not qualify to be called an *instructional* system. However, we see no reason why, in some circumstances, objectives should not be derived from previously defined content. This second approach is indeed the more common in the conventional educational context, and educational systems have worked more or less satisfactorily on this basis for centuries. The first approach has more relevance in the job-related or vocational training context, where there is always a *model of the desired post-instructional performance* to be analysed and thus specific objectives can be defined as the starting point for design. We call this model the 'master performer'.

In *Designing Instructional Systems* we referred to these two approaches as:

1. The 'outputs to inputs' approach.
2. The 'inputs to outputs' approach.

We showed that both could be used and, occasionally, one could be converted into the other in mid-stream as it were. In this book, we shall be following the outputs to inputs route as a general rule. However, in this section, we shall discuss both approaches as they are applied to the development of self-instructional materials. Each approach depends on a different form of analysis. We shall start with outputs to inputs techniques.

4.3. Techniques for objectives-oriented analysis: output to input

Skills analysis

Unlike Level 3 analysis, which is characterized by a more or less direct matching of objectives at the lesson level to instructional tactics, Level 4 analysis analyses the objective yet further, breaking it down into component behaviours or skill elements. It is broken down in this way to identify learning difficulties with greater precision and so to specify the instructional tactics, or rather to design the detailed steps of instruction more precisely.

The descent to such a detail of analysis must be justified, and the justification is that learning difficulties (which cannot be identified at a shallower level of analysis) really do exist. The overcoming of these learning difficulties must further be judged in the economic light of whether the cost, in time and effort, of the more detailed approach is justified by the importance of the objective.

We now consider three approaches, which we may classify as examples of Level 4 analysis. We shall describe two briefly and illustrate the third more fully by means of an example. It is worth mentioning all three because they relate to quite different types of learning.

Seymour's approach to skills analysis

Skills analysis is probably the oldest 'deep analysis' technique of a systematic nature to be widely used. It was developed out of time and motion study techniques and adapted to the design of training for complex operator skills as early as the 1940s by the Seymour brothers (Seymour, 1954; 1966).

Skills analysis takes over where the Level 2 type of anlaysis (eg listing the task steps and their level of difficulty) ends. It is particularly suited to complex semi-skilled operations in industry, especially if they are of a repetitive nature, involve high-speed operation or a high level of manual dexterity. In such tasks it is often very difficult to identify why the novice performs so much worse than the experienced worker. There is little knowledge content to learn and the novice soon masters that. However, he continues to perform way below the experienced worker's standard for weeks, months and, in some instances, years. Is it an inborn knack that is involved? It used to be thought so in some circles, but the skills analysis approach has done much to dispel this view, having been responsible for the creation of training schemes that have in days or weeks imparted the skill level that was previously considered to require months or years of on-the-job experience.

Separate the sensory and motor aspects of the task

The analysis procedure developed by Seymour is based on separating out carefully the sensory information and the motor actions for each minute step of the task. This was done by use of a multi-column form on which there is a provision for describing what each of the limbs and the senses involved actually does. This separation is important, as it helps the analyst to pinpoint the exact differences between the method of working of the expert and that of the novice. These differences may be in the sequence of actions performed, or in the way that certain actions are performed, or in the sensory information that is used to control the actions. This latter source of difference is most commonly the root of the different rates of performance of master and novice. The novice carries out each movement watching carefully what he is doing, while the expert has passed control of certain movements to other senses, particularly to touch or to the kinaesthetic sense of what it 'feels like' to be doing the job correctly. Thus, in many assembly operations that involve, for example, putting a screw into a hole, the novice flounders around trying to get it in by looking carefully at where he is putting the screw. The expert slides the screw from the side, preceded by the pad of his thumb. The touch senses in his thumb locate the hole and then his other fingers guide the screw into position, without any use at all of the sense of vision. The eyes may already be involved in planning the next stage of the operation or in controlling another action, happening in parallel, being executed by the other hand. It is this sort of level of detail that the analyst is looking for.

Once the analyst has identified the chief causes of the difference between master and novice (identified the knacks, if you like), the suggested procedures for instructional design are simple. As there are usually only a few knacks identified, even in the most complex operator task, Seymour suggests that one should teach these first. Often, this will require the preparation of special training exercises to remove these skill elements from the context of the job and to give concentrated practice in the knack. Sometimes this is possible in the context of the real job situation, but more commonly it requires the design of special training 'rigs' or 'simulators'.

Once the skill elements of outstanding difficulty have been learned up to a reasonable standard of performance, these elements are incorporateed into the practice of the whole task. For sequential tasks, made up of a series of definite steps, the preferred tactic is the 'progressive parts' sequencing of practice. The trainee practises the first step A, the next B, then A ɪ B, then C, then B ɪ C, then A ɪ B ɪ C, and so on. If the task is an integrated one, with no fixed sequence of steps, it is practised as a whole.

The mathetics system of Gilbert

The mathetics system of task analysis and instructional design was developed by Tom Gilbert from the behavioural analysis techniques already used by psychologists in the 1950s. He published his system in a two-volume treatise called *Mathetics: The Technology of Education* (Gilbert, 1961).

Mathetics is a complete system of analysis and design, replacing Level 2 and 3 analysis of the more usually practised type, but capable of descending to a deeper level of analysis as and when required. The main difference between the mathetics approach to analysis and most other approaches is that it pays equal attention to the input, or *stimulus*, aspect of behaviour as it does to the output, or *response* aspect. Most other systems of analysis tend to concentrate only on the responses – on what is done – and do not give equal attention to the stimuli that control these responses: the 'when to do' and 'when not to do' aspects of the task.

The basic unit of analysis is the operant, which is one stimulus/response connection. An operant can be large, for example:

Stimulus: **Response:**
Incoming telephone call ⟶ Connect it to the right extension
...or it can be small

Stimulus: **Response:**
Telephone rings ⟶ pick it up and hold to ear

The same notation of analysis can therefore be used at various depths of detail. Rather than using a standardized form, the analyst simply uses a large sheet of paper and constructs a behaviour map or, as Gilbert calls it, a prescription. The analyst prescribes, diagrammatically, how the trainee should be able to perform after training. He does this by observing a 'master performer' and noting down his behaviour. This, in principle, is what any system of task analysis tries to do. However, the mathetics system does it in a particularly thorough way, passing through a series of deeper and deeper analyses (if the problem requires them) until the analyst has a clear idea of how to proceed in designing the instruction. Coupled to the range of special analyses, there are a series of rules or tactics which are brought into play depending on the *patterns of behaviour* which the analyst discovers during his analysis.

Although this approach is elaborate (therefore not always justifiable in its full form on economic grounds) and very behaviour-centred (therefore inappropriate for the initial stages of analysing subject matter or information), the more detailed stages of the method can always be used when the detailed planning of the steps of instruction is envisaged. At this level, the analysis techniques are supplemented by detailed instructional tactics or rules.

The 'mental operations' approach of Landa

Lev Landa is a Russian psychologist who has performed much useful research on the learning and teaching of intellectual tasks. His work is concerned with the analysis of the instructional needs for such subjects as Russian grammar and problem solving in geometry. It is interesting, therefore, to examine his approach to the analysis of such 'school' subjects and compare it with the other two approaches we have discussed (which appear to be more concerned with the analysis of practical tasks).

Algorithmic and heuristic procedures

Landa refers to the 'mental operations' that must be performed in order to solve a problem, be it a standard rule-following (algorithmic) type of procedure, as is the case of applying the rules of grammar to sentence construction, or a more open-ended (heuristic) procedure such as the solution of problems in geometry.

These mental operations are what the behavioural psychologists have tended to ignore, preferring to concentrate on the observable behaviour only.

Landa's approach emphasizes the need to 'know' the specific operations that are required to solve a specific type of problem. He makes a point of stressing that competence in a given type of problem solving involves 'knowing the mental operations involved'. This 'knowing' is made up of knowing how to carry them out and knowing when to use them. He also emphasizes that 'one can know some of the operations involved and not know others' and that this can have a profound influence on performance in actually solving problems. (This reminds one of Gilbert's distinction between 'accomplishment' and 'acquirement' (Gilbert, 1969) and the profound effect that a quite small deficiency in acquirement can have on accomplishment.)

The need for a deep analysis

Landa argues (as does Gilbert) that effective instruction must be based on a previous identification of all the mental operations (Gilbert says 'operants') that are involved and their interrelation. The teacher who is serious about wishing to instruct his students in the solution of a particular category of problem must first know the detailed mental operations that the solution of such specific problems entails, and must then plan his instruction in such a way that he can 'see' whether the learner is performing the operations corrrectly. A deep level of analysis is required of any category of problem solving activity that in reality offers learning difficulties to students. The analysis performed for one type of problem solving may not necessarily help the teacher to teach another type, unless there are marked similarities in the mental operations involved.

The methods of analysis

The publications of Landa's which have been translated into English are not very explicit on how to perform this detailed analysis of mental operations, as they do not include any complete examples of his analyses of geometry and grammar problem solving, only isolated part-examples to illustrate certain points he is making. However, these are sufficient to illustrate the level of detail to which the analysis is taken. After all, one is attempting to analyse mental operations which are not directly observable. Therefore, the analysis must be a reflective self-examining process carried out by a 'master performer'. An instructional designer who is not an expert in the subject under analysis may help the expert, but he cannot alone do the analysis.

The final object is to map out or list all the mental operations and decisions that have to be made by the competent problem solver. These may turn out to be interrelated sequentially in a fixed manner and be of the type that can be transformed into specific instructions that, if followed correctly, guarantee the solution (ie an algorithm). Alternatively, they may turn out not to be related in a specific sequence, nor to guarantee a solution, even if followed correctly but nevertheless are the most effective way known to increase the probability of a solution being found (ie a set of heuristics).

Landa found that using Russian grammar correctly was a set of algorithmic procedures, whereas proving geometry theorems was a set of heuristic procedures.

4.4 The three models of analysis: similarities and differences

The differences

The major difference between the three models for Level 4 analysis that we have so far discussed is in their intended areas of application. Seymour's model was developed specifically for the analysis of industrial skills. It is easiest to apply this model to repetitive, high-speed, physical operations. It needs to be modified to be used for the analysis of tasks in which there is not a fixed sequence of operations, such as the tasks of supervising process equipment. In such tasks the difficulty rarely lies in the physical motions that are demanded by the operations, but rather in the decisions that the operator must make in order to choose between alternative operations. In a plastics production plant, for example, the operator does little actual physical work, other than the operation of some valves and switches. The difficultly lies in deciding when or whether to operate certain controls and by how much, in order to maintain the level of plastic being produced. He has no trouble in learning to operate the controls. His difficulty lies in interpreting the signals (or stimuli) which the various dials, meters or analysis sheets present to him.

In such tasks, Gilbert's approach pays off, as it forces the analyst consciously to look for all the stimuli that may occur and to analyse the desired responses for each one. All the alternative courses of action are explored and then mapped out (using stimulus/response notation). Important discriminations are identified, as are the generalizations (or classifications of situations that require the same response). This enables the analyst to specify the concepts that the operator should master. We are now in the cognitive domain, analysing what the learner should 'know' to be able to 'perform'. Seymour's methodology does not really get involved with the knowledge content of the task in a systematic manner. At most, some observations are made in the 'comments' column to indicate that the analyst has noticed the need for some specific knowledge. The analysis of this knowledge to ascertain whether it might not be the root of the learner's difficulties is better done using Gilbert's methodology.

However, despite Gilbert's original claims for mathetics, it becomes rather difficult for an analyst to map out a behaviour pattern when the intermediate steps of the behaviour are triggered off by internally generated stimuli, that is when the master performer is thinking rather than doing. The analyst may observe the initial stimulus (say a problem in mathematics) presented to the master performer. He may also observe the final response (the solution), but he must use indirect means (questioning or introspection) to identify the complex pattern of intermediate decisions that the master performer made to arrive at his solution. Experienced analysts will testify to the difficulty often encountered in extracting from a skilled performer the exact reasons for his actions, or the exact steps he takes in planning his decisions. There is no easy way around the difficulty. The skilled performer must become the analyst, must become introspective, must try to identify how he makes his decisions, what internally stored knowledge he uses and what are the thinking processes that go on in his head.

It is this type of introspective analysis by the expert that is the basis of Landa's approach. He attempts to identify all the specific mental operations that need to be performed to solve a particular category of problem, together with all the alternatives and the decisions which must be taken to choose between the alternatives. Thus Landa is mapping out the internal perceptions of an intellectual nature that are stimulated by certain aspects of the problem and the resultant intellectual decisions (or responses) that the expert makes. This perception-decision-operation sequence is akin to the stimulus-decision-response sequence used as the unit of analysis (the operant) in mathetics. The difference is that the latter is an externally observable event (at least the stimulus and the response), whereas the former may be a completely internal, not directly observable, unit of behaviour.

The similarities   Despite the differences in the three approaches, one can observe an overlap in the areas of application for which they are best suited and a basic similarity which links them and differentiates them from Level 3 analysis discussed earlier. The overlap in their levels of application can be visualized by drawing a graph (see Figure 4.1). Skills analysis is at its best when the task is composed of operations of high physical complexity (because of high-speed performance or of the need for special perceptual acuity or manual dexterity) and low intellectual complexity.

Landa's approach is needed for tasks composed of intellectual operations of high complexity (intellectual problem solving). Mathetics bridges the gap, being excellent for the analysis of tasks that have both physical and intellectual components of reasonable complexity, but the physical component does not involve high-speed repetitive operations and the intellectual component is not entirely made up of internalized reflective thinking, but is linked to external events (stimuli).

To complete the picture, one might consider the bottom left and top right areas

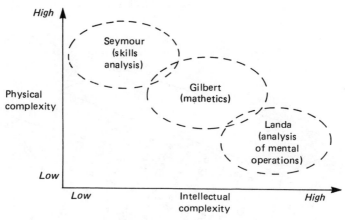

Figure 4.1 *Areas of overlap in the application of the Level 4 models analysed*

of the graph. At the bottom left, no Level 4 analysis should be required, as tasks of low complexity should be effectively taught on the basis of Level 3 or even shallower analysis. The expense of a more detailed analysis is not justified.

At the top right of the graph, there are probably few tasks that are highly complex both in their physical and intellectual aspects. Perhaps the creative arts qualify in some respects. The painter composing and painting an original picture and the jazz musician improvising freely on a basic melody use a high level of both physical and intellectual skill when performing. However, it is possible to separate out these two components, analyse and teach them separately (it is possible to appreciate a work of art and to criticize it without having the skills necessary to produce it – a conductor can 'manage' all the instruments of an orchestra without necessarily being able to play them). Thus such complex multi-faceted tasks can be analysed by a combination of techniques suited to physical operations and techniques suited to mental operations.

The basic similarity in the three approaches discussed (which incidentally differentiates them from most other methods of analysis) is that all three pay attention not only to what is done (motions, responses, mental operations), but also to what triggers off action (perception of relevant signals or stimuli or characteristics of the problem). This focus on the role of perception (both in the physical and intellectual sense) in any skilled activity assists the analyst in identifying the precise roots of learning difficulties that are typically experienced by novices. More often than not, the learning difficulty does not lie in the performance of the operation (mental or physical) but in the perception and interpretation of information in order to decide what operation to perform and when. The emphasis on both perception and resultant action, together with the minute level of detail at which the observations are made, are the chief characteristics of Level 4 analysis.

4.5 A practical example of mathetical analysis

Perhaps all this sounds rather vague, so let us illustrate the process of analysis with a practical example. To put the task into a real context, imagine that we are contracted to produce a training course for the telephonists of a large company. The performance of the telephonists is below the required standard. There are many complaints from other staff and from clients of the company that they are not getting the sort of service that they would like when they try to make phone

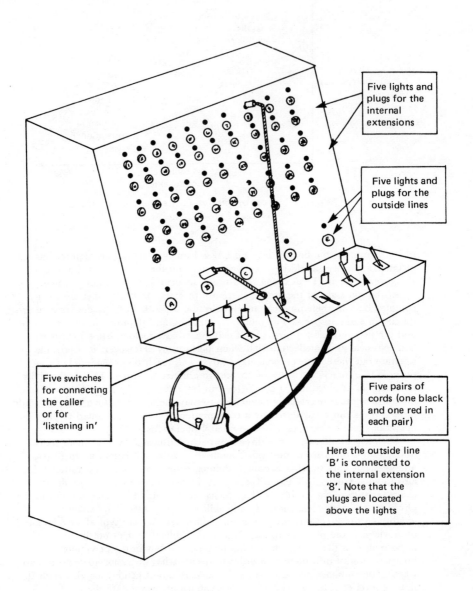

Five lights and plugs for the internal extensions

Five lights and plugs for the outside lines

Five switches for connecting the caller or for 'listening in'

Five pairs of cords (one black and one red in each pair)

Here the outside line 'B' is connected to the internal extension '8'. Note that the plugs are located above the lights

Figure 4.2 *The telephone switchboard*

calls. One serious problem is that incoming callers are often left waiting or are connected to the wrong extension.

We have started to do a job analysis and have identified the tasks that the telephonists perform. We are now focusing our attention on the 'incoming calls problem' and wish to analyse the task of 'answering and redirecting incoming calls'.

The telephone switchboard is of the rather ancient variety, and uses plugs and cords to make the connections between internal extensions and external lines.

Step 1: Describe the task in S-R notation

Using the stimulus-response notation, we can describe the task as one unit of behaviour, as was done earlier in this chapter. However, as we know that this task causes problems, we need to analyse it in more detail. Through observing a telephonist or indeed by reference to the manufacturer's manual, we identify the basic structure of the task as follows:

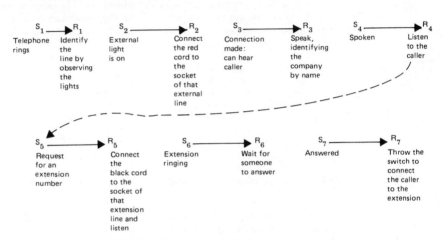

Note that we have expanded the task into seven units or 'operants'. This is as far as observing a perfect, trouble-free call, attended by a good telephonist, gets us. But naturally, this is the ideal situation. The telephonist must also be able to deal with non-standard situations. What variations from the norm can we expect? In most systems of analysis one hopes to find this out by asking open-ended questions like 'what can go wrong?' and by observing many telephonists in the hope of catching all the varieties. The mathetics system has a more foolproof way, called the analysis of stimulus variation.

Step 2: Analyse the possible stimulus variation

Taking each operant in turn, the analyst asks 'what are the other stimuli (consequences) which can result from making this response?'. For each stimulus identified, he asks 'if this does occur, what should the telephonist do?'. In this way, every operant of the original prescription is expanded to include all the variations that might be expected to occur, even in the most extremely rare cases. For example, take the operant:

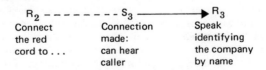

We can identify a number of variations of stimulus, leading to further responses that the telephonist should be able to make, which in turn lead to further possible variations of stimulus. In laymans's language, we are trying to map out all the various things that can happen at this point in the task.

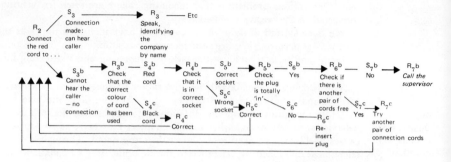

The analyst should initially consider the widest possible range of variation in the stimuli, to make sure he does not overlook the important ones that affect the standard of job performance of the telephonists. In doing this, operant by operant, he will come to consider stimulus number 5:

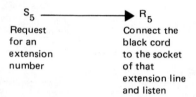

In analysing the stimulus variations possible in this case he will come up with at least the following variations:

* Request for an employee by name.
* Request for a certain department.
* A question or statement concerning some matter of concern to the company, but without any indication of extension, person or department that should deal with it.
* A question or statement which is of no concern to the company (eg do they manufacture beds, when in fact they manufacture fertilizer for bedding down plants).
* A wrong number (not always obvious initially).
* Heavy breathing.
* Silence.

Each of these variations leads to an appropriate response that the fully proficient telephonist should be able to make, often to a series or chain of further stimulus-response units or to even more complex patterns of behaviour. We shall see some of these in a moment. Of course, not all the variations create learning problems. Some occur so rarely that the analyst soon decides that it would not be effective to teach them as they would be forgotten before the chance to practise occurred. Others are so obvious or so well known to just about anyone that they do not need to be taught. The next stage of analysis therefore attempts to establish which of the variations in fact produce learning problems. The analyst tries to establish the 'training deficiency', in other words, the difference between master performer and the novice. Which are the operants that constitute this difference?

Step 3: Identify training deficiency

For example, the analyst might soon dispose of the 'heavy breathing' variant, as his evidence shows that it has hardly ever been known to occur in the history of the company. He may also dispose of the 'wrong number' variant, as most people know how to deal with that.

He might have some thoughts about the 'request for an employee by name' variant. He may observe that the best telephonists seem to remember most of the extension numbers of the employees and seldom refer to the internal telephone directory. He may enquire whether this practice leads to misconnections and find that in some cases it does, but generally it results in correct and quicker service. So should we teach the following behaviour? (Which is the way the 'best telephonists' behave after a year or so on the job?)

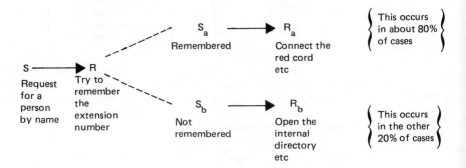

Our 'master performer' telephonist is using here the information that is stored in her memory, but we can observe her resultant behaviour as one of being able to differentiate, or discriminate, a whole series of possible incoming stimuli (names of employees) by matching to each the appropriate response (extension number). In S-R notation, such a multiple discrimination can be represented as follows:

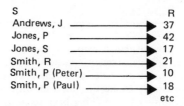

In effect, the total multiple discrimination is the memorization of the whole internal directory of about 50 extension numbers. This is not all that difficult to master (Morse Code, for example, is composed of some 37 symbols to discriminate and can be learned in a few hours). But is the benefit of mastering this behaviour going to pay off? How long before there are changes in staffing and extension numbers? Should we teach the above-mentioned multiple discrimination, or should we teach the following behaviour, which is a straightforward step-by-step procedure (or chain of operants), and which we know that anyone can master in a few minutes?

The answer to this question could depend on whether speed of response was a critical factor. The difference in speed between the use or non-use of a directory of a few (about 50) extension numbers is not very great, so the probable decision would be to rely on the directory in training. However, there are so many instances of requests for one of the six main departments by name, that the analyst decides to teach these six extension numbers formally during training. In stimulus-response terms the behaviour pattern looks like this:

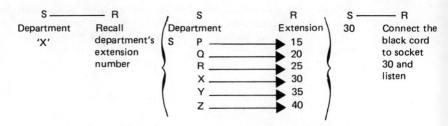

Finally let us consider the 'question of concern to the company but without indication of extension, name or department' variation. In following up whether this happens frequently, the analyst discovers not only that this is a frequent occurrence, but that it is the situation that leads most often to poor service. What happens looks something like this:

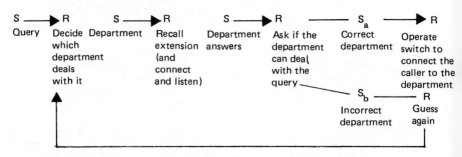

Sometimes the return loop of guessing again is repeated four or five times before the right department and person is located. Worse still is the case when the telephonist leaves the caller to do the finding out, by not listening in and asking advice but simply connecting the call to the department of her first guess, throwing the switch and going back to sleep until the department calls her up to transfer the call or the irate caller hangs up and tries again.

What should of course happen is that the telephonist should classify each incoming question correctly as pertaining to a certain department, and connect herself to that department's secretary in order to obtain the information necessary to connect the outside caller correctly first time. This means that the telephonist should have developed a behaviour pattern something like this:

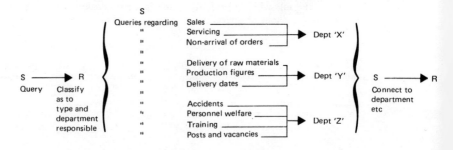

Our master telephonist should be able to generalize the response department X to a certain class of question types, and not to other questions that pertain to other departments.

Step 4: Carry out further analysis as required

As no one up to now has approached the telephonist's job from this angle of 'knowing what the various departments do', the analyst will not encounter readily available data on what types of queries are received by telephonists and with what frequency. However, it is by now patently obvious that the telephonists will need to be taught something about what the company does. This training should be based on the types of queries actually dealt with, and the training must be limited to the most commonly occurring and most troublesome types of queries. Otherwise, there would be no end to the training programme.

Our analyst sets about an error type and frequency analysis, keeping note over a period of all errors made in connecting calls and classifying them by type and department. He does this in such a way that he can eventually say, for example, 'if the telephone operator learns to classify these, she should deal with 80 per cent of incoming queries correctly'. The criterion of 80 per cent is arbitrary, and the analyst could aim higher or lower, depending on the criteria of good performance established by the client when stating his problem.

These analyses of frequency and errors are at this stage becoming quite deep. They may require the tape recording of a month's work on the telephone switchboard, a detailed analysis of all the calls into type, and an error count – perhaps another month's work. Whether the problem, as posed, warrants such depth is a moot point, but the example we are examining – based on a real-life case – serves to illustrate the various types of analyses possible in the mathetics system.

Step 5: Indentify the basic behaviour patterns

Before we go on, let us recap. By now, certain very distinctive behaviour patterns have emerged. The analyst is on the look-out for these as he goes along, because he has an armoury of specific instructional tactics to use on specific types of behaviour pattern.

There are three basic behaviour patterns recognized in mathetics. Gilbert claimed that all observable behaviour could be analysed into combinations of these three patterns. The three patterns are:

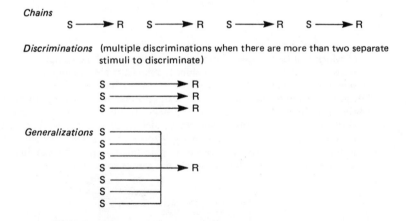

Chains

S ⟶ R   S ⟶ R   S ⟶ R   S ⟶ R

Discriminations   (multiple discriminations when there are more than two separate stimuli to discriminate)

S ⟶ R
S ⟶ R
S ⟶ R

Generalizations   S
S
S
S ⟶ R
S
S
S

We saw examples of all three as we performed our analysis. The original shallow analysis of the task appeared as a chain of seven operants, though as soon as we went to a deeper analysis (by analysing the stimulus variation) we found the chain was breaking up into a more complex pattern. We found another chain, however, in the behaviour of looking an extension number up in the directory.

On the other hand, the behaviour of pulling the extension number out of memory is a multiple discrimination. The telephonist who could do without the directory for 80 per cent of the extension numbers had mastered a complex multiple discrimination of about 40 separate stimuli. Whichever of these 40 names came over the line as a stimulus, her immediate response was to connect the correctly matched extension number. You will remember that we decided this was exceptional performance and that we would only teach a much simpler multiple discrimination of six separate stimuli, namely the numbers of the six main departments.

In the case of responding to queries of an undefined nature, we encountered an example of generalization, or rather of a group of six generalizations, one for each department. Indeed the behaviour called for in this case is to *generalize* the 'department' response to a certain set of stimuli and to *discriminate* these stimuli from others which belong in another department. This combination of generalization (or classification into a group) and discrimination (or keeping out of the group) is the behaviour normally referred to as 'having a concept'. What the telephonist needs is a set of concepts which will define the responsibilities/activities of each department.

We might easily have reached this conlusion by some other process of analysis or even by intuition. But note the very practical way in which the mathetics approach presents the discovery. By insisting on the stimulus, as well as the response, it forces the analyst to consider and clarify the types of telephone queries that illustrate the duties and responsibilities of each department. There is no chance of the instructional designer including hours of theoretical descriptions of the products, history or policy of the company. The only theory that goes into the proposed course is what is necessary to relate typical telephone queries to relevant departments.

Are we sure that there are only three basic patterns of behaviour? What about the situation often encountered in practice that a certain stimulus leads us to react in one of several possible ways? When the traffic lights go to amber as you approach them, you either brake or accelerate. The behaviour appears to be of the following pattern:

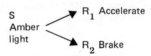

But this is not quite so. You brake or accelerate in response to your judgement of which will be the safer option, in the light of a whole complex of stimuli (the traffic conditions in front and behind). When the above pattern occurs it is a signal to the analyst that the analysis is not yet complete. He must identify the subsidiary stimuli to which the master performer is responding, in order to choose which of the alternative responses to make. In the above case, the real pattern is very complex, but would be something like the pattern below. The analyst must identify this pattern in order to teach it to the novice. His only alternative is not to teach it, but to leave the learner to gain this through post-training, on-the-job practice (the common, but obviously dangerous solution).

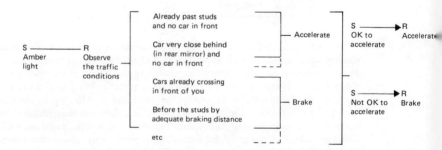

There are, in fact, some variations on the three basic patterns of behaviour, which are occasionally observed in some analyses. An example is the 'stimulus continuum' – a case such as responding to shades of colour in a painting, or indeed identifying colours by name. This is a discrimination, but there are no specific and separate stimulus situations that can be specified. Rather there is a continuous range of stimulus situations, one merging into the other. The learner must, as it were, learn to draw a mental line at some point in this range, responding 'red' to stimuli on one side of the line and 'orange' to stimuli on the other.

The precision of such a skill may vary enormously among people, but it may be necessary to develop a high level of precision for some specific tasks, say interior decorating or artistic painting. Further variations of this nature exist, and most of them have been matched to specific instructional tactics. Thus the analyst should be on the look-out for them.

**Step 6:
Analysis of
competition
and
facilitation**

But for the time being let us return to consider yet other types of analysis that can be performed if the problem warrants it. We have reached a stage of analysis when we have a pretty good idea of what should be taught. We have established:

* The variations from the normal behaviour pattern which should be included in a 'complete' prescription of mastery performance.
* How the typical trainees and 'master performers' differ in respect of this behaviour prescription.
* What subordinate or covert behaviour patterns are necessary to execute the main overt steps of the task.
* What type of behaviour patterns make up the mastery performance – chains, discriminations, generalizations.

We now come to an aspect of mathetical analysis which is probably unique among the commonly used systems of analysis – a formal way of looking for potential learning difficulties and aids. This is performed by searching the behaviour patterns for possible instances of competition or facilitation.

Everything that the learner has learned in the past may *possibly* affect the learning we want him to achieve now. It may help, or facilitate the new learning by providing analogies, involving the same skills, etc. It may hinder, or compete with, the new learning if 'similar yet different' things have to be learned – a frequent source of confusion for the learner.

Examples of Facilitation:

* Experience of using telephones in the home facilitates mastery of many of the behaviours required of telephonists in our hypothetical company. Imagine setting up a new company in an eskimo village and having to teach locals which end of the phone to speak into, etc.
* Good examples abound in language learning. For example, the knowlege of the term 'manual work' facilitates the learning of the French word 'main'. In S-R terms:

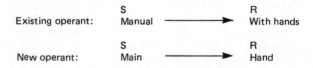

	S		R
Existing operant:	Manual	⟶	With hands
New operant:	Main	⟶	Hand

The instructional tactic in this second example would be to use the word 'manual' as a sort of bridge (or 'mediator') between the French word 'main' and the English word 'hand'. One can encounter more complex examples, when whole sets of existing concepts act as facilitators for a new set (eg they act as analogies). For example, I was taught the principles of electrical current flow – circuits, resistance, potential difference, etc – by an analogy to water flowing in pipes of various diameters, pumps, etc. These two sets of concepts are analogous, but the tactic of comparing them did not work in my case because I did not understand anything about water flow in pipes, at the time. In order to act as a facilitator, the analogous behaviour pattern must already be mastered by the student.

The analyst should therefore examine the behaviour pattern to be learned and compare it with behaviour patterns he knows the learners will have already mastered, in order to identify potential mediators or analogies, which he can later use in designing instructional exercises.

The behavioural characteristics of an operant which will act as a mediator are: similar stimulus and similar response.

Examples of competition:

* Prior experience as a telephonist on a switchboard of different manufacture may compete with the learning needed to master the new switchboard. For example the colour coding of the cords may mean different things, or the external and internal sockets may be in different positions, causing confusion.
* Turning again to language learning, an example experienced when learning Portuguese illustrates the point well. Study the S-R representation shown here:

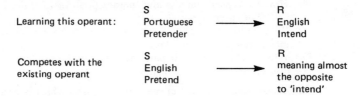

Imagine the confusion caused by statements taken to mean 'I will pretend to meet you outside the cinema at eight this evening', or 'I pretended to do my homework, Sir'.

* Once again, competition may also be encountered at the level of conceptual learning. For example, our telephonist who has now learned to conceptualize the operation of the company according to a model built up on the basis of typical phone enquiries gets promoted to the accounts department. Quite possibly, what she now has to learn about the structure and operation of the company will in some aspects compete with what she learned before.

Thus our analyst can search the behaviour prescription he has drawn up and compare it, operant by operant with what he knows that typical learners will already have learned, identifying sources of possible competition. The instructional tactic to adopt in the case of competing behaviours is to develop special exercises to discriminate the new learning from the old. The behavioural characteristics that identify possible competition are:
similar stimuli but different responses.

One should also mention that competition and facilitation can exist between the operants that make up a given task. For example, in using mathematical tables, the learning of one type of table (say logarithms) facilitates some parts of the learning of all the others in the book (sines, cosines, etc) because most of the steps of actually searching the table are the same for all the tables. However there are also some aspects which compete strongly between different tables. In some, you look up a final figure and add it to the main figure. In others, which look identical but for the title, you look up the final figure in the same way and then subtract it. Gilbert would call this an example of intra-domain competition to differentiate it from the examples involving previous learning, which he terms extra-domain competition.

Just to give some practice in identifying facilitation and competition, you might like to share an experience I had when teaching some educational technology courses in Egypt in 1970.

We in the West are very proud of our number system and of how efficient Arabic numerals are for mathematics compared with Roman numerals. But it so happens that the Arabic states do not use the same Arabic numerals as we do. All motor car number plates are, however, inscribed in both systems. Thus, from the moment of stepping into my first taxi at the airport, my re-training began, with ample opportunity during every journey to test performance. It took over three weeks to become comfortable with the new system, and even then occasional mistakes were made when attention strayed.

A glance at the comparative table of the two systems, shown below, will show why. The learning problem is bristling with examples of competition. The reader might like to try to identify some of the competing operants and suggest some ways of overcoming the competition:

Figure 4.3 *Two sets of Arabic numerals*

I found the most troublesome numerals to learn were the following (in order of frequency of error):

Egyptian numeral	read wrongly as:
zero	decimal point
five	zero
six	seven
four	three
seven or eight	confused with each other

Note that this last one is an example of intra-domain competition. The tactic to adopt is to develop special exercises which discriminate the old and new learning by some means. Here are some of the aids that I tried. Believe it or not, they worked.

- ☐ a point has no (zero) area
- ☐ Hawaii 5–0
- ☐ six is a drunken seven
- ☐ four is a backward three
- ☐ seVen

This is an example of extremely low-level learning (in terms of our taxonomy it would be knowledge of facts – the lowest category), but it serves to illustrate a technique of 'looking for trouble' before it actually occurs, which no other analysis technique does in quite such a thorough and formal way.

Step 7: Further possible analysis

Though demonstrated here with very simple examples, the mathetics technique of analysis can be applied to analysing tasks of any complexity, and can be supplemented as necessary by other types of specialist analysis. The 'error-frequency' analysis described in the telephonist example above is one such special purpose analysis which was not invented with mathetics but can be incorporated if necessary. Other such special analysis methods include:

- * Differential analysis: what is the range of differences among trainees' entry behaviours and how will it affect the instructional design?
- * Danger analysis; especially concentrating on the safety aspects of the task.
- * Efficiency analysis: is the mastery behaviour as now practised really the most efficient way of performing?

One can also revert to other techniques if the problem warrants it. For example, the discovery of difficult-to-analyse sensori-motor skills in the middle of a task would suggest the use of a Seymour-type skills analysis, as this is better adapted to identifying the overlapping coordinated actions and the role of each sense in the control and coordination of the actions. Alternatively, the discovery of no great learning difficulty may lead one to abandon the S-R approach at an early stage and revert to a simple 'task and key point listing'.

Conclusion: strengths and limitations of the mathetics system

Finally, if there are no tangible tasks to analyse, the problem under study being an 'information' problem, it is difficult to start the analysis by the mathetics approach. One is better off using one of the two routes described earlier. Either follow the inputs-to-outputs route, performing a subject matter analysis and topic analysis, finally deriving testable objectives or, I think preferably, follow the outputs-to-inputs route, by transforming the subject matter directly into a set of objectives. This is done by describing an imaginary master performer of an imaginary job. The latter was suggested by Gilbert in 1969 to complement his earlier mathetics approach. Once you have transformed your subject matter into detailed objectives, you can treat these objectives as units of behaviour, and subject them to further detailed analysis (eg of competition and facilitation) by the mathetics approach.

4.6 Techniques for content-oriented analysis: input to output

The techniques of analysis that we have looked at so far are all objectives-oriented or task-based, in that they start the process of instructional design by analysing the tasks or the general behaviour expected of the successful student after the instruction. The key concept that we have come across in all three techniques is that of the master performer – the person who can be observed in order to discover just what is expected as the *output* of instruction.

The reality of the educational context

However, it is common in education to have very little idea of what behaviours are expected after instruction. School subjects are not taught with a specific real-life model of mastery performance in mind. To a large extent, the interest of the learner, or of the teacher, dictates what *topics* should be learned. In other cases these topics are defined in some official curriculum, but generally as subject matter – a list of areas of knowledge to be explored.

Somewhere along the line objectives do get defined, but this is often late in the design process, when content, sequence, methods and media have already been established. At some point, the teachers say 'yes, but what shall we set in the final examinations to check if the course was effective?'. Thus, in a rather indirect manner, one can deduce the course objectives from an analysis of the last few years' exam papers.

The description we have given here is somewhat exaggerated, implying that no systematic planning has taken place, either in the selection of the course content or in the later definition of exam criteria. Although this is sometimes the case, we do not wish to imply that the inputs-to-outputs route of instructional design is necessarily asystemic or, indeed, inferior to the objectives-based approach. In many situations, one is faced with the reality of starting with the subject matter and selecting the 'topics worth teaching' more on the basis of what is interesting, or what may have some future value for the student, than on the basis of analysing the student's future occupations or roles. We hope to show, in this section, that this approach can be quite systematic.

Although we shall be dealing here with the detailed, Level 4 analysis of a topic that is to be taught, it is as well to review the earlier, more general levels of subject matter and topic analysis.

Level 1 analysis defines the subjects to be taught

Level 1: identifying the subjects worth teaching.
This is the socio-political level of decision making that often defines, in broad terms, what subjects will be included in a ciriculum and how much time should be devoted to each. The arguments for inclusion/exclusion may

be supported by reference to some very general 'aims' of education. Discussion may continue for years and slowly a particular viewpoint wins through, leading to a subject being included or excluded. Some 20 years ago, Latin disappeared from the range of compulsory subjects in the British secondary school curriculum, after more than a decade of heated discussion. At present, we are witnessing a discussion of the question of computer studies, which looks like becoming a compulsory subect soon.

Level 2: identifying the topics worth teaching.

Level 2 analysis defines the topics to be taught

This the strategic level of decision making that results in a definition of the specific curriculum to be studied. The curriculum designers, who are generally experts in the subject matter being analysed, take decisions on the parts or topics, of the total subject matter that have the most validity for inclusion in the course being planned. The basis on which these decisions are made varies. Generally speaking, however, the designers take their decisions on the basis of some model of the structure of the subject matter and some list of general aims in teaching it. In relation to the computer studies question, for example, the currently popular general objective seems to be 'computer literacy for all'. The problem for the experts is to construct a model of the structure of computer studies and identify which parts of the whole vast subject matter that *could* be taught best represent their concept of what a computer-literate citizen should know. Several conceptual tools exist to help in the construction of a model of the subject matter. We present three examples in the form of maps (see Figures 4.3, 4.4 and 4.5).

The analysis technique explained in Figure 4.4 is useful when dealing with subject matter that has a fairly neat and orderly structure, capable of being described in terms of a few basic 'organizing' or 'classifying' concepts. The diagram illustrated in the example could act as the basis for the organization of the topics in an individualized course of instruction. Indeed, it did act as the organizational basis for the chapters in Gibbons' (1971) book on the subject.

Figure 4.5 illustrates an approach more suited to a subject that has a complex structure with many interrelationships. The resultant diagram reveals key concepts or principles that must be included if the subject is to cohere and subsidiary concepts which may, or may not, be included, depending on the purpose for which the subject is being studied. It also suggests various possible sequences of instruction.

Figure 4.6 illustrates a more systematic approach to the mapping of a subject matter, which aims to present a visual, yet unambiguous, model of how certain concepts and principles interrelate. It is an attempt at visualizing the conceptual structure of a body of subject matter, presenting it as a type of schema. This schema could be used by the instructional designer as an aid to course and lesson planning and could also be used by the instructor as a visual aid to communication with the students.

Level 3: identifying the sequence and content of specific lessons.

Level 3 analysis helps us plan better lessons

The last example might be thought of as bordering on Level 3 analysis, as quite a clear picture of the content and the sequence of instruction is beginning to emerge. However, Level 3 design goes further than just the analysis of the interrelationships that exist between the concepts and principles that make up a topic. Figures 4.7 and 4.8 illustrate more clearly the decisions that are taken as the result of a Level 3 analysis of a topic. These two figures illustrate two alternative ways of analysing and then treating the same topic – Ohm's law. It may be worth taking a little time to compare these two figures carefully before reading on.

Procedure for the design of instruction by transforming topics into tasks

We may follow the logic of the examples in these figures by means of the numbers used to denote the steps. In Figure 4.7:

1. The instructional designer has selected the topic of Ohm's law as 'worth teaching', by means of some earlier, Level 2 analysis of 'electricity'.
2. He immediately transformed this topic into a 'task' (or rather a terminal

Introduction	An expert in a particular subject will form an opinion on 'what is worth teaching' largely on the basis of the *structure* of the information that composes the subject. Certain key concepts or topics that interrelate with many others, or which explain many phenomena, or which are a rich source of examples, would be considered as more worth teaching than rarely used peripheral information.
Mapping out the structure of the subject	In order to identify the key concepts or topics in a subject, one may attempt to map out the structure of the subject, using some form of pictorial representation. A visual representation of the structure is an aid both to the analysis of complex subject matter and to the communication of what is worth teaching and learning (to other teachers and also to students).
How to map out a hierarchical structure	One may use a family tree approach to represent the topics which compose the subject. One may descend to lower levels, identifying sub-topics, just as we saw in the case of job analysis. Jobs generally have a hierarchical structure, the job being the sum of a particular set of tasks and each task being the sum of a set of sub-tasks and so on. Subject matter structure can also be presented in this way. This is particularly useful when the subject contains a system of classification which is to be communicated.
Example	In the example below, several key concepts (eg active, direct, etc) are used to analyse the subject (individualized instruction) into a hierarchy of topics (instructional methods). This hierarchical structure may then be used to order the presentation of the subject in an organized manner.

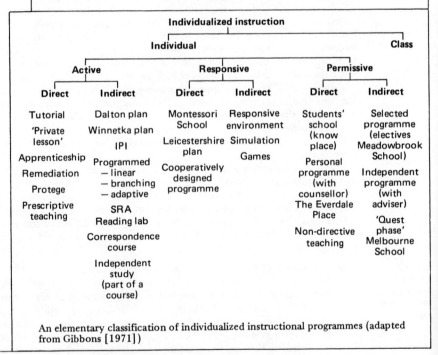

An elementary classification of individualized instructional programmes (adapted from Gibbons [1971])

Figure 4.4 *Map explaining the use of a hierarchy or family tree technique for the analysis and representation of the structure of a subject*

Introduction	Most subjects that are worth teaching have a more complex structure than do jobs. Not only do the topics add up to form the subject, but they also relate to each other in complex ways. Often the justification for teaching a particular body of information lies in the value to be gained from the investigation of these complex relationships. In such cases the analyst should attempt to visualize the complexity of the structure.
Example	This analysis by Clarke (1970) of a part of primary school mathematics clearly illustrates the importance of certain key topics (eg sets), and gives many insights into possible teaching sequences and revision points. 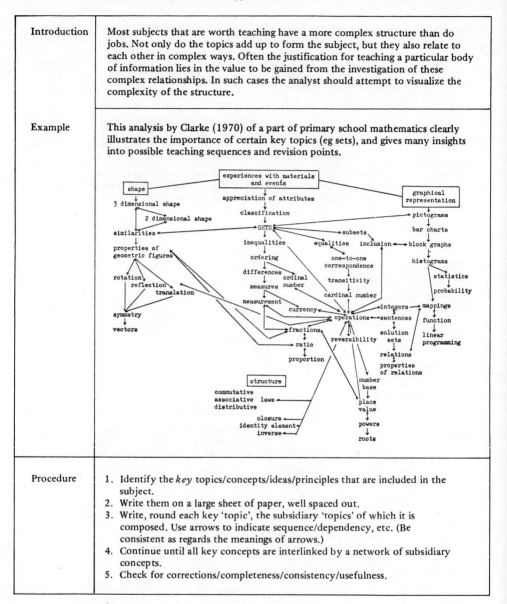
Procedure	1. Identify the *key* topics/concepts/ideas/principles that are included in the subject. 2. Write them on a large sheet of paper, well spaced out. 3. Write, round each key 'topic', the subsidiary 'topics' of which it is composed. Use arrows to indicate sequence/dependency, etc. (Be consistent as regards the meanings of arrows.) 4. Continue until all key concepts are interlinked by a network of subsidiary concepts. 5. Check for corrections/completeness/consistency/usefulness.

Figure 4.5 *Map explaining the use of simple networks to represent the structure of a subject*

Introduction	This approach is similar to the previously illustrated networks of key topics, but it is a little more rigorous. As Neil (1970) puts it, the 'stringing together of a series of topic headings by means of arrows is no more than a structured index or contents list'. Such a network often does not communicate much about the *types* of relationships that exist (except perhaps to experts or to the one person who drew up the network).
Procedure	Neil suggests the need for two refinements:
	(a) the need for a classification of the types of possible relationships, together with a language (notation) for representing them, and
	(b) the need for rules, for the clear visual presentation and organization of the resulting models (he calls this 'graphic topological design').
Example	The following example (Neil, 1970) illustrates the attempt at clear layout and the use of a notation for different types of relationships. A well-designed conceptual model should be read in the same way by anyone who 'knows the rules and the notation'.

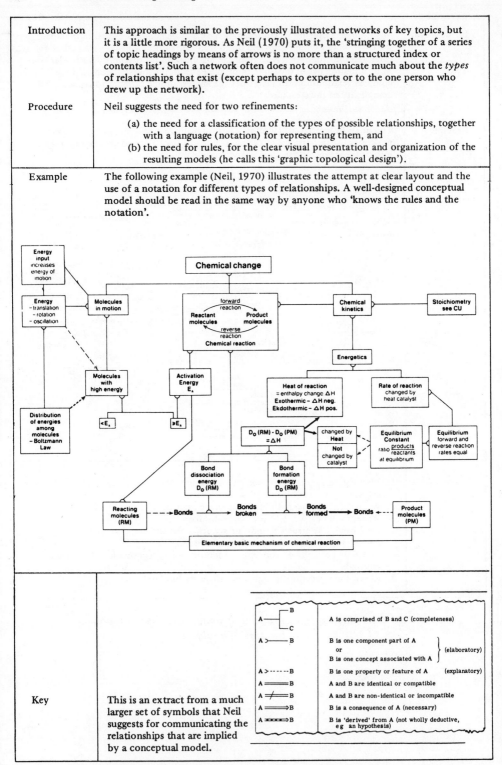

Key	This is an extract from a much larger set of symbols that Neil suggests for communicating the relationships that are implied by a conceptual model.

Figure 4.6 *Map explaining the use of more sophisticated network techniques for the analysis and representation of conceptual structures*

Introduction	This approach is akin to task analysis, but uses Gagné's technique of objectives analysis. The diagram below is similar in layout to Figure 5.4.
Procedure	1. Select the *topic* (eg Ohm's law). 2. Transform it into a *task* (ie a terminal objective). 3. Produce the *final test* item to match the objective. 4. Analyse the objective into a *hierarchy* of prerequisites. 5. Produce *test items* for each *sub-objective* identified. 6. Identify entry levels and produce an *entry test*. 7. Organize the objectives into a *logical sequence for learning*. 8. Select from the topic relevant *information/examples*. 9. Continue to develop an *instructional sequence* etc.
Example	The following example may be compared with the previous map. Note the similarities and differences.

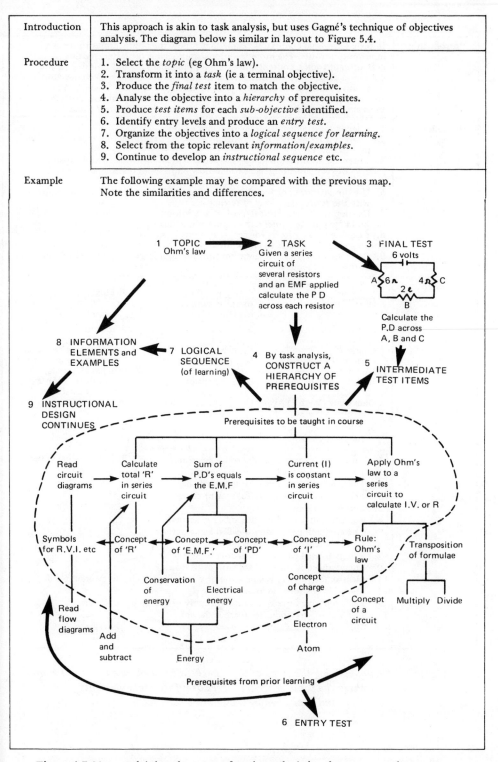

Figure 4.7 *Map explaining the steps of topic analysis by the output-to-input route of transforming the topic into a hypothetical task and using 'objectives analysis' techniques*

Introduction	In the educational context, we can either follow the output/input route or the input/output route. The more traditional approach is the latter, starting by the definition of topics, content and examples (information) and finishing by the definition of test instruments. Several technological tools have been developed to make this process more objective. Two of these (rulesets and Bloom's taxonomy) are combined in the methodology suggested in the example below. Both these tools are discussed in more detail in other chapters (see Chapters 3 and 12).
Procedure	1. Select a *topic worth teaching* (eg Ohm's Law). 2. Analyse it to identify the important *elements of information* (concepts, laws, procedures, etc). 3. Order the information in a *logical sequence for presentation* and expand it into a set of very short sentences which flow on (a set of *teaching points* or rules). 4. Agree on the *general aims* of teaching this topic. 5. Use the categories of Bloom's taxonomy to label each element of information with the appropriate *level of testing*. 6. Develop *appropriate test items* for each information element. 7. Build from these a *final test* of appropriate difficulty. 8. Examine the test items for necessary prerequisite skills or knowledge and prepare an *entry test* to check these. 9. Combine the logical presentation with the appropriate test items to develop an *instructional sequence*.
Example	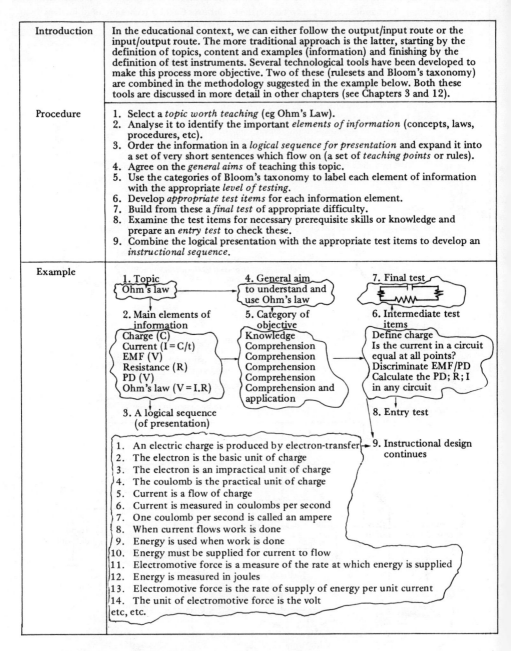

Figure 4.8 *Map explaining the steps of topic analysis by the input-to-output route of breaking the topic down into a set of information elements and transforming these into a 'rule set'*

objective) that suggests the form of the tasks that will be used to evaluate the learning.

3. The final test can therefore be derived immediately, by inventing some specific example problems of the task.
4. Using the form of task analysis (or objectives analysis) proposed by Gagné and other authors, all the intermediate or enabling objectives can be identified.
5. It is now possible to construct test items for all these intermediate objectives, to be used as part of the lesson evaluation system.
6. Any entry or prerequisite behaviours identified in the task analysis will, similarly, lead to the construction of an appropriate entry test.
7. An analysis of the interdependencies mapped out in the objectives hierarchy leads the instructional designer to select an appropriate sequence (or alternative sequences) for instruction.
8. This sequence of objectives, together with the basic informational content and suitable examples extracted directly from the topic itself, is all that the instructional designer requires in order to complete the instructional design.

In the case of a teacher-delivered lesson, he would proceed to develop a detailed lesson plan, perhaps following the model in Chapter 3. And in the case of opting for the preparation of self-instructional materials, and further analysis being necessary, the mathetics approach could be used, as illustrated earlier in this chapter, followed by the development of materials (see Chapters 4 to 6).

Procedure for the design of instruction by preparing a rule set In Figure 4.8, the instructional designer has adopted a very different approach. Rather than using the task-oriented approach, the designer continues his analysis by means of topic analysis. He proceeds as follows:

1. The 'topic worth teaching' is selected.
2. The designer analyses it to identify its structure and the main 'information elements' – concepts and principles – it comprises. This stage of analysis may be similar to that shown in Figures 4.4 or 4.5 earlier.
3. This topic structure is then further analysed and transformed into a logical sequence of teaching points, or rules as they are often called. One should note that the use of the term 'rule' is not quite the same as when Gagné uses the term to denote 'principles'. In the present context, a rule is any simple statement that communicates a single unique idea or item of information.
4. Backtracking at this point, the instructional designer starts to think in terms of aims and objectives.
5. He works more in terms of general categories of objectives, as defined, for example, by Bloom's taxonomy, or by his own knowledge and skill schemata, classifying the types of capability that the learner should exhibit in relation to every information element and every rule.
6. Suitable examples and problems are then generated for the testing of the recall, comprehension or application of each rule or group of rules.
7. A selection of these intermediate test items is used to construct a final test, if necessary.
8. An entry test is also constructed, if deemed necessary, through the analysis of information elements not to be included in the lessons on the topic, but necessary for its complete understanding. Finally, using the rule set and the intermediate test items as the prime ingredients, the instructional designer is ready to complete the instructional design.

In this case, a teacher-delivered lesson would appear in a much more conventional format than our three-column, objectives-based model. And a self-instructional text would appear very similar to currently popular forms of programmed or modular instruction. We shall see in the next chapter how this particular design would be completed.

Level 4: developing instructional materials.

Our discussion of the two alternative approaches to Level 3 analysis of the Ohm's law example leads us into our discussion of Level 4. As we saw, the task-oriented approach led us to produce a list of intermediate objectives (in behaviour terms), sequenced in an appropriate teaching order.

The topic-oriented approach, on the other hand, led us to develop a list of logically sequenced teaching points or rules. These may now be used for lesson planning, or for the development of plain textual hand-outs, or scripts for audiovisuals, or instructional modules, or programmed self-instructional materials. In the next chapter, we shall discuss the specific techniques for transforming a set of teaching points into a self-instructional programme. In later chapters we shall illustrate their use in developing other forms of instructional materials.

Before we go on, however, we shall analyse a little more closely how a set of teaching points is developed and how they may be organized into a logical teaching sequence. This sequencing of a set of teaching points or rules may be considered as a form of Level 4 analysis that follows the topic-oriented path to instructional design.

We have already mentioned that the term 'rule' has a special meaning in this context. Our aim is to break down the information that is to be taught into the smallest of units, so that all the possible alternative ways of sequencing and organization can be explored. Thus, our rules or teaching points represent just one item of information each. As a general guideline, therefore, it should be possible to write down a rule by means of a simple sentence, without ands, ifs or buts. For example the sentence 'all animals are born equal, but some are more equal than others' (from Orwell's *Animal Farm*) represents two distinct rules, linked together by the 'but'. Similarly, the statement 'isosceles triangles have two sides equal to each other and the two angles opposite these sides also equal each other' again represents two separate teaching points.

The exercise of breaking a topic into such atomic 'teaching points' transforms it into a very simple narrative that clearly explains the concepts and principles of which it is composed.

As an example, we present in Figure 4.9 the complete rule set that was developed for the teaching of Ohm's law by Ivor Davies (Davies, 1971). We may see that the law or principle is a relationship between five concepts – electrical charge, current, electromotive force, resistance and potential difference. The 'story' of this relationship is told in 29 simple sentences. Note that some of them have been repeated to emphasize how one concept interrelates with another. In Figure 4.9 these simple sentences are referred to as rules. However, as already mentioned, this may be confused with the meaning given to the term rule as a synonym for principle. For this reason, we prefer to use the term teaching point in the present context.

The use of a
matrix for
the analysis
of sequence
and
coherence
of a rule set

To get back to our 'story', the 29 teaching points listed in Figure 4.9 are organized in what seems to be a very logical sequence. They seem to tell the whole story, nothing seems to have been left out; continuity is excellent. But how did the analyst come to write such a perfect story? How did he decide that just those teaching points in just that order would be the most effective way of telling the story of Ohm's law? Was it instinct or systematic design? Well, of course, we are asking a leading question, for everything in this book is about systematic design. The technique used to check out the sequence and the completeness of the set of teaching points systematically is the 'matrix of interrelations'. This matrix is a graphical way of analysing the associations and discriminations – the links and breaks – that exist between the various teaching points. Figure 4.10 illustrates the matrix produced by Ivor Davies for the set of teaching points and sequence presented in Figure 4.9.

Note that the designer has shaded in various squares, on either side of the diagonal, to show the associations and the discriminations that he has identified between every teaching point in the set and every other teaching point. This

A topic analysis: Ohm's law and its five constituent concepts (Concept A + Concept B + Concept C + Concept D + Concept E = the principle of Ohm's law)

TOPIC: Ohm's law

(a) Concept of electrical charge

Rule 1. An electric charge is produced by friction.
" 2. The electron is the basic unit of charge.
" 3. The electron is an impractical unit of charge.
" 4. The coulomb is the practical unit of charge.
" 5. Current is a flow of charge.

(b) Concept of electrical current

*Rule 5. Current is a flow of charge.
" 6. Current is measured in coulombs per second.
" 7. One coulomb per second is called an ampere.
" 8. When current flows work is done.
" 9. Energy is used when work is done.
" 10. Energy must be supplied for current to flow.

(c) Concept of electromotive force

*Rule 10. Energy must be supplied for current to flow.
" 11. Electromotive force is a measure of the rate at which energy is supplied.
" 12. Energy is measured in joules.
' 13. Electromotive force is the rate of supply of energy per unit current.
" 14. The unit of electromotive force is the volt.

(d) Concept of resistance

Rule 15. Current meets resistance in flowing through a load.
" 16. The resistance depends upon the nature of the load.
" 17. Materials offering little resistance are termed conductors.
" 18. Materials offering very great resistance are termed insulators.
" 19. In overcoming resistance work is done.

(e) Concept of potential difference

Rule 20. The energy used in moving a charge is proportional to the size of the charge.
" 21. The rate at which energy is used is proportional to the rate at which charge flows.
" 22. The rate at which energy is used is proportional to the current.
" 23. Potential difference is the rate at which energy is used per unit of current.
" 24. The unit of potential difference must be the volt.
" 25. Potential difference is proportional to current.

(f) Principle of Ohm's law

*Rule 25. Potential difference is proportional to current.
" 26. Potential difference equals current multiplied by a constant.
" 27. The constant is resistance.
" 28. Resistance equals potential difference divided by current.
" 29. The unit of resistance is the ohm.

* Rules which overlap two related concepts.

Figure 4.9 *A well-structured set of teaching points (rules) prepared for the teaching of Ohm's law (adapted from Davies, 1971)*

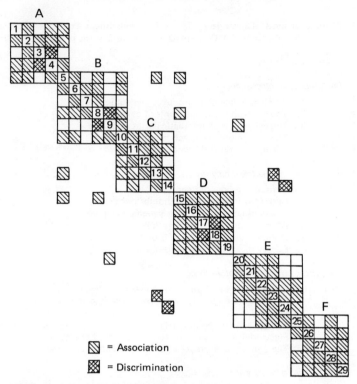

Matrix for Ohm's law (Concept A + Concept B + Concept C + Concept D + Concept E = Principle F of Ohm's law).

Note the clusters of associations between the rules that define a given concept and the relatively good continuity of associations between adjacent rules, all along the diagonal.

Figure 4.10 *Matrix showing the interrelationships that the instructional designer sees between the teaching points prepared for a lesson on Ohm's law (from Davies, 1971).*

checking has been done twice over, on one side of the diagonal and then on the other side. Thus each teaching point was considered in relation to each other point on two occasions. The fact that the pattern on one side of the diagonal is identical to that on the other is a sign that the designer has not changed his mind in relation to the associations which he sees between the items in the list.

It is a good idea to fill in one side of the diagonal, starting from teaching point number 1 and considering its relationship to each of the points that follow. Then take point number 2 and consider it in relation to points 3, 4 and so on. Then take point 3, in relation to 4, 5, 6 and so on. Having completed the consideration of all the points in relation to all the others, leave the matrix for a day or two and then, covering up the half already completed, have another go, starting this time from the last point, related to the penultimate point, etc. If you get a mirror image first time, you are lucky. Normally, there are some differences, and these serve to make you think again and resolve the discrepancies, clarifying your own thinking as you go along.

Look for a 'fully clothed' diagonal

Note also, that the continuity of associations is fairly uniform along the whole of the diagonal (there are no breaks in coverage of the diagonal), indicating that adjacent teaching points are always related in some way – a sign of good continuity and a satisfactory, logical sequence to the arguments being developed. This also does not generally happen immediately. Your first set of teaching points may give a fairly erratic pattern and be full of gaps, indicating that perhaps some points are not in the best order, or some key points have been omitted altogether. A study of the breaks and irregularities, and also of odd relationships way off the diagonal – indicating associations between items that appear in different parts of the list – lead the instructional designer to rewrite and restructure the sequence of the teaching points, in search of a better, more pleasing pattern. A good example of this procedure is given by Dean and Whitlock (1983), using a hypothetical example from bridge (see Figure 4.11).

Rewrite or re-order the teaching points to improve the matrix pattern
Figure 4.11 (A) shows the initial pattern of associations for the original version of a set of teaching points that describe the dealing procedures in bridge. Note that, although only one side of the diagonal has been used so far, it is already quite obvious that the sequence is not very good. Part B shows a later version of the teaching points, slightly rewritten and much reshuffled, that gives a much more satisfactory pattern of associations.

To close this section, we present yet one more example, which will be useful in our later work in the next chapter. In Chapter 5, we will study some examples of programmed instruction. One of these is entitled 'The Language of Computers'. Its content can be summarized in seven points, as follows:

1. Computer operators communicate with computers by means of controls.
2. Computers communicate with their operators by means of sets of lights.
3. The sets of lights indicate numbers that are stored in the computer.
4. One set of lights can, at any one time, communicate only one number to the operator.
5. The right hand light in a set of lights always has the value one.
6. Every other light in a set has a value equal to twice the value of the light immediately on its right.
7. To read the number being communicated by a set of lights, add the values of the lights that are lit.

The reader is invited to study the relationships that exist between the teaching points in this set (both the associations and the discriminations). Note that point 7 is strongly linked to points 6, 5, 4 and 3. Note that points 5 and 6 not only do not follow on from 3 and 4, but seem to enter into conflict – how is it that every light has a numerical value, when we were just told that a set of lights communicates only one number? Only when we read point 7 is this doubt dispelled: 'ah yes, we add the individual values of the lights to get the one value of the set'. Had we known point 7 before being introduced to points 6 and 5, we would not have had any doubts.

In Figure 4.12, we compare the matrices drawn from two alternative sequences – the original seven-point sequence and the new one with point 7 brought forward to precede points 5 and 6. Observe the Ds used to identify the aspects that need discriminating – the two uses of the word 'numerical value' – are now off the diagonal, while the diagonal is nicely 'clothed' with As – a sign of good continuity of ideas.

Beware: logical sequence and teaching sequence may not be the same
One might reflect, however, that from some viewpoints, the original sequence was more logical. After all, one cannot practise the reading of lights until one knows the values of each individual light. In other words, the knowledge expressed in teaching points 5 and 6 is a prerequisite to the execution of the procedure defined in teaching point 7. Thus the new sequence is more effective for telling someone the 'story' of computer lights, but the original sequence is more correct for the *teaching* and *practice* of the task that the computer operator performs. The example is very simple, but it raises questions that often are of the utmost importance to the instructional designer. We shall see, in the next chapter, how this apparently paradoxical sequence problem was resolved in one self-instructional programme.

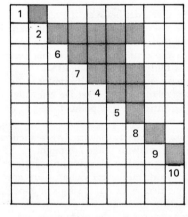

1. To choose partners, the pack is spread face down on the table.
2. Each player draws a card.
3. The dealer has the choice of seats.
4. The player with the second highest ranking card partners the dealer.
5. The other two players play as partners.
6. Cards rank in ascending order 2 to 10, then J, Q, K, A.
7. The player holding the highest ranking card is the dealer.
8. If two cards of like values are drawn, the higher suit rank wins.
9. Suits rank spades, hearts, diamonds, clubs in order of precedence.
10. This is reverse alphabetical order.

A first attempt at sequencing a set of teaching points
(from Dean and Whitlock, 1983)

1. To choose partners, the pack is spread face down on the table.
2. Each player draws a card.
6. Cards rank in ascending order 2 to 10, then J, Q, K, A.
7. The player holding the highest ranking card is the dealer.
4. The player with the second highest ranking card partners the dealer.
5. The other two players play as partners.
8. If two cards of like value are drawn, the higher suit rank wins.
9. Suits rank spades, hearts, diamonds, clubs in order of precedence.
10. This is reverse alphabetical order.

A later, revised sequence, showing the improved pattern
of associations (from Dean and Whitlock, 1983)

Figure 4.11

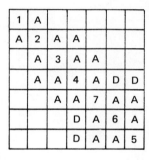

1	A					
A	2	A	A			
	A	3	A			
	A	A	4	D	D	
			D	5	A	A
			D	A	6	A
				A	A	7

Note the 'D' for discrimination — that indicates the points of apparent conflict between items of information presented to the learner. This results in a break in continuity of association.

(A) Matrix of interrelationships for the original sequence of teaching points on 'the language of computers'

1	A					
A	2	A	A			
	A	3	A	A		
	A	A	4	A	D	D
		A	A	7	A	A
			D	A	6	A
			D	A	A	5

Note that the only change is the position of point 7 in the sequence, but this has improved the pattern of interrelationships appreciably.

(B) Matrix redrawn for the revised sequence of teaching points

Figure 4.12

Other techniques of 'deep' topic analysis

The development of well-sequenced sets of teaching points is of great use for the subsequent writing of a variety of instructional materials – in general, those materials that are concerned with the communication of information and which follow a more or less linear sequence of presentation. This includes almost all textual material, most programmed instruction, audiovisual and video scripts, and even some forms of computer-assisted instruction. The content analysis and the sequence analysis techniques just described are therefore at least as important as the behaviour-based mathetics techniques described earlier.

The teaching point and sequence matrix 5 is not, however, the only technique of topic-based Level 4 analysis. We shall meet at least two other specialist techniques, which are not so linear in their communication philosophy, when we come to study structural communication in Chapter 7 and the structured writing techniques, based on information mapping in Chapter 8.

5. A Study of Change in Programmed Instruction: Analysis of the Topic-Centred and Task-Centred Programming Techniques*

5.1 The changing face of programmed instruction

In the early, heady days, when programmed instruction was still a new-born revolutionary, Skinner, its generally acknowledged father, predicted a rosy future: 'There is a simple job to be done. The task can be stated in concrete terms. The necessary techniques are known. Nothing stands in the way but cultural inertia.'

Looking back over 30 years later, it would seem that the job was not that simple, the tasks not all equally easy to state, and the techniques not as powerful as predicted. Or is it just that cultural inertia is the proverbial 'immovable object'?

On the one hand, programmed instruction has not fulfilled its early promise. On the other hand, the influence of the programmed instruction movement in education has gone much further and much deeper than many people care to admit. Programmed instruction was perhaps the first teaching movement to insist on clearly stated objectives. Now, this is the rule rather than the exception. Programmed instruction was one of the first teaching movements to lay itself open to evaluation at every stage. Now, we have accountability in education.

Of course, not all recent developments in educational thinking can be ascribed to programmed instruction. Rather, the necessities of the age fostered its growth and it, in turn, helped develop techniques to satisfy these necessities. Necessity is the mother of invention. That is why Pressey's excellent work on self-instruction, published in 1926, remained dormant for 30 years. That is why the principles already known to Quintilian and Rousseau and proved scientifically by Thorndike (1927) were not systematically applied earlier. That is probably why early attempts to install individualized systems of instruction at Winnekta and elsewhere failed (Washburne and Marland, 1963).

The two meanings of programmed instruction

Now necessity has caught up with invention. Progress in the design of individualized instructional systems has far outstripped the early versions of programmed instruction launched in the 1950s and 1960s. We have new forms of programmes and new forms of programming. Indeed, the very term 'programmed instruction' has grown to have two quite distinct meanings. To quote Biran (1974):

> In considering the limits of applicability of programmed instruction one must bear in mind the different levels on which this term is used. At the more general level of 'a systematic strategy of teaching, aided by technological aids and psychological tactics', programmed instruction merges into modern teaching methodology, and a separate discussion of its limitations becomes meaningless. At the most restricted level of 'structured, individual self-instruction', programmed instruction is a teaching technique in competition with others, such as independent learning in small groups, teacher-led instruction, project work, etc. Each of these has advantages, disadvantages and prerequisites on which a decision about the usefulness of a particular technique in particular circumstances can be based.

---

*This chapter was written in collaboration with Conrad Brunstrom, Dick LeHunte and Brian Slymon of Learning Systems Ltd, UK.

These two definitions – the 'process' and the 'product' definitions – highlight the difficulty of defining the characteristics of programmed instruction-based learning systems. Do we include only systems which present the learner with self-instructional, structured, individual learning tasks (usually programmed texts), or do we include all systems which have been developed according to the principles of programmed instruction?

The changing concepts in programmed instruction

One could argue that in the latter case, all individualized learning systems (excluding perhaps those based on small group dynamics) should be included. The concept of programmed learning held by most practitioners has changed much over the years.

George Leith in a pamphlet entitled *Second Thoughts on Programmed Learning* (1969), indicates the following changes:

Size of learning step

1. Small steps. Over-use of small steps usually leads to boredom on the part of the students. Current practice a step size as large as the student can manage. This is generally established by producing first drafts of programmes which are almost sure to be too difficult for the student, and then simplifying those sections which give students difficulty during validation.

Errors

2. Error-free learning. Skinner suggested the maximum permissible error rate on the frames of a linear programme to be in the region of 5 per cent. Subsequent studies have suggested that this figure is not very critical. Current programming practice pays much less attention to error rates within the frames of a programme, and concentrates instead on analysis of student performance on post-tests and on key criterion frames within the programme.

Responses

3. Overt responding. Again, studies have shown that, in most instances, students who do not write down responses to a programme's frames, but simply think them out, do just as well on final post-tests and generally take less time in studying the programme. Leith carried out a series of experiments on this factor, which seem to indicate that less mature students benefit more from responding overtly, and that the benefit is related to the type of subject being studied. For example, learning to spell is improved if students write their responses, learning concepts is not.

Pacing

4. Self-paced learning. This was held to be one of the main reasons for the success of early programmes – 'the learner can proceed at his own pace'. But the pace the learner chooses is often much slower than the pace he could proceed at if he were given some indication of what was expected of him. Today, programmes are sometimes designed with built-in pacing – a characteristic of some tape-slide programmes.

Grouping

5. Individual learning. Allied to the above point, if self-paced learning is abandoned, one can also abandon individual presentations and revert to group instruction, with a consequent saving of cost on hardware. Experiments have shown no loss of effectiveness for some programmes when used in the group situation, though this depends largely on the subject and the complexity of student responses demanded. However, even in complex, structured subjects such as mathematics, benefits have sometimes been gained from allowing students to work together in pairs or in small groups.

Programming methods

6. Programming methods. As already mentioned, programmers now tend to use both linear and branching sequences in the same programme, depending on the learning task they are tackling. There are also other programming styles in use. Some of these have developed their own jargon and sets of techniques. Examples of some notable current techniques (by no means a comprehensive list) are:

   (a) Mathetics: the creation of Tom Gilbert (1961), discussed in some detail in previous chapters and analysed further in this chapter and the next.
   (b) Information Mapping (T.M.): the creation of Robert Horn (1969). This

system attempts to obey all the established principles of good communication, transforming them into a complex set of rules for the layout of information in a book. Features include the titling, sub-titling and cross-referencing of all blocks of information in an attempt to make the purpose of the information clear and its retrieval easy (see Chapter 8).

(c) Structural communication: the creation of Anthony Hodgson (1974). This is a technique based on a quite different approach to learning from the previously described, mainly behavioural, models. It does not presuppose a correct answer or even a limited range of correct answers. Rather, it is designed to enable the student to respond to a block of information presented to him, in as open-ended a manner as is possible in a self-instructional exercise. The student composes a response from a set of 'building blocks', and the structure of this response is used to control the subsequent 'discussion' between author and student. Although textbook presentations in this format exist, it is better suited as a technique for computer-assisted instruction (see Chapter 7).

Media   7. Presentation styles. As already mentioned, teaching machines presenting only verbal information (plus occasional diagrams) are on the decline. There is also a decline in the production of 'traditional' programmed texts. However, audiovisual programming is on the increase, and one can also find examples of programmed film sequences, interactive video programmes and computer-controlled multi-media presentations. Group presentations are presented by some teachers, frame by frame on an overhead projector.

The most significant development is the concept of the instructional 'package'. This is an assembly of instructional materials designed around clearly defined objectives, complete with tests and other necessary controls, teacher notes, etc, and using the most appropriate instructional media.

5.2 Changing techniques of programming

In the early days of programmed instruction, the arguments for and against the
Linear or   linear or the branching format were the principal points for discussion between
branching?   programmers. We were all embroiled in the pros and cons of error-free learning and learning from one's mistakes. We tended to believe that practising erroneous behaviour was a 'bad thing' and were often ourselves led into the error of equating the *analysis* of erroneous behaviour with the *performance* of errors. We lost sight of the fact that errors in the conceptual analysis of a problem may simply be signs of incompletely formed conceptual schemata and might, indeed, be a necessary step to the fuller understanding of a given phenomenon or problem.

Slowly, programmers came to agree that there were times when a linear approach was more appropriate and other times when branching was a useful tactic to adopt. However, with very few exceptions, the programmes of early writers were based on *expositive* strategies. Such programmes as the 'Productive Thinking Program' of Covington and Crutchfield (1965), which aimed at the development of problem-solving skills by means of a discovery-type exploration of a detective story, formed a very small proportion of the total production of self-instructional learning materials in the 1960s.

Task- or   However, another important distinction in programming techniques appeared
topic-based?   in the 1960s – that between programmed instruction based on the analysis of the *content* to be learned and programmed instruction based on the analysis of the *objectives* to be mastered. We shall refer to these two alternative approaches as *topic-based* and *objectives-based*, or *task-based*.

Although from very early in the history of programmed instruction reference was made to the definition of instructional objectives, these were not always used as the basis for the design decisions then taken when developing the detailed material (or frames) that were to be studied by the learner. Many early approaches were based on the analysis of a topic into discrete items of information – variously called rules or teaching points. This was followed by the careful logical sequencing of these items into a programme of teaching steps composed of new information, question and response.

In many ways, this approach has much more in common with the traditional way of planning classroom instruction than with the principles of behavioural analysis and reinforcement developed in Skinner's psychology laboratory at Harvard University. The main point of programmed instruction, as a technique for the shaping of behaviour, seemed to get lost somewhere along the way, as more and more teachers became involved.

However, techniques of programming human learning more firmly based on Skinner's principles were available. One early exponent was Tom Gilbert, who developed the methodology of mathetics, perhaps the first attempt to prescribe specific procedures for the development of learning sequences based on the detailed analysis of the *behaviours* necessary to master the learning objectives.

The remainder of this chapter analyses two samples of programmed instruction materials, prepared at two points in the history of the technique's development. Both samples deal with the same topic and attempt to lead the learner to achieve broadly similar objectives, though some slight differences in the objectives, or rather the range of objectives, will be identified in our analysis. One of the samples was produced by a technique which is 'topic-oriented' and the other by the much more 'objectives-oriented' mathetics technique.

The two samples, presented as Exhibit 5.1 and Exhibit 5.2, are particularly interesting to compare, as they were used, at different times, as samples of 'good programming technique'. The first, originally written as a sample programme by Robert Mager, was used as a programming model in the 1960s, and the second, written by Brian Slymon and Dick LeHunte of Learning Systems Ltd, was written to illustrate how programming methodology had changed by the 1970s.

We recommend the reader to study the two exhibits carefully, not only as a student of the content being taught, but mainly to analyse and compare the programming techniques. In our later discussion we shall indicate that the principles we are analysing are of general applicability. We shall indicate some analogies to the design of expositive classroom instruction, and we hope that the reader will notice yet other analogies. We hope, by the end of this chapter, to build a general approach to the design of expositive self-instructional materials.

5.3 Analysis of Exhibit 5.1

The sequence of teaching frames

The programme in Exhibit 5.1 has a structure typical of many early linear programmes. This structure is illustrated in Figure 5.1. There is a main sequence of *teaching points*, or topic elements (A, B... L, M, N...etc), presented to the student in sequence by means of a series of small units called *frames*. Generally a number of frames are used to teach a given teaching point. Normally, at the beginning of a sequence, there are one or more *teaching frames* (M1, M2, etc), where information is presented and the student is asked to respond to show that he has understood. His responses are helped by the presence of various *prompts*, so that it is almost certain that he will respond correctly. As the student progresses through a sequence these prompts become progressively weaker, until at the end of the sequence there is what might be called a *test frame* which is equivalent in difficulty and in standard to the behaviour expected of the student after the completion of learning.

Overview and review frames

In addition to these teaching and test frames, one may have other frames which do not actually teach a particular point but which either introduce a section or summarize the previous section. These *overview* and *review* frames can have various functions in the programme. The overview frame, for example, often states not only what will be presented in the next section, but also the objectives which the student will be expected to reach at the end, the probable time of study that he

Exhibit 5.1 *An excerpt from an early linear programme*

The following sequence is an extract (frames 10 to 32) from a linear programme
entitled 'The Language of Computers', written in 1964 by Robert Mager. It is
reproduced here, by kind permission of the copyright holders, Learning Systems
Ltd, to act as an example for analysis. Please study this excerpt before continuing
this chapter.

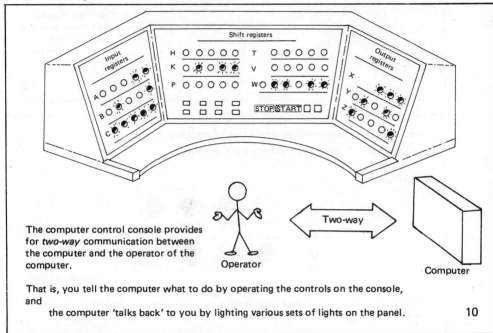

The computer control console provides for *two-way* communication between the computer and the operator of the computer.

That is, you tell the computer what to do by operating the controls on the console, and

the computer 'talks back' to you by lighting various sets of lights on the panel.

10

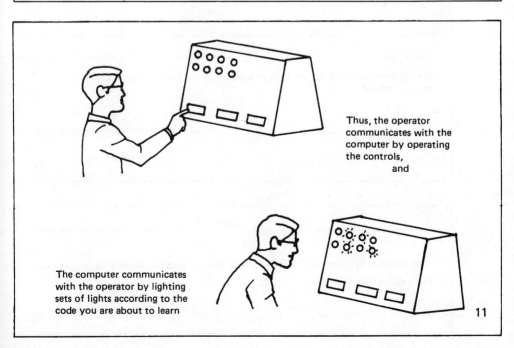

Thus, the operator communicates with the computer by operating the controls,
and

The computer communicates with the operator by lighting sets of lights according to the code you are about to learn

11

Exhibit 5.1 *(continued)*

How does the operator tell the computer what to do?

_____

How does the computer inform the operator?

_____

Answer these questions,
then turn to the next
page to check your
answers

12

The operator tells the computer what to do by operating the controls.

The computer informs the operator by lighting lights.

---

one set of lights

one set of lights

The <u>sets of lights</u> on a computer console indicate numbers that are being either stored or worked with in various parts of the computer.

13

The <u>numbers</u> located in various parts of a computer are reported to the operator by means of sets of

_____   Check your answer on the next page.

14

_____ *lights* _____

A <u>set</u> of lights on a computer control console tells the operator about a_____

that exists somewhere in the _____ .

15

Exhibit 5.1 *(continued)*

---

number

computer

Each <u>set</u> of lights represents only <u>one</u> number.

Therefore, a <u>set</u> of three lights indicates only <u>one</u> number at a time.

A set of five lights also represents only _____ number.

16

ONE

At any moment, then, a set of 12 lights can represent only _____ number(s).

<u>How many numbers</u> are represented by each of the following sets of lights?

Set 1   _____

Set 2   _____

17

ONE

ONE

ONE

To <u>read</u> the number displayed by a set of lights you need only:

   1. <u>know</u> the numerical value of each light,

   2. <u>add</u> the numerical values of the lights that are <u>lit.</u>

This is how you do it: _____

18

Exhibit 5.1 *(continued)*

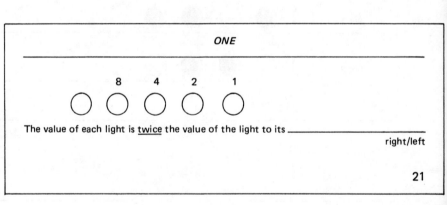

A set of lights

The light at the <u>right end</u> of a set is a always (ONE)

The light to its left is always a (TWO)

The light next to the TWO is
   always a (FOUR)

16   8   4   2   1

19

The light at the right end of a set always means _____ (what number?)

20

ONE

8   4   2   1

The value of each light is <u>twice</u> the value of the light to its _____

right/left

21

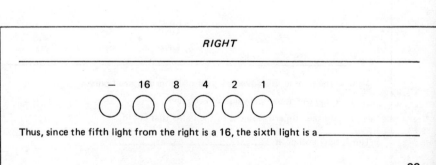

RIGHT

—   16   8   4   2   1

Thus, since the fifth light from the right is a **16**, the sixth light is a _____

22

Exhibit 5.1 *(continued)*

32

Complete the missing values in the set:

—   —   —   16   8   4   2   1

◯ ◯ ◯ ◯ ◯ ◯ ◯ ◯

23

---

128 64 32

No matter how many lights in a set, the light at the far right is a _____

24

ONE

Though each light in a set has a specific value, how many <u>numbers</u> are displayed by <u>one</u> set of lights?

_____

25

ONE

To learn the number displayed by a set of readout lights, we <u>add</u> up the values of <u>all</u> the lights that are <u>lit.</u>

For example,

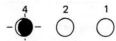

Since only the 'four' light is lit, the number shown by the set of lights is _____

26

Exhibit 5.1 *(continued)*

FOUR

In this example,

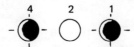

The '<u>four</u>' and the '<u>one</u>' lights are lit, so the number shown by the set is the <u>sum</u> of four and one, or _____

27

FIVE

The number shown by a set of readout lights is the _____ of the values of the lights that are <u>lit.</u>

28

SUM, or TOTAL

What number is shown by this set of lights?

_____

29

Twelve (12)

What number is shown by this set?

_____

30

Exhibit 5.1 *(continued)*

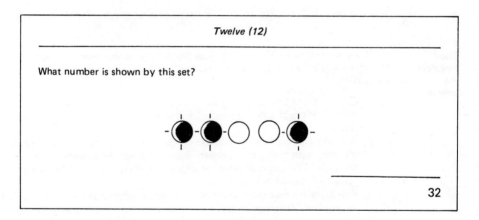

will require to complete the section, and possibly some hints on how to organize his study.

A review frame may simply summarize what has been presented in the last few pages, or may set an overall test or emphasize the relevance of what has been learned to other material that has been learned previously. However, we are here interested principally in the main teaching sequences – the teaching and the test frames.

The rules or teaching points

Let us now analyse a little more closely the example of the programme presented. We will refer again to frames 11 to 32. Taking as our definition of a teaching point the amount of new information that can be presented by one simple sentence without any 'ands' or 'buts', we can identify approximately six or seven teaching points in this section of the programme. (A suggested summary of teaching points is presented in Figure 5.2.) The first two points deal with the general context of the task to be taught, and then points 3, 4, 5, 6 and 7 summarize all the information necessary in order to perform the job. Note, therefore, that the route of analysis which this programme is following in its construction is the topic analysis route, starting with content or subject matter, which later is transformed into objectives.

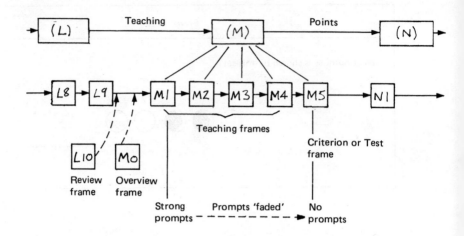

Figure 5.1 *The structure of a typical linear programmed sequence*

The functions of the teaching frames

If we now consider the functions of the frames in this example in relation to the seven teaching points, we can see that the frames perform one of three functions:

1. They present new information. The teaching point is presented for the first time. We might call this type of frame an 'information' or a 'demonstration' frame.
2. A frame can contain an exercise or question based on this new information, together with prompts, in order almost to force the student to respond correctly. We can call these 'prompted' frames, 'exercise' frames or 'practice' frames.
3. Finally, a frame can evaluate if the desired behaviour has been learned by the student. In other words, can he perform without the help of any prompts? We can call this a 'test' frame.

Of course one should note that a particular frame may sometimes have more than one function. For example, it might introduce new information while prompting or testing learning that has been introduced before.

Let us analyse a few of the frames in this programme. Frame 11, for example, introduces teaching points 1 and 2. Frame 12 tests both these points by means of a straight question. There are no prompts to help you. Thus, frame 11 is an information frame and frame 12 is a test frame. Further, frame 13 is a new information frame introducing teaching point 3. Frame 14 is a prompted frame for teaching point 3. Frame 15 is a test frame for the same teaching point.

If we continue with this analysis, we end up with a classification of our frames in relation to the teaching point they deal with, and the function which they perform, rather like the analysis in Figure 5.3. The reader may like to check our analysis of these frames, and may well come to disagree on some points, particularly whether certain frames should be classified as 'test' or 'prompted' frames. Sometimes the prompts are very obvious, but at other times they are rather weak. For example, in some cases one might consider a particular frame which asks a direct question (which is exactly the same as the question asked in the previous frame) as a prompted frame rather than a test frame because of the effect of the sequence of the frames.

1. *Operators* communicate with *computers* by means of *controls.*
2. *Computers* communicate with *operators* by means of *sets of lights.*
3. *Sets of lights* represent *numbers* stored in the computer.
4. *One* set of lights represents *one number* at any one time.
5. *To read the number* represented by a set of lights, *add the values of the lights that are lit.*
6. The *right-hand light* has the value *one.*
7. *Every other light* has a value equal to *twice the value of the light immediately to its right.*

Observation: The sequence indicated above is the sequence in which the rules are introduced in the programme. The practice of the reading of computer lights (application of rule 5) must wait until the values of the individual lights (rules 6 and 7) have been defined. This helps explain the sequence of frames in the programme. (Remember our comments regarding this example, in the latter part of the previous

Note: This is a topic analysis presented as a sequenced set of rules or teaching points that define the content of the lesson.

Figure 5.2 *Analysis of the content of the programme presented in Exhibit 5.1*

Teaching point	Classification of frame functions		
	Information	Teaching (prompted)	Test
1	11		12
2	11		12
3	13	14	15
4	16	16, 17	25
5	18	26, 27, 28, 29, 30, 31	32
6	19		20, 24
7	19	21, 22, 23	

Figure 5.3 *An analysis of Exhibit 5.1: frames 11 to 32 classified according to their functions in the programmed sequence*

The pro-
gramme's
objectives

We may make a further observation about this programme. The behaviour which the student is asked to perform can be categorized into two basic types. He responds by reading examples of read-out lights, which is the practice of a *skill*. But he also responds by completing or stating the rules, which have to be applied in order to read the lights. In other words, the programme is also teaching and testing *knowledge*. We can say, therefore, that this programme appears to have two overall objectives: first, that the student should be able to read read-out lights on a computer panel; and, second, that he should be able to state the rules verbally and explain how the read-out lights are read.

Consider the
relevance of
the
objectives

This raises an interesting question: are both these objectives necessary? If one were to consider a target population made up of computer operators who have to learn to read the read-out lights of the computers they work with, then one might argue that getting them also to practise *stating the rules* is really a waste of time. The programme could be just as effective, in terms of job performance, and considerably shorter if it concentrated only on the one objective of *reading the lights*. This does not mean, of course, that the programme would not state the rules. What it means is the programme would not test whether the *student* could state the rules, and therefore would not include practice or prompted frames that require the student to repeat the rules verbatim.

On the other hand, we might consider that the target population for this programme is a group of students in school who are studying this programme as a step towards other verbal learning. They may be tested in their final exam by questions which ask them to explain how read-out lights work. Or perhaps this programme is only a step in getting the student to understand the binary system of notation (which is, after all, the principle behind the read-out lights), so that the student can get an insight into an area of mathematics which he will soon study. Then we could argue that it is very important that the student should not merely read the read-out lights but also learn the rules, learn to state the rules and learn to explain the procedure to somebody else.

We shall come back to these considerations of whether the programme should teach both objectives or only one of them a little later. For the time being, let us return to considering the structure of the programme as it stands.

Consider the
instructional
strategy

Teaching points like the one which states 'in order to read a set of lights you must sum the values of all the lights which are lit' are called procedural rules. Such rules are learned by being stated and by being applied to examples. There are two alternative sequences: the *expositive* sequence, which would normally proceed from the statement of the rule to the giving of examples and eventually to the student applying the rule to new examples; and the *discovery* approach, which would usually start by the student working with examples and, by the solution of several examples, discovering the general rule which applies to them all. Thus, one could call the expositive approach a 'rule to example' approach, and the discovery approach an 'example to rule' approach. Both these two approaches can be used in the preparation of instructional programmes. The programme we are analysing follows an expositive strategy, first stating the rules and then applying them to examples. In the early 1960s, Glaser and others called this the Rul-Eg approach – rule followed by example.

The Rul-Eg
technique

The Rul-Eg approach developed its own system of notation to describe frames which *present* the rules (*information* frames), which ask the student to *state* the rule (*test* frames), and which ask the student to *complete* a partly stated rule (*prompted* frames). A similar three-stage notation for completely worked examples, partly worked examples and examples for the student to work on his own is also possible. This gives six component elements of rule or example from which a given frame can be composed. These are illustrated in Figure 5.4 together with an analysis of some of the frames of our programme showing the functions which these frames perform expressed in the Rul-Eg language.

Ref p602

Symbol	Meaning of the symbol
Rul	Communication of a rule to the learner as a complete statement (information).
R͡ul	Presentation of an incomplete statement of the rule, for the learner to complete (prompted).
R͡͡ul	A request for the learner to state the rule, without the help of prompts (a test).
Eg	Presentation of an example of the application of the rule (a full demonstration).
E͡g	Presentation of an example that is partly worked or solved for the learner to complete (prompted).
E͡͡g	Presentation of an example for the learner to resolve unaided (a test).

Note: The terms in brackets relate this classification to the one used in Figure 5.3.

Example of Rul-Eg analysis applied to a section of Exhibit 1.

This analysis deals only with the sequence of the programme that deals with teaching point 5 (frames 26 to 32).

Frame number	Function of the frame
26	Rul and E͡g
27	Rul and E͡g
28	R͡ul
29	E͡g
30	E͡͡g
31	E͡͡g
32	E͡͡g

Figure 5.4 *The Rul-Eg system of notation, useful for the analysis or design of instructional sequences for conceptual learning*

Planning the detailed instructional tactics

Of course, what we are doing in our analysis is exactly the opposite procedure to what the programme writer would have done in his design, or synthesis, of this sequence. He would have started by listing his teaching points, organizing them into what he considered to be the ideal teaching sequence, and then deciding exactly what frames he would write for each of the teaching points. In order to help him design the frames he would, before actually writing them, plan out a map of frames against teaching points, using the Rul-Eg notation to describe exactly what function each of the frames should have. Figure 5.5 illustrates the sort of plan which might have been produced as a step in the design of the programme that we have been analysing.

Applicability to other instructional methods and media

The Rul-Eg approach is interesting because it illustrates a fairly formal technique by which step-by-step instruction can be developed. Although it is not much used nowadays, it is in fact a technique capable of being applied not only to designing step-by-step, small frame, linear programmed instruction, but may also be applied, for example, to the development of a lesson plan which a teacher may give. Figure 5.6 illustrates the sort of lesson plan that might be developed using the general idea of Rul-Eg in order to teach the same computer read-out lights skills in a classroom instruction situation.

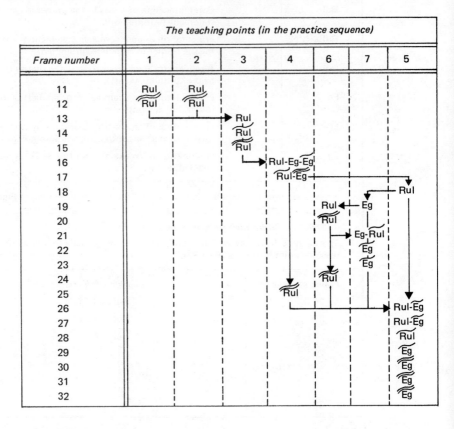

| | The teaching points (in the practice sequence) | | | | | | |
Frame number	1	2	3	4	6	7	5
11	Rul	Rul					
12	Rul	Rul					
13			Rul				
14			Rul				
15			Rul				
16				Rul-Eg-Eg			
17				Rul-Eg			
18						Rul	
19					Rul ← Eg		
20					Rul		
21						Eg-Rul	
22						Eg	
23						Eg	
24					Rul		
25				Rul			
26							Rul-Eg
27							Rul-Eg
28							Rul
29							Eg
30							Eg
31							Eg
32							Eg

Note: whereas teaching point 5 is taught by an expositive (Rul-Eg) sequence, teaching point 7 is taught by a discovery (Eg-Rul) sequence. Note also how the plan 'ties everything together' by revising points taught earlier before the final practice sequence.

Figure 5.5 *A plan for the programme in Exhibit 5.1,*
using the Rul-Eg Notation

We shall now proceed to the study and analysis of a second example of programmed instruction, based on mathetics. This example was developed from the first one, so readers will notice a good deal of similarity in the first frame or two. However, we recommend a close study of the differences in structure and programming methodology.

5.4 Analysis of Exhibit 5.2

Now let us analyse the second version of the programme to teach the skill of reading computer read-out lights (see Exhibit 5.2). You will notice that this programme is very short compared with the previous one, attempting to cover the same ground in four frames instead of more than 20. This programme has been written using the mathetics technique. One of the principles of this technique is that one should in the first version of a training exercise attempt to make the exercise as difficult as the student can take, while not allowing him to fail on the

Course unit:	Binary systems	Lesson number: 1	
Lesson title:	Reading computer read-out lights		Time: Half an hour

Equipment:	A simulated computer console, showing three or four sets of lights which may in some way be marked as 'on' or 'off'. This may be a real console or a simple OHP transparency/magnetic board/flannel board representation.

Materials:	Some slides or photographs of real computer consoles and of computer operators.

Step	Teacher activity	Learner activity	Time plan (mins)
1. Motivation and introduction	**Motivate:** Introduce the topic of man-machine communication and the 'language of computers'. Use slides or photos.	*Observe* and *attend* (should show observable signs of *interest* in the topic).	0 5
2. Rules 1 and 2	*Explain* by visuals the two directions of man-machine communication.	*Recall rules 1 and 2.* Man⟶computer controls Computer⟶man: sets of lights.	
3. Rule 3	*State* that sets of lights signify numbers. Give *examples* of some activated read-out lights.	*Recall rule 3.* What do the lights communicate? Numbers stored in the computer.	10
4. Rules 4 and 5	*State* that one set communicates one number at a time. *Test comprehension* by reference to both large and small sets of lights on the simulated panel. *Show* how the value of the set is found.	*Apply rule 3 by discriminating* the individual values of the lights and the value of the set. Show *comprehension* that the number of lights in the set does not relate to the 'number of numbers' that can be communicated.	15
5. Rule 6	Present *examples* of the individual values of the lights.	*Recognize rule 6.* The right-hand side is always ONE.	
6. Rule 7	Give further *examples* of values of individual lights, until learners generalize the rule.	*Discover* the generality of *rule 7.* 'Each light has value twice the light on its right-hand side'.	20
7. The whole procedure	*Revise* rules 4, 5, 6 and 7 by *questions.* Present a range of *examples for practice* graded from easy to difficult. *Prompt* the learners if this proves necessary. Measure accuracy and response rate.	*Recall rules 4, 5, 6, 7. Apply* them to reading sets of read-out lights presented on the simulated panel/console. Continue practice until demonstrating required skill and confidence.	30

Figure 5.6 *A lesson plan for teacher-led group instruction based on the same analysis and content as the self-instructional programme presented in Exhibit 5.1*

Exhibit 5.2 *Mathetics programme*

The excerpt which follows is an attempt to write a programmed sequence
equivalent in content to Exhibit 5.1 but employing the techniques of
mathetical exercise design, as originally outlined by Tom Gilbert and
modified by Imbucon Learning Systems Ltd. Please study this exhibit before
continuing this chapter.

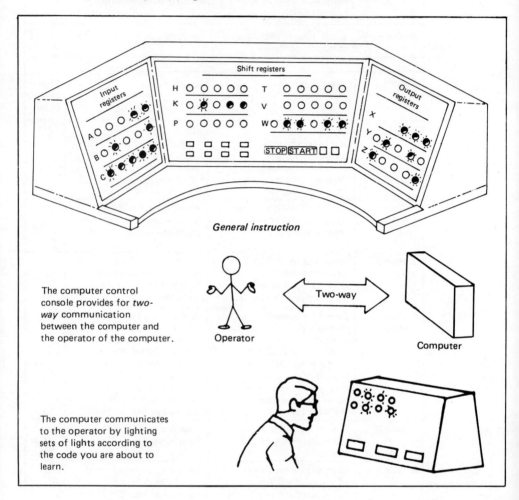

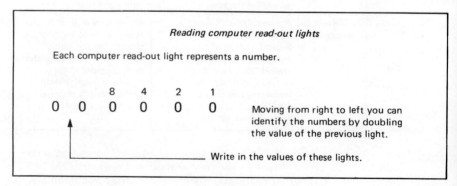

Exhibit 5.2 *(continued)*

32	16	8	4	2	1
0	0	0	0	0	0

Reading computer read-out lights

To find the number represented by a set of lights, sum the values of the lights which are lit.

Example:

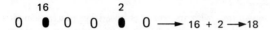

16 + 2 ⟶ 18

(a) What are the values of the last two unlit lights of the set above?

(b) What number is represented by the set of lights below?

0   0   0   0   0   0

(a) 32 and 64   (b) 49

Reading computer read-out lights

Write the numbers being communicated by the sets of lights illustrated below.

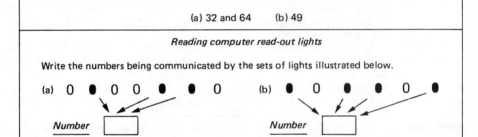

(a) 38   (b) 45

Reading computer read-out lights

What are the numbers represented by the set of lights below?

Set 1	●	●	0	●	●	0
Set 2	0	0	0	0	0	0
Set 3	●	●	●	●	●	●
Set 4	0	●	●	●	●	0

final test. Another principle of the mathetics approach is that only behaviour that is directly related to the terminal objective is worth practising during the training exercise. We note, therefore, that this second version sets out to teach only one objective (reading the computer read-out lights). The programme states the basic rules but in a much more condensed fashion, and nowhere does it require the student to write or state the rules. On the contrary, the student immediately passes to the stage of applying the rules to practice problems.

It may be difficult for the reader to evaluate objectively the two versions of the computer read-out lights programme that we have presented, having first seen the long and detailed 1964 version and then the very short 1970 version. However,

Comparison of the two versions of the programme

tests of the two programmes with groups of students who had not seen the other version beforehand have shown that there is not much difference in the initial *effectiveness* of these two programmes. Obviously, the *efficiency* of the second shorter version is much greater, since the student takes much less time to complete this version. It is possible, however, that a day or two later a retest of students' ability to read the lights may show in favour of the first version, which gives rather more practice and takes great pains to explain and to get the student to memorize the principles which have to be applied.

However, is this really necessary? Do we need to ensure that our students remember the content and the technique of this programme over a period of days without further practice? Of course it all depends on the context in which the instruction takes place.

Exhibit 5.1 was prepared for use in schools

The first programme, with its contents-based approach, seems to have been written more for a school audience which is studying computers as part of a more general educational course. A study of the whole programme from which this small section was extracted confirms this. The chapter we have presented is the first of an introductory programme to binary arithmetic to be used as part of a mathematics course. As such it is quite a clever approach, since the learners (who may perhaps not be very keen on mathematics and who might be put off by a new method of counting such as the binary system when they first meet it) are introduced to this through an interesting exercise with content that seems fairly remote from the mathematics they will learn. When they come to learning the term binary code, they have already learned indirectly, to use it. They simply need to abstract and generalize what they have already learned from the specific case of computer read-out lights.

In the school context we have a situation where the students may need to remember over long periods, and without practice, the technique of counting in binary numbers. They will need this in the final examination, and may not practise it very regularly in the meantime. Hence, learning to verbalize the rules may act as a very useful aid to remembering, or to deriving again from first principles, the procedures for binary counting.

Exhibit 5.2 was prepared for training computer operators

The second programme, on the other hand, has only one objective: to get trainees to read computer read-out lights. If we postulate that this had, indeed, been prepared to train future computer operators in this particular task, then it would be logical to assume that immediately after training these people will pass to real-life practice of the new skill they have learned. They will no doubt read computer read-out lights every day thereafter, so there is no great likelihood that they will forget how to read the lights. Nor is there a high probability that the skill will be lost through lack of use, and thus there is no need for over-training in the skill during the learning process. Also there is no need for practice in verbalizing the rules for the reading of computer read-out lights, as they will never be called upon to restate those rules as part of their job.

There are other aspects to this second version of the programme which are quite similar to the first. We have, in the first frame, an overview or introduction putting the skill in its context. Then we have three frames, the first of which presents new information – the rule which gives the value of each of the individual lights. It also provides practice in applying the rule. Then, frame 3 presents the second main rule – how to compute the value being shown by an activated set of lights; that is, summing the value of those which are lit. Once again, it immediately gives practice in applying the rule. It also gives further practice in applying the previous rule. The fourth frame is a test frame, which expects the student to apply what he has learned to the reading of several read-out lights (both rules in combination).

The main differences in approach

It is interesting to examine the theoretical basis for the difference in approach to programming in these two examples. The first exhibit was based on an analysis of the subject matter or topic. It started by establishing the rules (the knowledge content) and then sequencing them in what appeared to be a logical order. The second version started in a different manner. It started from the objective (from the desired terminal behaviour) and tried to determine the *minimal* explanation and the *minimal* practice necessary to establish that behaviour in the trainee.

Thus, the first version has followed what we earlier called the 'input to output' route of instructional design, whereas the second version has followed the 'output to input' route. Let us analyse more closely the stages that one would follow in designing such an instructional programme.

Stimulus-response analysis of the objective

We start with the objective – a behavioural objective. Behaviour has two components: a response, emitted by the person or the organism that is behaving, and a stimulus that has activated that particular response. It is very useful to think of the stimulus and response as a unit, as a totality, because behaviour which is intelligent behaviour does not occur without a reason; a response does not exist without the presence of some stimulus that has activated or triggered it. It is therefore very useful to analyse behaviour in terms of stimulus and response. We saw in Chapter 4 how this can be done in the analysis of the problems experienced by a trainee telephone operator. The stimulus-response analysis revealed several typical behaviour patterns, which could be identified in the task of answering a telephone call and connecting it to the right person. The three behaviour patterns were *chains*, *multiple discriminations* and *generalizations*.

The objective as one operant

For our present example, the stimulus-response pattern of the behaviour that our computer read-out light reader has to learn is very simple. It is:

$$S \longrightarrow R$$

Activated set of read-out lights | Read (and note down) the number being communicated

We *could* attempt to teach our trainee this unit of behaviour in one step, simply explaining and demonstrating how to do it and getting him immediately after that to practise to perfection. This would be, in fact, very similar to the TWI approach, a method based upon demonstrating the whole task, explaining key points, and then getting the learner to practise the whole task (and possibly explain the key points back to the trainer).

The objective split into two sub-objectives

In the case of version 2 of the read-out lights programme, the programmer decided to break this behaviour pattern down into two stages. The first of these, shown below, is the behaviour of stating the correct value of a given light.

$$S \longrightarrow R$$

Any set of read-out lights | State the value of each light in the set

The second behaviour which follows on the first is shown below, and is concerned with totalling the values of the activated lights in order to get the total value of the set.

$$S \longrightarrow R$$

Activated set of read-out lights | Total the values of the lights which are lit (and note down the result)

Analysis of the structure of Exhibit 5.2

Now let us analyse exactly what each frame of our programme does in relation to these two units of behaviour, treating them as two sub-objectives of our terminal objective. The chart below shows an analysis of the second version of the programme in a way similar to that in which we analysed version 1 at the beginning of this chapter, classifying the functions of our frame in relation to the objectives in terms of whether they are demonstrating new behaviour, giving practice in a particular behaviour or testing a particular behaviour.

We note that, just as with the previous programme, there is a sequence of frames which go from demonstration to practice to test. We notice also that the frames have multiple functions. Frame 1 is demonstrating the first behaviour and

	The sub-objectives (or operants)	
Frames	Operant 1	Operant 2
1	Demonstrated and practised	
2	Prompted practice	Demonstrated and practised
3	Tested indirectly	Prompted practice
4	Tested indirectly	Tested directly

Notes:
(a) 'Demonstrated and practised' means that the frame states the rule, demonstrates its application to an example, and asks the learner to demonstrate his understanding by also applying the rule to a similar example.
(b) 'Prompted practice' means that the frame requires the learner to perform as is specified in the sub-objective, but with the help of 'prompts' of some form.
(c) 'Tested indirectly' means that the question asked requires the performance of this operant as a prerequisite to the behaviour being asked for directly.

Figure 5.7 *An analysis of the four frames of Exhibit 2*

including a little practice on it. Frame 2 is demonstrating the second unit of behaviour, giving a little practice in that and also giving some practice on the first unit of behaviour. Frame 4 is testing all the behaviour together.

The basic 'demonstrate practice and test' strategy

We notice that whereas the first of the two programmes (Exhibit 5.1) was based on an analysis of the content and teaching of rules, memorization of rules and hence application of rules to problems, the second version (Exhibit 5.2) concentrated only on the *application* of the rules to problems. However, both the programmes apply the same instructional strategy – that is, a basically expositive strategy of demonstration, practice and evaluation.

One of the programmes includes many small steps minimizing the practice in each step and giving a lot of clues and prompts. The other programme expects the learner to do more for himself, giving him less practice steps, but more practice in each step, thus getting him perhaps to think more, and to restructure for himself internally the rules which have been presented to him, which he is never asked to restate but simply to apply.

The difference between the two programmes is therefore not at the overall instructional strategy level, but rather in terms of viewpoint as to the relative importance of verbal knowledge as a step towards performance of a given task, and in the tactics of programming adopted. In Exhibit 5.1, the knowledge content is emphasized. It is stated, practised and tested, and is used as the basis for analysis and sequencing of the programme. In Exhibit 5.2, the required behaviours are analysed and sequenced, and the minimum requisite knowledge is stated quite clearly, but is then practised and tested indirectly, as part of the skill.

Applicability to other instructional methods and media

In Figure 5.8 we show a lesson plan developed from the same analysis that was used to construct the programme presented in Exhibit 5.2. We have allowed the same lesson time (30 minutes) as that for the previous example (Figure 5.6) though our experience suggests that the second, mathetics-based lesson plan would probably require less time to achieve the same objectives. However, we have enlarged the introductory part to 10 minutes, in order to give a more complete overview of the context in which computer read-out lights are read.

Note that the lesson plan in Figure 5.8 has fewer distinct steps, a result of not asking for separate verbal definition by the learners of the basic rules used to read the lights. Note also the care with which the planner has defined four transparencies, to be used at specific points in the lesson. The content of these

Course unit: Computer operation		*Lesson number:* 1	
Lesson title: Reading computer read-out lights		*Time:* 30 minutes	

Equipment: A simulated computer console, presented as diagrams of sets of lights, in a sequence of OHP transparencies

Materials: Some photographs of real computer consoles and operators

Step	Teacher activity	Learner activity	Time plan (mins)
1. Motivation and introduction	Introduce the topic of man-machine communication and the 'language of computers'. Use *slides and photos* to *explain* the two directions of man-machine communication. Define and show examples of computer read-out lights.	*Observe* and *attend* to the presentation. The learners should show observable signs of interest in the topic to be developed during the lesson.	0 10
2. Demonstrate operant 1	Demonstrate a set of read-out lights (use transparency 1). Explain the rules that define the values of the individual lights.	*Demonstrate comprehension* of the rules by identifying the values of some of the lights presented in the examples on transparency 1.	15
3. Demonstrate operant 2 and practise operant 1	Present several sets of lights (use transparency 2). Explain the rule for reading the number being communicated by adding the values of the lights that are lit.	*Recall* prior learning by naming the values of the lights that are lit. Show *comprehension* of rule 2, by reading the numbers being displayed, on some of the examples presented.	20
4. Prompted practice of both operants	Present further sets of activated lights (use transparency 3). Guide and prompt the learners as necessary.	Respond by *reading the numbers* being communicated by the sets of lights illustrated in the transparency.	25
5. Test of the new behaviour	Present the final test, as a hand-out or on the OHP, using transparency 4.	Respond individually, *writing down the numbers* being communicated. Hand in to teacher.	30

Figure 5.8 *A lesson plan for teacher-led instruction based on the same analysis used for the programme presented in Exhibit 2*

transparencies could well be exactly identical to the visual content of frames 1 to 4 of the programme shown in Exhibit 5.2. The transparencies do not, of course, need to present exactly the same examples used in the programme (although they could do so), but they should be identical in general structure. This is because the visual content of the demonstrations made during instruction is derived from an analysis of the stimuli which the trainee will receive when performing the task in reality.

Although at first glance the two lesson plans presented in Figures 5.6 and 5.8 may appear quite similar, a deeper analysis reveals substantial differences in approach.

6. Tactics for Producing Expositive Self-Instructional Materials *

6.1 The basic model for expositive instruction

Where we are

In Chapter 5 we analysed two examples of self-instructional materials. These were based on different approaches to programming, but both employed an expositive strategy. We identified the three principal component parts of an instructional sequence which follows an expositive strategy, namely *demonstration* (and/or illustration and explanation), *prompting* (provision of guided and graded practice, with corrective feedback if necessary), and *testing* (some authors call this the 'release' stage).

Where we are going

We shall now proceed to a more detailed study of how to plan and develop *expositive instructional materials and exercise sequences*. Later, we shall contrast these techniques with those applicable for the development of *experiential learning materials and exercises*. Throughout this chapter, we will illustrate the principles of the techniques suggested by means of examples of written exercises. It is more difficult in a book to present examples in other media, though we do occasionally describe other applications in outline. This stress on paper-based examples should not be taken to imply that the techniques suggested are limited to the design of print-based programmed instruction. The reader should attempt to generalize from the examples presented in the exhibits to other types of media.

We have already shown that the analysis prepared to plan a sequence of self-instruction may also be used to plan a conventional classroom lesson. It is true, of course, that conventional lessons may be planned on the basis of a less thorough analysis (a Level 3 analysis). In such cases, we leave more of the detailed decision making to be taken on-line by the teacher. However, the basic principle holds good – there is largely no difference between a well-planned classroom lesson and a self-instructional sequence aiming at the same objectives and using the same basic strategy.

We now hope to show that the basic expositive strategy can be adapted to various types of instructional objectives, and may be applied to produce various types of instructional materials.

We shall start with a simple example of one objective taken from the knowledge domain: the rule for the rounding off of a decimal number to a given number of decimal places. This constitutes a simple procedure (or algorithm) which involves the discrimination between two possible cases and the performance of a simple chain for each of the cases. In Exhibit 6.1 we present a detailed account of how to construct a mathetical exercise for this example. Please study this exhibit carefully before continuing with this chapter.

Please study Exhibit 6.1 before continuing

6.2 A discussion of Exhibit 6.1

The description and example presented in Exhibit 6.1 were first published by Conrad Brunstrom, in an article for the 1974-75 *APLET Yearbook of Educational and Instructional Technology* (ed: Romiszowski, 1974). The article outlined the approach

---

*This chapter was written in collaboration with Conrad Brunstrom, Dick LeHunte and Brian Slymon of Learning Systems Ltd, UK.

Exhibit 6.1 *Extracts from* Developing in Training Design *(Brunstrom, 1974), illustrating the mathetics 'demonstrate-prompt-release' exercise design model based on the suggestions of Tom Gilbert (1962) and further developed by Learning Systems Ltd*

Instructional Design Rules

Behavioural analysis is the basic analytic approach in training design. It is complemented by instruction design rules or models which provide a completely formalized approach for deriving instruction materials from the training prescription. The rules can be applied to any instruction method (and at different levels of detail). This section illustrates the application of the rules to the construction of programmed learning materials, since programmed learning uses the rules in their most detailed and explicit form.

The instruction design rules can best be described under three separate headings:

1. Demonstrate-Prompt-Release Model
To teach trainees a particular behaviour unit (ie to teach them to make the response to the stimulus) you can first *demonstrate* the response to the stimulus, then *prompt* the trainee to make the response, and finally *release* it, by presenting the trainee with the stimulus and asking him to make the response without any prompts. This is the basis of the D-P-R model, but it is not a rigid three-step process — you can also use combinations like DR, PR, DPPR, etc.

2. Instruction tactics decisions
In order to translate the training prescription into training you have to make a number of decisions on instruction 'tactics', such as:
- ☐ instruction sequence for behaviour units
- ☐ emphasis on DPR, DR or PR
- ☐ layout and format of instruction materials
- ☐ feedback method and amount of feedback
- ☐ step by step 'D-P-R' map for construction of training*
- ☐ style of prompts
- ☐ overview requirements.

All these decisions should be made on a formalized basis, relying on 'rules' relationships between criteria for the decisions and the decisions themselves. *An important bridge between the final prescription and the construction of the training is the D-P-R map — a plan showing D, P and R intentions for each behaviour unit in each section of the training.

Let us take this example of a discrimination for rounding off numbers to three decimal places:

S——————————R

| Number in fourth | Omit all figures |
| place less than five | after third place |

S——————————R

Number in fourth	Add one to figure in
place five or more	third place and omit all
	subsequent figures

Suppose that you decide to teach this with an overview page (telling or showing trainees what they will learn) followed by three teaching pages — 1D, 1P and 1R. Starting with the *demonstrate* page you take a large sheet of paper and follow these materials construction rules:

1. Record *stimulus* centrally on page — in this case we show each of the stimuli which can occur —————	3.1624 3.1625 3.1627
2. Record *response* close to stimulus —————	3.1624 rounds off to 3.162 3.1625 rounds off to 3.163 3.1627 rounds off to 3.163

Exhibit 6.1 *(continued)*

3. Write an *instruction* for the trainee — telling him what you want him to do on this page

> 3.1624 rounds off to 3.162
>
> 3.1625 rounds off to 3.163
> 3.1627 rounds off to 3.163
>
> Follow these rules to round off:
> 10.8022
> 0.0385
> 5.6648

4. Write an *overall* context statement. This is usually a title or continuity statement — probably not necessary in a short piece of instruction

> *Rounding off to three decimal places*
> 3.1624 rounds off to 3.162
>
> 3.1625 rounds off to 3.163
> 3.1627 rounds off to 3.163
>
> Follow these rules to round off:
> 10.8022
> 0.0385
> 5.6649

Demonstrate page

5. Finally, write any *observing information* needed to draw attention to the stimulus and response, and to explain or clarify how the response is made

Only include the minimum of necessary information

> *Rounding off to three decimal places*
> To round off to three decimal places, drop any figures after the third place
>
> 3.1624 rounds off to 3.162
>
> But if the figure in the fourth place is *five or more* ADD one to the third figure
> 3.1625 rounds off to 3.163
> 3.1627 rounds off to 3.163
>
> Follow these rules to round off:
> 10.8022
> 0.0385
> 5.6699

This completes the *demonstrate* page. To summarize, you construct the page in this way:

1. Stimulus
2. Response
3. Instruction
4. Overall context
5. Observing information

To construct the *prompt* page you follow the same sequence, but this time:
You do not record *the response* at all. You want the trainee to make the response, with prompting. You put in less *observing information*, since you are only prompting the trainee to make the response, not showing him how to do it.

Prompt page

> 3   *Rounding off to three decimal places*
>   Remember, when you round off to three
> 4   places, to check whether the fourth
>   place figure is five or more
> 2   Round off to three decimal places:
>   4.3726
>   1.59006
> 1   0.0744
>   28.2915

Exhibit 6.1 *(continued)*

The numbers show the sequence in which the page is written:
1. Stimulus
2. Instruction
3. Overall context
4. Observing information (reduced)

Finally, on the release page you require the trainee to make the response with no prompting, so this time there will be no observing information.

Release page

3	*Rounding off to three decimal places*
2	Round off to three decimal places:
	105.1871
	0.622501
1	0.33333
	8.1997

In this case the sequence is:
1. Stimulus
2. Instruction
3. Overall context

The materials construction rules for content and sequencing can be shown in a table:

	Demonstrate	*Prompt*	*Release*
Stimulus	✓	✓	✓
Response	✓	✗	✗
Instruction	✓	✓	✓
Overall context	✓	✓	✓
Observing information	✓	[✓]	✗

In addition, you need to include exercise answers in the programme. The correct answers to the exercise on each page would be given on the following page, say in the top left hand corner.

As you complete each page, following the materials construction rules, you should then 'stand back' from it and consider whether the presentation can be improved or the text reduced. You may have to shuffle pieces of information about, or highlight key points — you can use boxes, arrows, large print, sketches, etc to do this.

used by Learning Systems Ltd at that time for the design of learning materials and exercises. This approach was a development of Gilbert's mathetics methodology (Gilbert, 1961). Whereas Gilbert's original writings are rather long and technical, and thus not too easy to read, Brunstrom's article presents the basics of the methodology very clearly and succinctly. It is for this reason that we decided to reproduce the extract presented in Exhibit 6.1. Several interesting observations may be made in respect of the example and the exercise design technique presented in this exhibit.

The sequence of steps in exercise design

In the first place, observe that the instructional designer is, as it were, 'painting the exercise'. He is not setting pen to paper in the conventional sense of beginning at the top of a page and writing one paragraph after another. He is developing the message in a somewhat unorthodox sequence of steps, commencing with the *stimulus* situation to which the learner should respond (this may be an example of a problem or a simulation of a real-life situation) and the type of *response* that the learner should make to this stimulus. Other aspects, such as the rules, the general context, instructions on what to do, improvements in the layout and sequencing of the information on the page, are added on, like paint to a canvas, in order to produce the most effective and most efficient message that the designer can invent.

Importance of the behaviour prescription in controlling exercise design

Second, note that all the design stages are controlled by the initial analysis of the behaviour that is to be taught by the exercise. Every one of the pages, or frames, of the exercise contains the same stimulus situation (including all the variations of the stimulus that occur in practice) and the learner is expected to supply appropriate responses. (Each of the pages presents examples in which the fourth decimal place is less than five, is equal to five and is greater than five.) This is a characteristic of a simple mathetical exercise.

In a more complex case, the objective may be split into two or more sub-objectives (component behaviours), which would then be taught in separate exercises, before the total behaviour is practised and tested. These exercises may, however, be superimposed on each other, so that the learner is practising one component behaviour while receiving initial demonstration of another. This was done in the second version of the computer read-out lights programme presented in Chapter 5 (Exhibit 5.2). In this instance, the two component behaviours of naming the values of individual lights and reading the number being shown by a set of lights were taught in tandem.

Importance of the basic expositive exercise design model

Third, note that the author has used the terms demonstrate, prompt, release (or D-P-R for short) to describe the basic expositive design model illustrated. This is Gilbert's original terminology. The words make sense when we remember that we are always referring to behaviour. First, the new behaviour is *demonstrated* to the learner, then the learner practises under the *control* of *prompts* and other help supplied to him, and, finally, the newly mastered behaviour is *released* from the control of the teacher and practised freely by the learner, just as in the real-life task being taught.

However, this third (release) stage is invariably used as a means of formative evaluation, to decide whether the amount of prompted practice was sufficient, or whether the amount of theory and explanation given was sufficient. We therefore feel that the more common term 'test' may be substituted for all practical purposes, and thus refer to the steps of the model as 'demonstrate, practise, test'. However, this difference in terminology does not imply any divergence from the basic methodology described.

Theoretical and practical implications of the model

Fourth, although the methodology described is so closely based on an analysis of the objectives to be achieved, in terms of stimulus and response, we do not believe that the adoption of this methodology necessarily implies that the instructional designer is espousing all the tenets of behaviourism. This might be the case if he were to apply this slavishly, and only this, model of design for *all* the instructional design tasks he may be called upon to perform. But the adoption of a stimulus-response notation as a way of analysing the end results of training is not necessarily behaviourism. Nor is the strategy of attempting to simulate, during training, the end-of-training situation, necessarily out-and-out behaviourism. It is no more than the systematic application of the age-old tenet of learning by doing.

Versatility of the model

Fifth, the basic model may be varied, as Brunstrom indicates, to fit better the needs of a specific training task. A task which is very difficult to learn may require a more complex approach, involving several practice sessions, gradually increasing in difficulty, before the finally desired behaviour may be expected from the learner. Such a design may have the form of DPPT (DPPR if you prefer) or even DPPPPT (DPPPPR). This is typical of the spaced practice designs used for teaching complex, high-speed skills in industry.

A task which does not form part of the job of a given trainee, but which he should have some notion of, could be treated by means of an exercise comprising a demonstration and some limited practice. There is no need for all trainees to reach a final, predetermined standard. This model could be described as DP and may often be encountered in education when people are learning something for its own sake, rather than for vocational reasons. It is also applied in the training of managers and supervisers, who should know enough about the jobs that they are supervising to detect errors, inspect performance standards, and so on.

A typical conventional textbook on, say, mathematics consists of chapters on specific topics that include many worked examples and (sometimes) a few comprehension tests. At the end of each chapter there is a selection of problems from past papers that serve to illustrate the level of knowledge and skill expected by the examiners. Such a chapter is essentially a series of demonstration pages, followed by a comprehensive test page (after all, the exam problems do not contain direct and deliberate prompts or simplifications designed to enhance the chances of success). If a chapter deals with, say, four principal learning objectives, then we might conceptualize its structure as DDDDTTTT. The analysis of existing materials by means of these three basic components of our expositive instruction model, can be very illuminating and may lead to the identification of possible, and very obvious, improvements in the chapter's structure.

A skills practice session, such as typing at a particular speed, may not need any new demonstration or any theoretical explanation. The session may start by paced typing, at a speed above current performance standard, aided by musical accompaniment, a metronome, or even a programmed exercise tape of the type used in the Sight and Sound system. After an appropriate period of such paced (or prompted) practice, the trainee transfers to the real typing situation, trying to maintain the new speed, without making errors, and without the aid of pacing. The trainee's behaviour is now released from the control of the instructional system. It is being tested and evaluated. The basic model here is therefore PT, although in some cases, where the trainee fails to maintain the new peformance standard and must return to more paced practice, it may be PTPT or even PTPTPT. Variations of this model are involved in all instances of drill-and-practice exercises.

We believe that all expositive instruction may be described as a combination of the three basic units of exercise design that we have introduced here. For this reason, we believe that the D-P-T (or D-P-R) model is a very useful tool for the design of expositive instructional exercises in any subject area and in any medium of instruction.

Use of the model for the design of classroom instruction

Sixth and last, to illustrate the use of basically the same design model for the design of other methods and media, we present another excerpt from Conrad Brunstrom's article, in which he illustrated the design of classroom instruction for the same rounding-off objective, based on the analysis presented in Exhibit 6.1. The approach to the design of classroom instruction, shown in Exhibit 6.2 is essentially the same as that we have been using since Chapter 3, for lesson planning. It shows, however, in a very clear manner, how the stimulus part of the behaviour operant being taught largely determines the teaching activity – what Brunstrom calls the 'instructor support' given to the learner – and how the response part of the operant largely determines the learning activity (the column named 'trainees' in the exhibit). It also illustrates how the steps of the lesson follow the steps of the model adopted.

Exhibit 6.2 *Further extract from* Developments in Training Design
(Brunstrom, 1974) showing the application of
the basic expositive design model to classroom instruction

As before, you wish to teach the 'rounding off' discrimination with an overview followed by three steps — 1D, 1P and 1R. After the overview, follow the materials construction rules to write in notes for the demonstrate step:

1. Record *stimuli* in *instructor* column

Instructor	Trainees
3.1624	
3.1625	

2. In this case you need not record the response (unless the instructor lacks technical expertise in the task) since the instructor should know it. Next you record on *instruction* for the instructor — and one for trainees if you want them to do something in this 'demonstrate' step.

Instructor	Trainees	
Round off 3.1624		
Round off 3.1625	Round off	0.8022
		5.6649

3. Finally, include notes on any necessary *observing information*

D	Round off 3.1624 explain cut off Round off 3.1625 explain addition and cut off	Round off	10.8022 5.6649

This is the completed demonstrate step. The trainee instruction could be omitted if you think it unnecessary for the trainee to make a response at this step (other than observing what the instructor does).

The whole lesson plan, including the *prompt* and *release* steps, will look like this:

	Instructor	Trainees	
	Introduction — state aims and show example		
D	Round off 3.1624 explain cut off. Round off 3.1625 explain addition and cut off	Round off 3 d.p.	10.8022 5.6649
P	State to check whether fourth place five or more	Round off 3 d.p.	4.3726 1.59006 0.0744
R		Round off 3 d.p.	105.1871 0.622501 0.33333 8.1997

Comments

1. The prompt and release steps are constructed in accord with materials construction rules — note that in the prompt step the stimulus is recorded in the *trainees* column, and only the observing information is recorded in the instructor column.

2. This is a short and simple illustration. In practice, for a prescription with several behaviour units, the final release step is the session end exercise.

3. The example shown contains detailed specific information for the instructor (eg stating the actual examples to use). The training designer can construct a lesson plan at a lesser (or greater) level of detail — this will depend on the technical and training skill of instructors, and the stability of the training organisation. If the lesson plan shown is constructed for highly skilled instructors it should contain less details, and examples should be left to the instructor.

**Applicability
of the model
for the
design of
video for
training**

We shall conclude this section of the chapter by quoting an experience that illustrates the value of the approach outlined here (and above all the need for a systematic approach to instructional design) in all media. Some years back, I was responsible for the installation and implementation of a videocassette production facility for a large telecommunications organization in Brazil. Once the basic equipment was installed, I set about planning a training course for the people involved in the planning and production of training through video. Because of the past history and recruitment policy of the organization, the target population for this training was made up of approximately equal numbers of 'education' experts (graduates in education or psychology), with some experience in the training context but no experience of the use of television, and television experts (graduates in engineering and communications), with some experience in the operation of studios and equipment or in the planning and production of commercial television, but with little or no work experience in education or training. The course was given by a team of international experts.

In the first module of the course, an expositive approach was adopted, and the special characteristics of video for *training*, as opposed to commercial, documentary or even broad educational uses, were carefully explained. At the end of this module, four teams were created to produce training videocassettes for given objectives. The result was four programmes of varying artistic quality (some quite imaginative), varying technical quality (some quite professional), varying programme design strategies and tactics (some quite strange), but none of them really likely to achieve the proposed instructional objectives (at least, not without the help of an awful lot of extra teaching, preparation and follow-up).

The trainees, when invited to analyse each other's productions, demonstrated a bewildering variety of criteria. As these were analysed by the team of instructors, the trainees were gradually led to the opinion that video cannot be a training medium in its own right, but must always be used for 'enrichment', or as the demonstration component in a more complex, multi-media package. Whereas this may be true in a great many applications, it is not always so. There are cases, particularly in the training of assembly or machine operation tasks, where the video medium may be the most important, or indeed the sole, component of an effective and efficient training system. There is the example of Ford in South Africa, which successfully replaced almost all of its training centre-based live instruction by on-the-job video instruction.

In the second module of the course a different strategy was adopted. All four groups received the same project – to produce a videotape that alone would teach an illiterate who had never worked in a factory or workshop context to take apart and correctly re-assemble a domestic telephone. The instructional strategy was now experiential. Each group had to try to meet the project requirements, analyse any weaknesses encountered in its initial attempts, replan and reshoot the tape, analyse, remake and try again as often as was necessary in order to reach the project objectives. The groups worked separately, without interchange of ideas. All groups finally achieved the project objectives, but no group took less than three remakes to achieve the target of teaching a range of specially selected 'naive trainees' to assemble a telephone quickly and accurately.

A study of the successive remakes was most illuminating. Each group started with substantially different general approaches to the design problem and completely different shooting scripts, dialogues, camera positions and movements. As they progressed to successive remakes, all four groups sooner or later cut down camera movements until the camera stayed in the position of the operator's eyes (though some zooming in, to show detail, persisted), cut down dialogue to the bare minimum, while improving the visual clarity of the demonstrations on video and providing built-in 'practice breaks' to allow the trainee to develop skill in one operation before building it into another. All four groups ended up with the same design model – $D_1P_1D_2P_2D_3P_3D_4P_4T_{(all)}$.

We shall return in Chapter 14 to a more detailed analysis of this case study. For the meantime, it is interesting just to note that the same approach to analysis may be used for both printed and audiovisual materials design.

6.3 Sophistications of the basic expositive model – special tactics for the teaching of multiple discriminations

The teaching of multiple discriminations

The exercise we studied in the earlier part of this chapter set out to teach a very simple discrimination between two alternatives. The decimal number we were expected to round off to *three* decimal places has a *fourth* decimal digit either less than five, or five or more. If it is less than five, we simply omit the other digits; if it is equal to or greater than five we add one to the third decimal place and omit the others. Let us now look at a slightly more complex discrimination. Exhibit 6.3 reproduces an extract of four frames from a classic example – a programme written by Tom Gilbert to illustrate the mathetics technique in 1961. The objective of this programme is to teach students to read the colour code used on electrical resistors. It is an example of a relatively complex multiple discrimination of ten items which have to be distinguished one from another and which are quite easily confused. The programme is self-explanatory. We suggest you read it as if you were really trying to learn the content, reading one frame at a time, not looking ahead to the other frames, covering the responses until you have made an honest attempt at answering and evaluating your performance on the exercise.

Most people do quite well on this exercise. Indeed, it has been tried in groups of many hundreds of trainees, and has given results in three or four minutes of study equivalent to 10 or 15 minutes of study of a straightforward, unstructured list of the 10 colour-number associations.

Study Exhibit 6.3 before reading on

When studying the examples, note the special prompting techniques employed by the author, the amount of new information presented right from the start of the exercise, and the sequence in which this information is presented.

It is obvious that the structure of this programme also follows the expositive 'demonstrate, practise and test' model.

The use of mnemonics as mediators

However, perhaps the most characteristic feature of this programme is the use of mnemonic reminders built into the ten sentences describing the colour-number associations. This technique has been known and practised for many centuries. However, there are good mnemonics and mediocre ones. This one is particularly good and bears a little analysis.

We have before us the task of learning a set of ten colour-number associations, which are relatively new and without particular meaning for us at the moment. Any one of these associations is a stimulus-response link as shown below.

$$S \longrightarrow R$$

Red band   'Two'

The technique the programmer has used is to look for two other stimulus-response links which he believes already exist as a result of the trainee's previous learning. For instance, in the case of 'red' and 'two', he has dug up the associations *Red-Heart*, and *Heart has two parts*, thus transforming the direct stimulus-response relationship 'Red-Two' into a two-stage or triangular process, as shown in the diagram below. Thus the key word heart becomes a link between the two items to be associated. The technical term for this sort of link is a 'mediator'.

Mediators and analogies

The use of mediators for the teaching of associations is a very useful technique when teaching factual information, where there is no meaningful relationship between the two items to be associated. Of course, in the case of conceptual learning there are other techniques that would be used. An equivalent of the mediator in the learning of concepts and principles is the analogy, for example the analogy mentioned earlier of water running through pipes or cars running through street networks to electrical circuit theory.

The planning of sequence within the demonstration page

There are a number of other interesting points to observe about this particular example.

The units of behaviour, or operants, are presented in a random order, and this order changes in every frame. A non-sequential presentation is preferred because, when the trainee practises the reading of coloured bands on resistors, he

1a. Some electrical resistors have a colour band that tells how much they will resist electric current. On small resistors you can see colours better than numbers. Each colour stands for a number.

THE FIRST THREE COLOUR BANDS ARE READ AS THE NUMBER OF OHMS RESISTANCE

THE FOURTH COLOUR BAND IS READ AS THE PERCENTAGE OF ERROR IN THE RATING

1b. Each of the **First Three Colour Bands** can have one of 10 colours. Read through this list twice. Learn the **Number** for which each **Colour** stands.

a **Five** dollar bill is Green
One Brown penny
a **White** cat has **Nine** lives
Seven Purple seas
a **Blue** tail fly has **Six** legs

Zero: Black nothingness
a **Red** heart has **Two** parts
Three oranges
a **Four** legged **Yellow** dog
an **Eighty** year old man has Grey hair

---

2. List the number for which each **Colour** stands:

Red (heart)	White (cat)	Purple (seas)	Brown (penny)	Black (nothingness)
Green (bill)	Grey (hair)	Blue (tail fly)	Orange (oranges)	Yellow (dog)

---

3. List the **Number** for which each colour stands:

Black	Brown	Yellow	Grey	Green
White	Purple	Red	Orange	Blue

Exhibit 6.3 *An exercise for teaching a multiple discrimination.*
An extract from a programme on reading the colour code
of electrical resistors (Gilbert, 1962)

encounters these colours in a random sequence. An ordered numerical list would therefore be an imperfect design. The learner might be unable to do the exercise without the help of the sequence built into the material which he is studying. He would recite the colours in the order from zero through to nine until he learned them in that sequence, and henceforth would always have to recite to himself the sequence in order to pick out the colour he wants. This is the sort of behaviour that many people display throughout their lives because they learned multiplication tables by heart in this way.

While on the question of learning aids and sequences that may or may not be helpful in learning, it is interesting to note that whoever invented this colour code built it around a sequence. The colours are, in fact, sequenced in the same order as the colours of the rainbow, with the addition of what one might call the 'dirty shades' of black and brown as zero and one (next to red) and the 'clean shades' of grey and white, representing eight and nine, built on the upper end of the scale (after violet). That is fine, except that it is not a very useful learning aid in practice. There are two reasons for this.

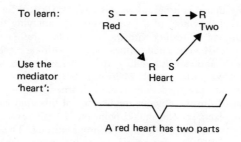

To learn: S – – – – – →R
Red   Two

Use the
mediator   R   S
'heart':   Heart

A red heart has two parts

Figure 6.1 *The use of mediators in the teaching of an association*

First, as mentioned above, it can result in the sequence being an important element in memorizing the various associations, so that one has to follow through from zero every time one has to pick out a colour-number association.

Second, the particular learning aid built in would only be useful to a student who already knows the sequence of the colours in the rainbow, and this is probably quite rare among the sort of people who need to learn the resistor colour code.

If the colour code were being taught to physics students in a secondary school as one item of knowledge that they required within a much wider course that also happened to include the colours of the spectrum, there might be some sense in teaching the colour code of resistors by the order of the colours of the spectrum, because the learning of one would reinforce the other. However, we would still not overcome the first objection, that the colour code would be remembered as a sequence, when in practice use of the colour code requires random-access.

The planning of relevant responses

Throughout this programme the student only practises one behaviour pattern:

S ——————→ R
A number   A colour

He never practises the opposite – given a number, state what colour represents it.

The point of interest here is that many educators would consider it well worth giving practice in both behaviours – matching numbers to colour and colours to number. This would obviously take more time, unless one cut down on the practice of the important behaviour, ie matching numbers, given a colour. So a further principle in this mathetics programme is 'only practise the behaviours that are relevant in the final job situation'. Any other practice inevitably reduces the amount of time which one can give to the really relevant practice.

Comparison with the Rul-Eg approach

One of the first instructional programmes written in England dealt with the colour code of resistors. It was a typical linear sequence, though perhaps a little less well written than the good linear programme we showed in Exhibit 5.1. It had more or less the following structure:

Responses	Frames
	1. Brown represents 1.
	2. Red represents 2. What is the meaning of a red band followed by a brown band?
2 and 1	3. Orange represents 3. What is 2?
Red	4. What is brown? What is orange?
Brown is 1 Orange is 3	5. Yellow represents 4. What is red?
Red is 2	6. Green represents 5. What is yellow?

This sequence continues for a total of about 64 frames.

Figure 6.2 *Extract from an early linear instructional programme for colour code of resistors*

Note that this programme was based on a Rul-Eg type of design, starting with the verbal statement of the rules and then applying them to examples. Note also that, instead of dealing with the topic in four rather fat frames, it split the topic down into 64 very small frames dealing with each colour-number association separately and slowly building up the behaviour on what might be called a 'progressive parts' model, where first one is learned, then two is learned, then one and two are practised, then three is learned, and so on.

Present discrimin- ations 'all at once'

Apart from the obvious inefficiency of this approach, leading to a much longer learning exercise, one should point out a basic error which again illustrates one of the important rules of mathetical programming. The learner, in studying this sequential Rul-Eg programme, goes quite a long way through the sequence before he is really comparing various colours together in the way that he will have to when he is faced with a real job situation (ie many resistors and many coloured bands on each one). In other words, it isn't until well into the programme that the learner really starts discriminating between a large number of alternatives, and it is indeed not until the final two or three frames of the programme that he is dealing with all the ten colour-number associations together.

By contrast, the mathetical programme starts off immediately with all the colour-number associations presented to the student. It presents the complexity of the final job situation right from the beginning. The reason for this is that, as the learner is required to learn to discriminate, he must be given a variety of stimuli to discriminate between, and he must be given these as soon as possible in the training sequence. The splitting down of a discrimination into step-by-step learning in the old-fashioned example is directly opposed to this principle. It is anti-instructional. It is a technique which does have its applications, but not in the teaching of discriminations. It is applicable to the training of step-by-step procedures, or chains.

6.4 Sophistications of the basic expositive model – backward chaining

The teaching of chains and simple procedures

Exhibit 6.4 deals with the calculation of the income tax that a person should pay (in a hypothetical country). The rules are quite simple, and there are just three steps to the calculation process. We would suggest that the reader studies the programme as if he really had to learn the procedure being taught, and we shall then analyse the structure of the exercise.

Analysis of Exhibit 6.4

Let us go quickly through some of the analysis of this particular programmed sequence. The behaviour pattern that we have learned is a chain made of three links, or operants, as shown in Figure 6.3.

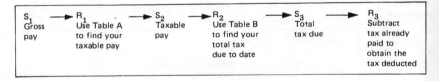

| S_1 Gross pay | R_1 Use Table A to find your taxable pay | S_2 Taxable pay | R_2 Use Table B to find your total tax due to date | S_3 Total tax due | R_3 Subtract tax already paid to obtain the tax deducted |

Figure 6.3 *The behaviour prescription*

The functions of the five frames which were presented to teach this chain are summarized in the table presented in Figure 6.4.

Backward chaining

Perhaps the most unusual aspect of this particular presentation is that it has broken with the traditional custom of teaching a procedure by starting from the beginning, and has in fact begun by *first* demonstrating the *last* step. Note that although the last step was the first to be demonstrated, the actual practice that the student performs goes through from the beginning to the end, but in early frames he practises only the last step or last two steps, and it is not until the last frame o

Frame	Operant 1	Operant 2	Operant 3
1.	–	–	Demonstrate
2.	–	Demonstrate	Prompt
3.	Demonstrate	Prompt	Test
4.	Prompt	Test	Test
5.	Test	Test	Test

Figure 6.4 *Analysis of the frames in Exhibit 6.4*

two that he practises all the steps together. This is an example of the application of the principle of 'backward chaining', which was suggested by Gilbert as the most appropriate sequencing technique for linear procedures, or chains of behaviour.

Origins in animal training

The justification for this approach springs from the animal laboratory. If we wish to train an animal in the performance of some trick which involves two or three stages (as for instance pressing a button which may or may not give a green light; and if a green light appears, pressing a lever which may or may not open a door; and if a door opens, going through it to get food), we will not succeed unless we start with the last step in this chain. A multi-stage behaviour must be built up step by step, starting from the last step and working back to the beginning. This is the only way in which a complex procedure can be developed without verbal communication. It is interesting to note that the instructional strategy that is being applied in this example is not really expositive, because there is no verbal communication or practical demonstration. It is in fact a guided discovery approach, where particular behaviour is interpreted as being the right behaviour because of the results it produces.

Relevance to the human learner

A quick glance at the income tax example will show that the instructional strategy used is not a discovery strategy. The learner does not 'discover' what the steps of calculating his income tax are. He is told. This is an expositive strategy. Indeed, we are using the 'demonstrate, practise and test' instructional model which was applied in the previous examples. One might ask, therefore, why are we following a backward chaining approach? After all, it is obvious that this sort of skill can be learned by the normal A, B, C method.

The answer depends to a certain extent on the reinforcement (or reward) principle mentioned in the animal example. In this case, however, we are suggesting that the sort of person who wants to calculate how much tax he will have to pay at the end of the year is interested principally in one thing, and one thing only: the amount he has to pay. This is the bit of information which is motivating him to study the exercise. Therefore, the sooner he calculates this, the more likely he is to continue to remain motivated to learn.

Furthermore, things slip into place and take on relevance which sometimes they do not do when one follows the more traditional sequence. In this example, there are certain technical terms, such as gross pay and taxable pay, which may be new to the learner. If he is not very well prepared, and not particularly interested in mathematics and the intricacies of accountancy and income tax, he may even have a built-in aversion to such technical jargon. If one starts in the normal A, B, C sequence, one of the first things one has to do is to define some of these terms. Furthermore, the learner has to start by using a fairly complex table. This is not motivational, nor is it the easiest part of the procedure to learn.

However, when starting at the end, the last step is obvious, and obviously relevant to the learner's interest. He learns simply that what he has to pay is the difference between what he owes for the year and what he paid during the year. That's obvious: he knows what he's paid during the year, because he's kept records. So it is suddenly of vital importance to know how to calculate the amount

Exhibit 6.4 *An exercise for teaching a simple procedure (chain) —
an excerpt from an exercise to teach how to check your income tax
deduction. (The procedure has been simplified
for the purposes of this Exhibit.)*

---

1. Calculating the tax deduction

The tax deduction that is made at the end of the year is calculated
as follows:

total tax due — tax already paid = tax deduction

Example:	Total tax due	70
	Tax already paid	45
	Tax deduction to be made	25

Now you calculate the tax deduction to be made if

| the total tax due is | 80 |
| and the tax already paid is | 15 |

Response
to frame 1

65

2. Calculating the total tax due

Your total tax due depends on your taxable pay.
Use Table B to read off the total tax due as follows:

	Table B	
	Taxable pay	Total tax due
(a) Find your taxable pay in column 1	50	10
	60	12
	70	14
	80	16
	90	18
(b) Read off the total tax due in column 2	100	20
	110	22
	120	24
	130	26
	140	28

Now complete this example:

Your taxable pay is 100.
1. Use Table B to find the total tax due and
   write this down here _____
2. Now, supposing you have already paid 15,
   subtract to find the value of the tax
   deduction that will be made _____

Exhibit 6.4 *(continued)*

Responses to frame 2	3. Calculating your taxable pay

1. 20
2. 5

Your taxable pay depends on your gross (total) pay and on the number of dependants that you have.
Only wives and children under 18 count as dependants.
You may use Table A to find your taxable pay as follows:

Table A							
Gross salary	Number of dependants						
	0	1	2	3	4	5
100	100	90	80	70	60	50
200	200	180	160	140	120	100
300	300	270	240	210	180	150
400	400	360	320	280	240	200
500	500	450	400	350	300	250
600	600	540	480	420	360	300

(a) Find your gross salary in column one.
(b) Find the column for the number of dependants (eg three).
(c) Read off your taxable pay.

Now complete this example:

(1) Your gross salary is 200. You have a wife and three young children. You have already paid 20 in tax this year.

Find your taxable pay in Table A. Write it here _____

(2) Use Table B to find your total tax due. Write it here _____

(3) Calculate the tax deduction to be made. Write it here _____

Responses to frame 3	4. The whole procedure – a practice problem

1. 120
2. 24
3. 4

You have a wife and one son aged 16 and one daughter aged 24.
Your salary is 800. You have already paid 25 in tax this year.

(1) Use the real (complete) book of tables (A and B).
   (a) Find your gross salary in the first column (Table A).
   (b) Check how many *eligible* dependants you have.
   (c) Read off your taxable pay in appropriate column.
                              Write the result here _____

(2) Now find your total tax due. Write it here _____

(3) Calculate the current tax deduction. Write it here _____

Responses to frame 4	5. The whole procedure – a test

You have earned 1,200 this year and have already paid 150 in tax.
Your eldest child is 14 years old. Your wife is expecting a fourth child early next year. You are keen to pay any back income tax before the new baby is born, so you wish to calculate what you owe.
What is the currently owing tax deduction that you should pay?

owed for the whole year. That then leads to the penultimate step. If we know our taxable pay, we can use Table B quite simply to read off what is owed for the whole year. 'Right,' says our learner, 'that is obvious; I can do it. But what is my taxable pay? How is that calculated?' That, of course, leads naturally to the first step of the procedure, rendering the use of Table A relevant and obvious, and in fact leading in a direct line through to the desired outcome by steps that the learner has already mastered.

As in previous examples, the same basic instructional design can be applied to other plans and in other media. One may for instance prefer to adopt a teacher-led group instruction method, rather than a print-based self-instructional method. In that case, the lesson plan could follow an exercise sequence similar to the programmed sequence of Exhibit 6.4.

A possible lesson plan, which also adopts the backward chaining sequence, is shown in Figure 6.5. This plan differs from the programme in Exhibit 6.4 only in two aspects: it is a teacher-led instructional plan and it does not specify the exact examples to be used in the teaching, leaving this choice to the teacher concerned and in some cases the learners themselves. In this latter respect, the lesson plan differs from the one for rounding off decimals, presented in Exhibit 6.2.

6.5 Sophistications of the basic expositive model – conceptual learning

Concepts and principles

Space does not permit us to continue analysing the further applications of the expositive 'demonstrate, prompt, test' model of instructional design in the detailed way adopted in the examples in this chapter. A whole book could be devoted just to the design of such expositive instructional exercises. They are quite generally applicable, and are easily developed for the teaching of concepts.

The use of examples in teaching concepts

In the latter case, the exercise design is similar to that for a multiple discrimination, except that what the learner discriminates are examples and non-examples of the concept being taught. Care must also be taken in the selection of the examples. One must be sure to use examples representative of the whole range of instances that constitute the concept. One must include sufficient 'near-examples' to delineate the boundary of the concept to the required level of precision. And care must be taken to vary the examples so that other, non-relevant properties do not appear to be relevant. For example, in teaching a shape, the size, colour and so on must be varied in a random fashion. Finally, one must use different examples at each stage in the exercise, so that the learner learns to generalize to the whole class of examples (ie he has learned a concept) and not simply to identify certain specific examples (ie he has learned a discrimination).

Teaching concrete concepts

In the case of a concrete concept, such as the concept of red, the instructional designer must present many examples of red, non-red and near-red objects, varying in size, shape, etc. The explanations will be restricted to pointing out which of the objects presented are normally considered red and which are not. The programming in this case would probably take the form of initially presenting obviously red and non-red examples and then, as the learner's proficiency develops, gradually narrowing down the difference between the examples and non-examples, presenting shades of orange and pink that are ever closer to the accepted concept of red. In this exercise, there is no clear distinction between the 'prompted practice' and 'test' stages. They have merged into one continuum of progressively finer discriminations between reds and non-reds.

Primary school teachers are well acquainted with this sort of programming. It takes place over a period that might be weeks, months or, in the case of some concepts, years, accompanying the child's developing ability to discriminate ever finer shades of difference. Such a programme of conceptual development still follows the steps of our basic expositive instructional design model. It would be different if the teacher were not to structure his teaching of this concept, but were simply to provide the learner with a colour-rich environment and leave him to discover by interacting with this environment and with the people in it, just what

Lesson: *Calculating your tax deduction*		
Step	*Teacher activity*	*Learner activity*
1.	Introduce the topic. Motivate interest in checking one's tax deductions. Refer to the learners' job situation. Who deducts tax? Could they err?	Find and examine data on their own salaries and the tax they have paid. Express interest in learning to check their tax deductions.

2.	Explain the formula:
	   Tax deduction =
	   total tax due —
	   tax already paid
	Illustrate with an example.
		Perform the subtraction in one or two examples.

3.	Define taxable pay. Demonstrate the use of Table B by an example.	Use Table B to find tax due for a given amount of taxable pay. Then proceed to calculate the tax deduction.
4.	Explain how taxable pay depends on the number of dependants. Define 'dependants'. Demonstrate the use of Table A by means of one or two examples.	Use Table A to find taxable pay for a given amount of gross pay and number of dependants. Then proceed to calculate the tax deduction.
5.	Give one or two more practice exercises, prompting or helping individual learners if necessary.	Perform problems under teacher guidance until mastery is achieved.
6.	Ask each learner to calculate the tax deduction he expects on his next pay cheque.	Check own pay cheques, recalculating the tax deductions. Calculate future tax deductions.

Figure 6.5 *A possible lesson plan for teacher-led instruction of the same content as that in Exhibit 6.4 and employing the same basic instructional design*

Teaching defined concepts

constitutes the normally accepted concept of the colour red. This could be called a discovery strategy, though not a particularly effective or efficient one.

In the case of a defined concept, the instruction may be accomplished by an exercise resembling those presented earlier. A defined concept depends on previously learned simpler concepts and this prerequisite learning should first be tested. After that, an expositive exercise can be developed according to the rules of our basic model. Figure 16.6 presents an example of an exercise designed to teach the concepts of acute angled triangle and obtuse angled triangle. Prior learning includes the concepts of a triangle, a right angle, degrees, etc. Prior knowledge of the concepts acute angle and obtuse angle is not assumed, as this exercise is seen as a means of indirectly teaching these concepts also.

The exercise shown in Exhibit 6.5 illustrates clearly the designer's care in trying to give a sufficiently varied selection of examples and non-examples of the concept to be taught. As the non-examples of acute angled triangles are examples of right angled or of obtuse angled triangles, it makes sense to teach these concepts together. The exercise thus becomes a concept formation exercise for the two concepts 'acute' and 'obtuse' angled triangles and a discrimination exercise *between* these two concepts. In reality, the third concept of a 'right angled triangle' is also involved, as the limiting case between the other two concepts.

As it was given that the concept of a right angle was already known from previous learning, we used this limiting case as a learning aid, incorporating it as a sort of mediator, or rather a standard for comparison, in order to prompt the performance of the learners in frames 1 and 2. In frame 3, we no longer present,

Exhibit 6.5 *Example of an instructional sequence
designed to teach two related defined concepts*

1. Demon-
stration
frame

Acute angled and obtuse angled triangles

Triangles in which all three angles are *smaller than a right angle*, are
called:

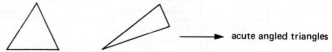

Triangles in which one of the angles is *equal to a right angle*, are
called:

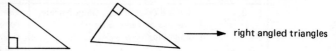

Triangles in which one of the angles is *greater than a right angle*,
are called:

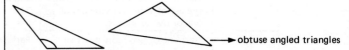

In the five examples shown below, mark the acute angled triangles
with an A, the right angled triangles with an R and the obtuse angled
triangles with an O.

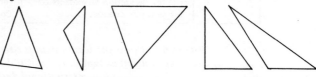

2. Prompted
practice
frame

Acute angled and obtuse angled triangles

Study the size of the angles in the six triangles shown below.
Identify the *one* right angled triangle.

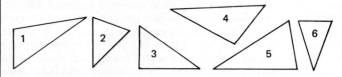

Which (numbered) triangles are acute angled? _____

Which (numbered) triangles are obtuse angled?_____

3. Test or
release
frame

Acute angled and obtuse angled triangles

Name each of the triangles illustrated below.

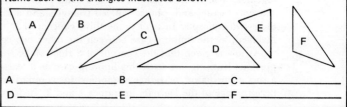

A _____ B _____ C _____
D _____ E _____ F _____

or mention, right angled triangles, as the final performance required is that students should correctly identify acute and obtuse angled triangles in *any* context and not just when presented alongside a right angled triangle, for comparison. The student must compare the examples with a right angle that is 'in his head'.

The number of examples presented and the amount of variety in size shape and position necessary to establish this concept in a young learner is something that the instructional designer must judge from past teaching experience, or else must discover by testing out the initial versions of his instructional materials on a suitable sample of the target population. If in doubt, it is better to err on the 'hard' side initially, as it is difficult to discover when one has included too many examples.

The limits of the D-P-T model Any relatively simple, or discrete, concept (or any group of closely related concepts) may be taught by exercises similar in structure to the example shown in Exhibit 6.5. As the concepts to be taught become increasingly complex, abstract and verbal, the examples and non-examples become increasingly lengthy statements, rather than neat collections of diagrams. The observing information or rules that define the concept also become longer and more complex. Before we know it, our instructional exercises become blocks of quite complicated text, made up of (yes, you've guessed) rules and examples, not to mention a few non-examples to define the concept's limit. This is the area of programming in which the Rul-Eg approach to programming really should be applied. It is in this area of verbally expressed, defined concepts and principles that a topic analysis is called for in order to identify the examples, non-examples, explanations and analogies that will lead the learners to a full understanding of the knowledge to be learned.

And the real value of Rul-Eg Thus, to summarize, while one is dealing with the acquisition of skills in which there is little knowledge-content or the knowledge-content has already been learned, the basic expositive instruction model may often apply. Also, in the knowledge domain, this is the most efficient model for the teaching of factual knowledge and procedures, made up of chains and discriminations, or simple concepts, both concrete and defined. As we proceed to more complex verbal learning, the mathetics model transforms itself into the Rul-Eg model, though usually presenting much more substantial units of verbal information in a frame than was the custom in the early examples of linear programmed instruction. The misuse of the Rul-Eg approach resulted from its application to subject matter made up of simple concepts, discriminations and chains and to the development of simple reproductive skills. This resulted in programmes with frames that were too small to be interesting, yet too verbal to be relevant to the real objectives of the instructional exercise being designed.

New wine in old bottles The reaction from the over-programmed small frame approach fostered in the early days of programmed instruction produced a number of 'new' methodologies. These were, nevertheless, still based on the old PI principles. Among these were:

1. The 'task card' or 'learning activity card' approaches, which differ principally in presenting a unit of information somewhat bigger than a frame, capable of being presented in one page, or on one large card, followed by a question or reinforcement task, to be responded to and handed in to a monitor or the instructor in exchange for the next card.

2. Various structured learning-manual systems, which generally present one or two full pages of demonstration or information, followed by one or more questions or tasks to practise skills or evaluate learning. Perhaps the most sophisticated of these techniques is the information mapping structured writing system developed by Robert Horn, which we study in depth in Chapter 8.

3. Various systems of instructional module writing, which are characterized by presenting a number of pages of information (perhaps four to eight) to the learner before presenting him with a self-assessment test or exercise. This test allows the learner to practise the application of the new knowledge to specific tasks.

4. Systems, such as the Keller Plan, which present the learner with a whole chapter of ten or more pages of new material, or a series of references to sections of books or journals in the library, followed by self-assessment and skill development exercises. These are usually handed in to a tutor or monitor for evaluation and discussion before proceeding to a new unit.

All these varieties of instructional systems are derivatives of programmed instruction, applying most of that methodology's basic principles, but adapting the structure of the materials to the needs and characteristics of specific types of target population, specific types of course content or specific characteristics of the overall training structure adopted by an institution.

All are examples of basically expositive instruction, based on sequences of demonstration and/or explanation, followed by practice and/or test. Most have specific advantages over the small frame style of programme writing in specific situations, but may not be quite so applicable in others. Some require a more detailed objectives or topic analysis than others. It is a fallacy to think that the larger the unit of information presented in one chunk to the learner, the less need there is for a detailed analysis. Rather, the factor that indicates the depth of analysis that is necessary is the amount of difficulty that typical learners experience in mastering the subject and its objectives. A detailed analysis of a difficult subject might, nevertheless, lead to the presentation of large units of learning material to the learners. However, these large units may involve a great deal of detailed design. And this design will be based on analyses and models similar to those presented in this and the two preceding chapters.

6.6 The limitations of expositive programmed techniques and some of the possible alternatives

As one proceeds to the teaching of more abstract concepts, or complex conceptual schemata, or explores heuristic strategies and general theories, one is less likely to select basically expositive strategies. The theoretical reasons for this were mentioned in Chapter 2 of this book and more extensively in *Designing Instructional Systems*.

Discovery methods for concept and principle learning

One would be more likely to adopt discovery learning strategies as the concepts become more abstract. This is not to say that it is impossible to use programmed, expositive instructional methods at this level, but simply that the discovery approach is often preferred. Once again, space precludes a full treatment of the design of discovery learning exercises. The methods that may be used include small group instruction, or tutorials, a host of laboratory or project-based methods, games and simulations and occasionally self-instructional programmed exercises. The latter may take the form of branching programmed instruction, adaptively branching instruction (as in computer-assisted instruction) or conversational tutorial methods, as employed in structural communication. We shall discuss some of these below, and shall devote the next chapter to structural communication.

Limitations of written, expositive instruction

There are still other factors which may limit the applicability of written, expositive, programmed instructional materials. Many of these were examined in an article by Biran in the 1974 *APLET Yearbook* (Biran, 1974). Most of these limitations may however be overcome by the inventive adaptation of the basic programming principles to the production of different instructional products. Some of Biran's list of limitations, together with ways of overcoming them, are presented in Figure 6.6.

6.7 The basic model for experiential instruction

Experiential instruction defined

To some people, the juxtaposition of the terms experiential and instruction may seem strange. However, our definition of instruction, given in Chapter 3, allows such a juxtaposition, so long as the experience that leads to learning is pre-

Limitations of programmed self-instruction (from Biran, 1972)	Some creative solutions which have overcome the limitations.
Unsuitability: For teaching manual skills. (Dexterity or high speed.)	'Programmed simulators' (Wheatcroft, 1973; Romiszowski, 1968, 1974). See also the 'RITT' system.
For poor readers. (Young children or adults in industry.)	'RITT' system of paced audio-programming in industrial training (Agar, 1962). Language Master (Leedham and Unwin, 1971). Audio Tutorial (Postlewaite, 1972).
For open-ended subject matter. (Management, literature, etc.)	Programmed group instruction (BOAC management training). Structural communication (Hodgson, 1974).
When the exact needs and objectives of the target audience are unknown. (Orientation, reference, revision or learning.)	Information mapping (Horn, 1969, 1973).
When the content is subject to frequent changes, or the methods of work vary.	'Learner cue sheets' (Le Hunte, Learning Systems Ltd).
When the learners have other study habits.	Oral group response to OHP programme (Blake, UNESCO, 1975). Keller Plan for undergraduates (Keller, 1968, Keller and Sherman, 1974).

Figure 6.6 *Some limitations of expositive programmed self-instruction and some ways of overcoming them (from Romiszowski and Atherton 1979)*

planned and is in some way guided to ensure that specific learning objectives are achieved. Thus the guided discovery method described by Gagné (1965) is *experiential learning which is nevertheless instruction*, whereas the free discovery approach favoured by some educators (based on the provision of a rich learning environment in which the learner may play freely without undue concern about the objectives to be achieved) is an example of *experiential learning which is not instruction*. The essential characteristics of an experiential instruction model are:

The basic experiential instruction model

1. A definition of the situation, problem or case to be studied, and (sometimes) of the general aims of the study. It is not usually necessary to inform the learners of the specific learning objectives, as these will be discovered as part of the learning experience.
2. The organization of appropriate situations, real or simulated, which will allow the learners to act out reality, to observe the consequences of certain actions, and to play with or manipulate real systems (whether mechanical, conceptual or organizational) or models that simulate these real systems.
3. The verification of the learning that has taken place. This includes: first, the understanding of the cause-effect relationships discovered in the real or simulated situation being experienced; second, the formation of general principles or theories that explain or predict the behaviour of the systems being studied, and, finally, the transfer of these newly formed principles to explain or predict other events or actions that might occur in similar related systems.

These three stages may be referred to as: briefing; experience (or action) and debriefing. We have borrowed these terms from military parlance, and, indeed, they have crept into the jargon of games and simulations by way of war games and their close kinship to management games. Of the three stages, the briefing is perhaps the least critical, as it is usually obvious what initial explanations are necessary once the action stage is planned. The action stage is the one that usually gets most attention during design, attempts being made to design experiential exercises that are capable of providing the basis for the insights we hope the learners will make. The debriefing stage is, however, at least as important and should get as much attention during design, as it is in this stage that the really important conceptual learning occurs and is evaluated. Unfortunately, however, the debriefing stage is often neglected by the instructional designer, being left to the instructor to carry out (or not) as best he thinks fit.

The importance of the debriefing stage, both as a means of checking that the experiences are being correctly interpreted by the learners, and also as a means for developing cognitive schemata that will enable learners to generalize to other situations, has led many writers to subdivide the debriefing into two separate stages. As an illustration of this thinking, in the field of mathematics teaching, we present the views of two authors, one involved with the teaching of young children (Dienes) and the other mainly concerned with older students and adults (Polya). Figure 6.7 summarizes their descriptions of the discovery approach applied to mathematics teaching.

Dienes	Polya
The preliminary or play stage corresponds to rather undirected, seemingly purposeless activity usually described as play. In order to make play possible, freedom to experiment is necessary.	A first, exploratory phase which is close to action and perception and moves on an intuitive, heuristic level.
The second stage is more directed and purposeful. At this stage a certain degree of structured activity is desirable.	A second, formalizing phase ascends to a more conceptual level, introducing terminology, definitions and proofs.
The next stage really has two aspects: one is having a look at what has been done and seeing how it is really put together (logical analysis); the other is making use of what we have done (practice). In either case this stage completes the cycle, the concept is now safely anchored with the rest of experience and can be used as a new toy with which to play new games.	The phase of assimilation comes last: there should be an attempt to perceive the 'inner grounds' of things; the material learned should be mentally digested, absorbed into the system of knowledge, into the whole mental outlook of the learner. This phase paves the way to applications on one hand, to higher generalizations on the other.

Figure 6.7 *The experiential approach to mathematics teaching, seen as a cyclic process of 3 stages (from Servais and Varga, 1971)*

The cyclic nature of experiential learning

Another way of describing the experiential learning process was presented by Stadsklev (1974). He contrasted the 'information processing' (or reception learning) process involved in expositive instruction, with the 'experience processing' that occurs during discovery learning:

1. Acting in a particular instance. One carries out an action and sees the effects. The effects may act as rewards or punishments (as in operant conditioning) or may simply supply information about a cause-effect relationship that exists.
2. Understanding the particular case, so that if the same set of circumstances reappears, one can anticipate the effects.
3. Generalizing from the particular instance to the understanding of the general principle, under which the particular instance falls. This may require action over a range of instances before the general principle becomes apparent.
4. Acting in a new circumstance, to which the principle applies and anticipating the effects of the action. (This last step closes the loop and leads to yet further learning.)

6.8 Some examples of experiential programmed instruction

The role of conversation in debriefing

All of these descriptions emphasize the need for understanding, introspection and restructuring of the learner's cognitive schemata, in order to make sense of the particular experience and also to learn from experience, in the sense of being better equipped to understand and act in new situations that may occur in the future and to which the same general principles apply. One should emphasize the important role played by the instructor during both the action and the debriefing stages, in observing, analysing, evaluating and correcting the cognitive schemata being formed by the learners. There is often a limit to what can be learned about the internal state of the student's mind by simply observing his actions. It is often necessary to 'get inside the learner's head', and the only readily available technique of doing this is to engage the learner in conversation. Hence, the extreme importance of planning and executing the debriefing stage – the analytical/evaluative conversation – with due care.

The need to penetrate the learner's thoughts may be considered a factor that limits the applicability of pre-packaged, self-instructional materials for experiential learning. Without a doubt, it is more difficult to develop appropriate self-instructional exercises that really lead the learner to introspection and restructuring of his cognitive schemata. However, examples of such materials do exist and have proved very effective in promoting just the kind of learning that we expect from an experiential strategy.

The productive thinking program

Even in the early days of programmed instruction, some materials appeared that proved to be capable of developing complex cognitive strategies and higher order problem-solving skills. A classic example is the 'productive thinking program' developed by Covington and Crutchfield (1965) at Berkeley University. This set out to 'teach students to think logically' by means of a programmed text in which they solve a series of crimes. Gradually, by exposure to a series of instances in which the students act as detectives, they come to understand the cause-effect relationships that exist in logical deduction and finally generalize these relationships to form a powerful schema or problem-solving strategy. Covington and Crutchfield found that this strategy transferred effectively to other categories of problems in school subjects and in real life, where logical problem solving was required.

Learning strategies from programmed instruction

Another specific instance from the early days of programmed instruction was the programmed text 'The Battle of Waterloo', written by Patrick Thornhill (1965). This text was designed to put the reader into the shoes of Napoleon and the other principal commanders at Waterloo. Written in branching text style, the programme would present a particular stage in the battle, giving details of troop disposition, lines of communication, supplies, terrain and so on. The reader was then presented with a multiple-choice question of the type: 'If you were Napoleon, how would you act in this situation?'. This was followed by four or more alternative courses of action. Depending on one's choice of response, one would then be directed to pages that analysed the advantages, disadvantages and implications of the decision taken.

Indirectly, one learns what Napoleon and his adversaries actually did, but the important learning occurs in analysing what *might* have been the outcome of the battle if other decisions had been taken. The book was intended as a novel approach to the teaching of a school history topic, aiming at cognitive objectives seldom included in conventional school history courses. It ended up being used, apparently with great effect, in officer cadet schools, as part of the standard training in military strategy.

The 'Eg-Rul' model for discovery programming

There are many other examples of programmed materials, especially in the sciences, that promote a kind of experiential learning, by inviting learners to analyse given experimental results in order to form their own hypotheses, which, in later stages of the programme, are put to the test in further simulated experiments. Some of these programmes are linear, but most adopt the branching format.

Many of the early attempts at 'programmed experience processing' were developed on the basis of a topic analysis, followed by a design based on rules and examples, in which the analysis of given examples precedes and leads to the identification of the general rule that they illustrate (the Eg-Rul approach). Programme writers favouring the topic analysis approach would sometimes use an Eg-Rul sequence of frames and sometimes a Rul-Eg sequence, almost indiscriminately. Only the more analytical writers seemed to identify the connection between the different types of objectives to be achieved, the different general strategies that should be used to achieve them, and the effective use of the rules and examples methodology of topic analysis and instructional materials design. As a result, very many poor examples of programming appeared, in which the true meanings of rule and example were obscured by the preoccupation with the subdivision of information into the smallest possible units.

More sophisticated models of experiential programming

The effect of this tendency was a growing reaction against programmed instruction and against programming in general on the part of many educators, especially those more interested in the development of higher order learning, cognitive schemata, problem-solving strategies and so on. This was unfortunate, as the result was to relegate programmed instruction to the domain of task-oriented training (where it still thrives today) and to the simpler levels of reproductive skills and factual knowledge (procedures, etc). Thus the task analysis-based methodologies continued to evolve (witness mathetics), while the topic-based approaches stagnated. There were, of course, some notable exceptions, for example the work of Landa (1974, 1976) in the Soviet Union on the analysis and programming of mathematical problem solving (geometry) and language structure (Russian grammar). Another very promising development was Hodgson's structural communication programming methodology, so important that we shall devote the next chapter to it.

6.9 Experiential programming and current educational trends

We are now witnessing a rebirth of interest and activity in the area of programmed experience processing. One reason for this is the advent of the microcomputer, which has made computer-assisted instruction an economic possibility. Whether it becomes an innovation that saves money will depend on the uses to which CAI are put. Many educators already criticize the current trend, as a mere revival of 1960's programmed instruction in a more automated format. Many existing CAI programmes are indeed this: simple expositive programmes that use the computer as a page-turning device. But the real potential of CAI is that it opens up possibilities for the development of experiential programmes of a much more sophisticated type than was possible with the technology of 20 years ago.

Micro-computers and the new interest in experiential programming

One development is the use of computer-based simulations. At present, these are generally limited to the action stage of the experiential learning process, and the debriefing stages are generally left to discussion sessions between students and tutors. But the debriefing can be effectively programmed, turning the simulation package into a complete experiential tutorial. Other approaches exist, for example the 'Dialogue CAI' technique advocated by Gordon Pask (1976). We shall examine such techniques more closely in Part 3 of this book.

Topic analysis — the many types of rules

As part of this rebirth, we are witnessing the resuscitation of such analysis techniques as Rul-Eg and Eg-Rul, and associated techniques such as the matrix analysis of a sequence of rules. We should take care not to commit the same mistakes that we made 20 years ago. The indiscriminate and indeed erroneous use of Rul-Eg/Eg-Rul as a programming technique contributed to the apparent 'death' of programmed instruction in education during the 70s. Let us not kill CAI by a similar misapplication of a basically useful methodology. Let us realize, for example, that there are many types of rules: rules that specify procedures (algorithms); rules that define concepts (defined concepts); rules that explain the phenomena we see around us (principles or rules of nature); rules that guide us to act in appropriate ways in novel situations (rules or principles of action, also called heuristics); and complex rule sets that govern our approach to specific categories of situations or to life in general and that are often difficult to analyse into their component simple rules (theories, strategies and philosophies).

Each of these different categories of rules requires different treatment in terms of the planning of instruction. We have already mentioned these differences in Chapter 2 when considering the conditions for learning for each of the types of objective that make up our knowledge and skills classification schemata. A topic analysis of sufficient depth to enable the development of adequate experiential learning materials must take into account these varieties of learning categories and identify the most appropriate instructional tactics for each.

The future importance of effective experiential program-ming

With increasing use of distance education, in both developed and developing countries, increasing use of computers as a means of instruction and the probable increase in *combined* systems (television distribution of computer-based learning materials), it is probable that educational systems of the future will depend to ever increasing degrees on pre-prepared, interactive self-instructional materials. The problem of 'courseware' design is therefore of increasing importance. The development of powerful techniques for the production of experiential self-instructional materials is a particularly important area, if we are not to limit the use of the new communication technologies to the simpler aspects of education and training or, worse still, to distort future educational systems by the inappropriate application of expositive strategies for nearly all instruction.

For these reasons, we shall be devoting the next three chapters to specific techniques of programming which break away somewhat from the early methodologies of programmed instruction.

7. Structural Communication: A Cognitive Approach to Self-Instruction*

7.1 The rationale for structural communication

Origins

This technique was developed by the Centre for Structural Communication, Richmond, UK, notably by A.M. Hodgson and his colleagues (Zeitlin and Goldberg, 1970; Hodgson *et al*, 1971, 1972, 1974; Egan 1976). The roots of this technique are in cognitive and field psychology, with a touch of cybernetics. Hodgson sees it as the harnessing of process control methods to the assistance of higher order cognitive or reflective learning.

To relate these two apparently inconsistent disciplines, Hodgson constructs a model of types of thinking. Egan (1976) describes this model as follows:

> The model distinguishes four more or less discrete levels or kinds of intellectual activity. I will try to indicate the distinctions by reference to common experience of reading.

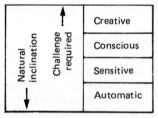

Figure 7.1 *A model of four kinds of intellectual activity (Egan, 1976)*

> Automatic thinking is the lowest level and it may be recognized in that frustrating experience of reading a paragraph and then realizing one has not taken in a word of it. Nevertheless, it is proper to say that one has mechanically or automatically read the words.
>
> Sensitive thinking is that of commonest awake experience; in reading, it is recognizable in the typical condition of reading and understanding the meaning of what is read.
>
> Conscious thinking represents that stage of heightened critical awareness that we recognize in reading and synthesizing what is read with other ideas and areas of knowledge.
>
> Creative thinking is marked by spontaneity, in which one's reading sparks new insights that are not accountable for by any evident lines of association. This rare kind of thinking is perhaps most vividly described as the 'Eureka!' experience.

Hodgson sees behavioural approaches to instruction, such as programmed instruction, as relevant to the automatic and sensitive levels – involuntary habits and reproductive thinking. He sees the need for a cognitive approach at the conscious and creative levels (reflective thinking).

---

*This chapter was written in collaboration with A.M. Hodgson.

Lack of practical techniques based on cognitive theories

Hitherto, cognitive field theorists have been in an awkward position when trying to combat behaviourist claims about the proper method of studying and promoting learning. Behaviourists have been able to shrug off criticism by pointing to linear and branching programmes and experimental results showing their efficacy in improving learning, while field theorists have lacked any technique drawn from their theories that contributes anything like as directly to pedagogy. They have been restricted to reasserting in a variety of forms that the most effective way of teaching subject matter is 'with due regard not only for coverage but also for structure'. Without techniques that can implement these recommendations, such arguments have little more effect on behaviourists than calling them names.

J.S. Bruner, who is the most visible proponent of a 'structural' approach to learning, has given us *Man: A Course of Study*, but this is a whole curriculum (whose efficacy is enormously difficult to measure reliably) and not a technique that teachers can readily use in order to encourage effective learning of all kinds of subject matter. Structural communication is, however, just such a technique.

The principles of structural communication

Structural communication is based on the concept of a 'guided dialogue'. Hodgson (1974) defines this as follows:

> In-depth communication, the 'pre' factors in judgement, conception and disposition are worked on and loosened up by a process of reciprocal action in which aspects of the message are explored, contradicted, negotiated and correlated between the people concerned. This process is known in classical terms as 'dialogue'. When there is present in the communications process insight and expertise in the discipline of dialogue itself, then we can refer to the process as 'guided dialogue'. The basic philosophy of structural communication as a technique, is that it is the medium of 'guided dialogue' which needs adding to the currently available technologies of communication and learning. This is a richer concept than feedback, since feedback can operate a process that is nothing but conditioning.

In a guided group dialogue we can discern the following elements:

1. Each learner is challenged to think for himself.
2. Each has to face the real facts of the case.
3. Each is making skilled inferences from the facts.
4. Several interrelated perspectives are considered.
5. Alternative solutions to each problem are raised.
6. The positive imagination of each learner is brought into play.
7. The coherence and consistency of the propositions of each learner are analysed.
8. Exchanges are adapted to individual differences.
9. An optional consensus is reached with commitment to act.
10. No learner is left out of, or dominates over, the group.

7.2 The structure of a typical SC study unit

We can simplify the process of dialogue to a cycle of directed challenge ('Will you take this up?'), responsive environment ('Tell us how you see it'), and reality testing ('Is this really what you intend, mean, feel etc?'). This cycle is shown in Figure 7.2.

To be effective, a guided dialogue should set a challenge which is seen to be pertinent, motivates and appeals, and has just the right level of difficulty to 'stretch' the learner without overwhelming his capacity. The responsive environment should be related to the theme of the dialogue, comprehensive towards that theme and rich in the scope for expressions of different reactions to the challenge. The reality testing should be based on analysis of relevance, consistency and as far as possible on an impartial view of the alternatives for viewing the theme.

This concept of a to some extent interactive dialogue brings SC close to the cybernetics view of dialogue or conversational programming. It is perhaps not

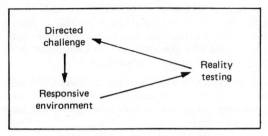

Figure 7.2 *Cycle of dialogue*

quite so ambitious as the attempts to automate the instruction process, but is quite consistent with contemporary CAL programming languages (eg Sperry Univac's ASET). Certainly, Hodgson does not accept the 'must be better than a live tutorial' criterion which Pask sets for CAI. He states that a study unit written in SC 'approaches the flexibility of a well conducted live tutorial'.

The basic element of SC is called a study unit, and each of these consists of six interdependent parts. The simplest kind of study unit may be contained in a booklet of about 12-20 pages. Typically, study units are designed for about one hour's work. They may be used individually, or they may be designed for group work. Below we name each section and describe its role in the communication:

Intention. This is self-descriptive. The author uses it to describe what the study unit is about, and may use it to specify certain 'behavioral objectives', provide an 'advance organizer', or, if it is one of a series, identify its context and role.

Presentation. Here the student first comes into contact with the subject matter of the study unit. In the simplest units, this is a written text, rather like a chapter of a book, though rather more condensed than is usual for a book that lacks the following sections of a study unit.

Investigation. This section usually comprises between three and five problems about the subject matter of the presentation. Each of these problems takes a different perspective on the material, and the student attempts to resolve them by composing a response from the next section.

Response matrix. The matrix is a randomized array of between about 12 and 35 'items' that restate in a concise form the significant elements of the study unit theme outlined in the presentation. Usually, these items will be statements of facts, theories, formulae, parts of strategies and so on, depending on the subject of the study unit and the type of question asked.

The student's task is to compose a combination or sub-set from the matrix which represents an appropriate response to the challenge posed by each problem. There is nothing on the face of any item that will tell him to which problem or problems it is relevant or in what combination of items it should fit. Each item may be considered a signifier of semantic content more fully expressed in the presentation. Thus, each item should carry a fairly rich semantic load. From his reading of the presentation, which is designed to create conditions of readiness, the student will be drawn to see complex sets of relationships emerge from the random matrix as he concentrates on it with a particular problem in mind.

Discussion. This is in two parts. First, there is the guide that analyses the student's responses and directs him to the appropriate discussion comments, which form the second part of the discussion section. The guide, in its printed form, poses diagnostic 'tests' as, for example: 'If you included in your response items 10 and 18, read comment P', or, 'If you omitted from your response items 7 and 11 read comment K'. The student will thus be directed to those comments that form the author's response to the student's response.

The possible variety of comments is hardly less than is available to a teacher in a classroom, and because of the author's control over the construction of the problems and response medium, it is possible to discuss his responses with the student in great detail and focus on his difficulties with great precision. One may design a dozen or more comments for each problem, of which any typical student might be directed to only two or three.

Viewpoints. At the conclusion of the discussion, the author addresses the student about the material of the study unit. He may use this section to indicate his own assumptions, and suggest further reading that will present other views on the subject under study.

7.3 An example of a typical SC study unit

A study unit may take many different forms, diverging in some respects from the structure outlined in the previous section. In his book on structural communication, Egan (1976) mentions several atypical applications. These include:

1. The use of a set of problems (investigation) and a response indicator as a means of testing and evaluating (formatively) a body of knowledge that should have been learned previously (by whatever means).
2. The substitution of the written presentation by an experiment (to be performed in a laboratory, for example), followed by an investigation aimed at uncovering the conceptual schemata that the learner has formed as a result of his experience.
3. The substitution of the printed presentation by a bibliography, or set of references, to be studied in the library or obtained from other sources, before attempting to respond to the problems in the investigation.

There are many other possible variations of the technique and, especially with the use of computer-based presentations, further variations will no doubt be invented. We shall, however, present a typical, printbased study unit, dating from around 1970. We have chosen an example from history, as most readers will find it easy to analyse the structure of the unit without getting lost in the intricacies of the content. Other units on, say, mathematics or management may have been even more illustrative of the technique's ability to analyse and evaluate complex cognitive schemata, but may be alien subject areas for a large proportion of readers.

We suggest that the reader now studies the example unit presented in Exhibit 7.1. We have, for reasons of space, shortened the unit somewhat, omitting about half the presentation. It will not, therefore, be possible to use the unit as a study exercise, but what is reproduced is adequate to illustrate the basic structure and the mechanics of its use by the learner. We suggest that the reader imagines other possible applications of the technique in his own subject area.

7.4. Designing and writing structural communications

A close study of the example unit presented in Exhibit 7.1 should convince the reader that the design and production of a unit of this type is not as easy as it might at first appear. On first sight, some readers may feel that the style of the material is much more familiar than the mathetically designed sequences of programmed instruction presented in the previous chapter. However, on closer analysis, the same readers come to realize that the production of a structural communication requires much more skill and effort.

It may be fairly easy for some gifted writers to produce clear and concise intentions or presentations that include all the information necessary for the formation of desired conclusions, without going so far as to form these conclusions for the reader. However, the design of a useful and challenging set of problems for the investigation section, and the selection of the most appropriate and useful

Exhibit 7.1 *The Merchant Adventurers: An example of a study unit from a history course, written in the structural communication style (reproduced by kind permission of A.M. Hodgson)*

Intention

In this study unit we leave the political scene and look instead at English trade, and how it affected, and was affected by, the economy of this country and of Europe. While the economic life of the country followed its course alongside and continually interconnected with the political story, it has different highlights, crises and disruptions. The deaths of kings, which tended to be climacteric moments in the political history of the country, might be quite irrelevant to the flow of the economy.

This study unit has the title 'The Merchant Adventurers'. The Fellowship, or Company, of Merchant Adventurers was formally created in 1486 and consisted of those London merchants who controlled most of the profitable cloth trade with Flanders. We will see how the fluctuations in trade affected this powerful and wealthy group and how they were reflected in its attitudes and activities.

We will follow the story of English trade into the turbulent days of the 1550s, and take a brief look beyond at the great expansion of markets which took place during the reign of Elizabeth I.

Introduction

Throughout the Middle Ages the main English export had been wool. English wool was the best in Europe and had been highly prized on the Continent, where it had formed the raw material for the great cloth industries of Flanders and North Italy. But during the century and a half before Henry VII came to the throne, England had itself been developing a cloth industry of its own. By Henry VIII's reign the English sold about ten times more cloth than raw wool in the great twice yearly shipments to the fairs in and around Antwerp.

To feed this cloth industry, it has been estimated, there were three sheep to every person in England. The trade in cloth was an easy and profitable one and consequently it grew steadily, until the whole economy was heavily dependent on the sheep. A large proportion of the population relied on the state of the wool and cloth trade for their livelihood: from shepherds to those involved in the cloth industry, from cloth dealers to the merchant adventurers.

The wool travelled from the backs of the sheep to be made up into cloth by either workshops in the towns controlled by the guilds, or as was more and more frequently the case, by individual workers in the country outside the guild restrictions. The cloth dealers would then travel round and buy the made-up cloth, and deliver it to Blackwell Hall in London – the headquarters of the merchant adventurers. From there it would be ferried across the North Sea to Flanders and be bought there by the men who would have it 'finished', ie dyed and made up into articles of clothing. It might end up on the back of almost anyone in Europe.

The trade might have continued in its dull, routine and profitable way had it not been for the dramatic effects of the price rise, which was being felt all over Europe, and Henry VIII's debasements of the currency, which brought the English economy to the brink of chaos.

Presentation

A merchant adventurer was one who traded with foreign parts. The Company of Merchant Adventurers... consisted of those men who controlled most of the cloth trade. Their aim in forming the company was to control the whole of the cloth trade, and ensure that no merchant not belonging to their company should get any of its profits. Besides this motive there was also the need for a strong organization to back up commercial ventures. The individual merchant stood little chance of surviving against the trading organizations – like the German Hanseatic League

Exhibit 7.1 *(continued)*

in the Baltic or the Venetians in the Mediterranean – and a trader putting into a port which a trading organization felt was 'theirs' might find himself negotiating by cannon.

At the end of the fifteenth and beginning of the sixteenth centuries everything seemed to be going well for the merchant adventurers. In 1496 Henry VII had negotiated a favourable trade treaty, the 'intercursus Magnus', with the Archduke Philip of Flanders which was to benefit the company enormously. Any port besides London was known as an 'outport'; and all the outports together probably handled only about one-tenth the quantity of goods which passed through the capital. The method the merchant adventurers chose to preserve their exclusiveness was to charge so high a fee for the right to trade with the Netherlands that few of the 'outport' merchants could afford it. And though Parliament forced the company to lower the fee, it recognized their right to levy one, thus making the London/Antwerp trade in cloth a virtual monopoly for the merchant adventurers.

A further advantage the merchant adventurers enjoyed during Henry VII's reign was the smallness of the Customs tax on cloth. During the Middle Ages, kings had continually increased the tax on wool, till by 1485 the tax amounted to about one-third of the value of the wool itself, whereas cloth which was a relative newcomer to the Customs was taxed a barely noticeable 3%.

During the first decades of the sixteenth century England experienced a gradual price rise. The price of goods depends on the relationship between money available and goods available. If there is not much money but plenty of goods, prices are low. If there is a lot of money but not many goods, prices are high. The causes of the sixteenth-century price rise, like any large-scale economic event, are very complex and even yet not fully understood. One clear cause, however, is traceable. That was the influx of silver to Spain from her newly won territories in the Americas. There was no increase in the production of goods in Spain, so the increased amount of money available led prices to rise to balance up with it.

This situation made trading with Spain very profitable. During the 1520s the rise in prices was affecting Spain before other countries. So an English merchant could buy goods cheaply in England and sell them at the higher price normal by then in Spain. Thus the English merchant made his normal profit and also the difference between prices in Spain and England.

Because of the heavy trade between Flanders and Spain it was not long before the price rise hit Flanders. (The way prices rose generally seems to have been something like this: when producers of materials saw the profits the merchants were making, they increased the prices at which they sold to the merchants, and then as they had money the people who sold goods raised their prices, and so it went on.) The profits of the Flanders merchants soon increased the supply of money in relation to goods there too, helped also by trade with Germany, where the increased output of German silver mines contributed to the same effect. Again the English merchants stood in a position of advantage. As the price rise in England was still less rapid than in Flanders, the English merchants could continue to buy goods relatively cheaply in England and sell them for the higher prices prevailing in Flanders.

This state of affairs continued throughout the 1530s when, despite some early trouble with the Emperor, Charles V, ruler of Flanders, the merchant adventurers enjoyed smooth and profitable business, helped by the policies of Thomas Cromwell. The growing prosperity and wealth of the English merchants, and most of those connected with cloth and wool, began to drive English prices up ever more rapidly – to the consternation of those who were gaining no profits to compensate for the increases in prices.

After the execution of Thomas Cromwell in 1542 Henry VIII took upon himself the responsibility for guiding the policies of his realm. The economic policies he pursued were short-sighted and disastrous. Involving himself in costly

Exhibit 7.1 *(continued)*

wars, he quickly exhausted the fortune Cromwell had made available to him, and so adopted the plan of debasing the coinage. He reduced the amount of silver in the coins, keeping the silver thus saved in his Treasury. The coins were now less valuable and there was consequently less real money available while the quantity of goods was steadily increasing. So, suddenly prices in England shot up sharply causing confusion and havoc throughout the country.

One immediate result of the debasements, however, was that the merchant adventurers gained enormously. In 1522, the English pound was worth 32 Flemish shillings, but by 1551, after the devaluations, it was worth only 13s 4d. This meant that if an English merchant paid £1 for goods in England he would, in 1522, have asked 32s (plus his profit) for them in Antwerp, whereas in 1551 he asked only 13*s* 4*d* (plus his profit). So while the devaluations hit the English currency at home, it made exporting much easier, as the dealers at Antwerp were able to buy the same goods at half the 1522 price.

Because of the apparent cheapness in price, English merchants found they could sell as much cloth as they could carry, and there were complaints from abroad that short sizes and inferior cloth were being sold: an indication that English merchants were taking across everything they could lay their hands on, sure of a ready market.

But in 1551, after social unrest and riots, the Government took steps to reform the value of the English coinage. Once this had been done, it became much more difficult for the merchants to sell their goods, because they now had to ask higher prices to recoup the cost of the more valuable money they had paid out in England. During the boom years, the merchants had encouraged the expansion of the cloth industry, and it had responded by increasing its output enormously – only to discover in 1531 and the following years that there was suddenly no market for the increased output. Producers were angry, many independent cloth makers ruined and even the merchants, finding themselves with more cloth than they could sell in the normal markets of Antwerp, were desperate to find new outlets.

Largely because of this crisis we see in the 1550s the first really adventurous voyages from London in any quantity. The ease and ensured profits of the LondonAntwerp trade made the London-based merchant adventurers reluctant to try further abroad. Earlier adventurous trading voyages, like those of the Bristol merchants to Newfoundland, or Plymouth merchants to South America, were nearly all undertaken by 'outport' merchants who were excluded from the cloth trade to the Continent by the self-protective policies of the Company of Merchant Adventurers.

Earlier in the century, English merchants had for political reasons been discouraged from seeking new markets around Africa to the south and to the Americas in the west. The Tudor throne had needed the support of foreign monarchs, and so the newly found 'empires' of Spain and Portugal had been left alone. But by the 1550s relations with Spain were growing worse, despite Queen Mary's marriage to Philip II of Spain, and the need for new markets made English seamen less inclined to respect Spanish and Portuguese 'property'.

With the trade collapse after the reforming of the coinage came a spate of voyages along new routes. The money to pay for these more risky adventures came from the tremendous profits made in the years between the debasements and the reforming of the coinage.

(Observation: For reasons of space limitation, the remainder of the presentation – about half – has been omitted.)

Investigation
Problem 1.
During the first half of the sixteenth century, the Company of Merchant Adventurers became one of the most powerful and influential groups in England.

Exhibit 7.1 *(continued)*

They came to control the important cloth trade with Antwerp on which much of England's growing prosperity depended. Their relationship with the government was usually close, and they were respected and considerately treated by even the highest in the land. State occasions were made even more magnificent by the splendour and display put on by the members of the company. Using the response indicator, explain why all this should have happened. What factors were favourable to the growth of their power and wealth?

Problem 2.

Which statements in the response indicator do you feel suggest that the merchant adventurers were not really very adventurous.

Problem 3.

When the trade in cloth became more settled after the disturbances of the early 1550s, the merchant adventurers' profits continued to come in steadily, but the central theme of the story of English trading history moves to the broader stage of the world's oceans.

With the extension of English trade – in area and variety of goods exported – the Company of Merchant Adventurers' fortunes were imperceptibly but surely in decline. The late 1540s mark the highpoint of the company's profits, and their trade never again came near to the level of those years of boom. Using statements from the response indicators, explain why their fortunes declined during the latter half of the century.

Problem 4.

As mentioned in problem 3, after the 1550s, the central theme of the story of Englis' trade moves away from the single contact with Antwerp and its great fairs, through which English cloth had travelled to all parts of Europe. English ships were soon calling into ports around the Baltic, the Mediterranean, Africa, the Americas, and later India and the islands of the Pacific. Which statements in the response indicator help to explain why there should have been such a burst of new trading activities after 1551?

Response indicator

Increasingly often, interlopers ignored the merchant adventurers' monopoly and traded in cloth with the Continent 1	English seamen hoped for the good luck of the Spaniards in finding silver and gold 2	Parliament backed up the merchant adventurers' claim to control the sale of cloth abroad 3	Capital was available in London to back risky expeditions 4
Henry VIII debased the coinage 5	Some foreign trading organizations were weakening 6	Trade with the Continent was disrupted by wars of religion, and the Spaniards' destruction of Antwerp 7	Henry VII negotiated the 'Intercursus Magnus' 8
The merchant adventurers' monopoly was withdrawn 9	Some of the new trading ventures reaped enormous profits 10	The merchant adventurers introduced 'stints' 11	It was 'outport' merchants who first traded with the Americas 12

Exhibit 7.1 *(continued)*

In 1485 the Customs tax on cloth was only 3 per cent of the cloth's value **13**	The price rise affected the Continent more strongly than England at first **14**	Steps were taken to reform the value of the coinage **15**	The merchant adventurers had to pay heavily for the privileges granted to them in Elizabeth I's reign **16**
The formation of the joint stock companies offered a new means of financing trading expeditions **17**	The merchant adventurers continually appealed to the Government for protection against competitors **18**	By trying to sell undersized cloths during the boom years the merchant adventurers damaged their reputation abroad **19**	The merchant adventurers controlled nearly all the cloth passing through the port of London **20**

Discussion guide

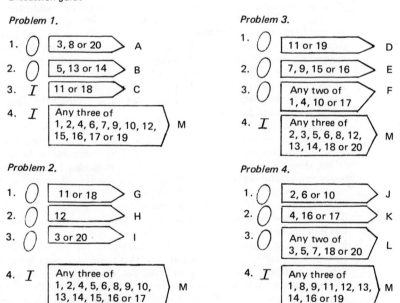

Problem 1.

1. ◯ [3, 8 or 20 ⟩ A
2. ◯ [5, 13 or 14 ⟩ B
3. I [11 or 18 ⟩ C
4. I [Any three of 1, 2, 4, 6, 7, 9, 10, 12, 15, 16, 17 or 19 ⟩ M

Problem 2.

1. ◯ [11 or 18 ⟩ G
2. ◯ [12 ⟩ H
3. ◯ [3 or 20 ⟩ I
4. I [Any three of 1, 2, 4, 5, 6, 8, 9, 10, 13, 14, 15, 16 or 17 ⟩ M

Problem 3.

1. ◯ [11 or 19 ⟩ D
2. ◯ [7, 9, 15 or 16 ⟩ E
3. ◯ [Any two of 1, 4, 10 or 17 ⟩ F
4. I [Any three of 2, 3, 5, 6, 8, 12, 13, 14, 18 or 20 ⟩ M

Problem 4.

1. ◯ [2, 6 or 10 ⟩ J
2. ◯ [4, 16 or 17 ⟩ K
3. ◯ [Any two of 3, 5, 7, 18 or 20 ⟩ L
4. I [Any three of 1, 8, 9, 11, 12, 13, 14, 16 or 19 ⟩ M

Discussion comments

A. In these times, when peace and order were easily and often disrupted, trading was a hazardous occupation. For regular trading relations between countries it was thought necessary to have powerful and wellorganized companies in charge of the operation. And as trade also benefited the king, whose subjects grew wealthy from it, wise monarchs were careful to help their merchants. Henry VII was particularly noted for the help he gave to the merchant adventurers, which, with the cooperation of Parliament, ensured that they were able to control the valuable English cloth trade, on which the company's fortunes depended.

B. In the early days after their formation, the Company of Merchant Adventurers

Exhibit 7.1 *(continued)*

was helped by the low tax on cloth, as opposed to the highly taxed wool with which another company traded. An important factor which favoured their growth during the first half of the century was the steady price rise which, though it also affected England, was at first more marked on the Continent: thus leaving English traders at an advantage. When the debasements of the English currency followed, causing chaos internally, the merchant adventurers made enormous profits on foreign trade.

C. It is difficult to assess how far Government 'protection' for the merchant adventurers was effective, especially at a time when the Government largely lacked the means to enforce its rulings. There were court cases against men who trespassed on the company's monopolies, but such cases do not seem to have been a great discouragement, in that shortly afterwards the same men can be found 'interloping' again.

The company's attempts to protect their own markets by introducing 'stints' to ensure a steady but not too great supply in the markets (thus keeping the prices up) was again of questionable value. The markets were capable of considerable expansion, as they had proved during the boom years, and it is possible that with a more adventurous policy they might have been able to sell a lot more cloth.

D. During the boom years, after the debasements and before the devaluation, the English merchants found they could sell everything they could carry across to the Antwerp fairs. Everyone connected with the cloth industry and cloth trade worked hard to expand production, but in the process a lot of shoddy material passed across to the markets. The goodwill lost might well have played its part in the drop in sales over the following years. The company tried to prevent flooding of the market and lowering of prices by introducing 'stints' in the 1560s. This ensured them steady profits by keeping the prices on what they did take over to the markets high, but this policy also ensured that their trade would not expand further.

E. The 1550s and 60s were not good times for the company. While their power and influence seemed to be still growing at the time, we can see that they had passed the highpoint of their trading years. The reforming of the coinage made selling their wares doubly hard after the low prices they had asked during the boom years. The continental markets became disrupted by religious wars and the staple town of Antwerp was destroyed by the 'Spanish Fury' – the Duke of Alba's army – in 1576. Elizabeth, herself short of money, granted privileges to the company only at a heavy price, and in 1585 their monopoly on white cloth was withdrawn.

F. It is difficult to assess what effects the 'interlopers' had on the merchant adventurers' trade. They showed at least that there was room for more trade than the company was carrying on itself, but whether they had any serious effect in taking trade from the company is doubtful, despite the clamour for Government protection. The joint stock companies too, by demonstrating a new and more flexible way of organizing trading ventures, and by attracting so much of the available capital for risky but often very profitable expeditions, may have had the effect of hindering the company in some way, but again it is not at all clear.

G. Particularly in the second half of the sixteenth century, the merchant adventurers seemed content to consolidate what they had gained and seemed quite uninterested in taking any risks to extend their markets. They introduced 'stints' for their own members to ensure that the cloth they exported would be scarce enough to keep the price high, thus ensuring regular and steady profits by the easiest means. During this period too, as had been the case since their foundation in 1486, they continually appealed to the Government to defend them against English 'interlopers' or foreign competitors.

Exhibit 7.1 *(continued)*

H. In the first half of the sixteenth century, the company struggled to exclude 'outport' merchants from the profitable and relatively trouble-free trade with Antwerp. The company ignored appeals from Henry VII to trade with North America, and it was left to Bristol merchants to brave the Atlantic to get to the rich fishing grounds of Newfoundland. Plymouth sailors were trading with Brazil and Central America, despite the Spanish ban, while the London merchants were too content with the Antwerp trade to take risks on far-flung adventures.

I. The main ambition of the Company of Merchant Adventurers was to get a monopoly over the cloth trade with Antwerp. With Government help it gained a virtual monopoly during the first half of the century, and came to control all the cloth that passed through London. The merchant adventurers' defence, that it was necessary to have a strong, well-organized company to control a steady and largescale trade with a regular market, is one that deserves respect, but even so, having won their monopoly, the company certainly did not do all it could to expand the cloth trade and help the cloth industry. Their policy was one of 'safety first', and they were more concerned with their own interests than the country's, or the workers who supplied their cloth.

J. With the weakening of foreign trading organizations and the fact that it was no longer politically necessary to respect Spanish claims to exclusive trade with the Americas, English seamen were eager to try to emulate the huge profits the Spaniards and Portuguese were able to gather from their newly won 'empires'. The vast profits reaped by successful expeditions drew the adventurous like bees to a honeypot.

K. The reforming of the coinage, combined with political problems, was making trade with the traditional continental markets more difficult. With the capital made available during the boom years, many merchants, businessmen, politicians, etc were ready to invest in adventurous expeditions, in the hope of the large profits which they saw Spanish merchants and, as time went by, English merchants bringing home.

L. Any of the statements which brought you to this comment could be seen as stimulating the expansion of the area of English trade after 1551. The hold of the merchant adventurers over the biggest single item of English trade, backed up by the Government, acted as an inducement to other merchants to try their hands elsewhere – especially as political disturbances on the Continent made the more traditional markets less secure. The debasements made the fortunes of many traders, and so provided much of the capital which financed the voyages to all parts of the world during the next decades. The influence of these factors on the expansion of English trade is very difficult to trace or measure, but their influence is not necessarily insignificant because of that.

M. All these statements are considered more or less irrelevant to the problem under discussion. It might be that you have seen a connection which is considered too indirect to offer hints about, or perhaps you have not understood exactly what the problem is looking for. It may be that you have confused the time-scale, or included causes rather than effects, or effects rather than causes, or it may be that the connection you have seen is a perfectly good one which has escaped the writer. Whatever the reason, if you re-read the problem and the relevant part of the presentation you should in most cases be able to work out why the statements you included were too indirect to be discussed in the above comments.

items for inclusion in the response indicator, require a much deeper analysis of the subject being taught than is usually required for the design of expositive programmed instruction. Most of all, the decisions on what comments to make to the learners, and the writing of these comments in the appropriate conversational style, are specialist skills that most writers only develop after initial training and a good deal of practice.

We might add that, in contrast to expositive programmed instruction (where generalist programmers can usually tackle almost any assignment, learning the subject as they go along), structural communication writing requires a deep and thorough knowledge of the subject to be taught and a considerable amount of prior experience in teaching it to students representative of the study unit's target population.

At this point, we shall therefore turn over the account of how to design and write structural communications to the original inventor of the technique, Anthony Hodgson. The remainder of this section is based on his article 'Structural Communication in Practice' (Hodgson, 1974), first published in the 1974-75 *APLET Yearbook* (Romiszowski, 1974), and reproduced here with the author's permission and with modifications specially prepared by the author for this chapter.

7.4.1 The analytical phase

The first and most important step is to verify that the right kind of communication problem has been chosen. The criteria to bear in mind are:

- (a) Communication or learning difficulty which needs an act of judgement to overcome it.
- (b) This difficulty involves rather more people than can be handled by face-to-face groups in the normal way.
- (c) The availability of a broad-minded expert in the content area.
- (d) An opportunity to field test the unit on a pilot scale.
- (e) A time-scale of one to three months before the unit is needed.

These five conditions provide the best situation for the beginner to try his hand at a structural communication.

The second step is to study and observe the difficulty as seen by the expert and as seen by the target population. This is best done by seeking personal interviews with individuals and groups who will all be involved in using the unit. As well as developing a 'feeling' for the problem and helping to involve and commit the different parties to the communication, the writer has several technical tasks. These are:

- (a) Taking note of the different viewpoints and perspectives on the problem which emerge.
- (b) Recording samples of the kind of language and vocabulary which different groups of people use to talk about the same thing.
- (c) Identifying the prevalent attitudes of the different groups both towards the problem and towards each other.
- (d) Identifying any organizational and logistical constraints on the situation.
- (e) Mapping out the areas of expertise involved and studying their interrelationship as a dynamic structure.

All of these activities together will enable some sort of picture of the situation to emerge as a *definition of the communication gap*.

The communication gap will be identified by the following parameters:

- (a) The intentions of the source of the communication.
- (b) The present experience of those to be involved in the communication.
- (c) A definition of the step to be made in terms of achieving congruence of attitudes, perceptions, intentions and responsibilities.
- (d) The new learning which will be needed to overcome the gap.

Having identified in some way the communication gap, the writer brings to it his own knowledge of structural communication and how it might be applied. This stage aims to make clear the intentions of the exercise, the strategy to be adopted and the criteria for satisfaction which the different parties to the communication will accept. At this stage the writer has to anticipate the kinds of ideas he is likely to get and his own competence in communication design. He also has to weigh up what the situation will stand – personally, technically and economically. At this stage he puts his proposition and negotiates its acceptance by the sponsors or clients.

Of course, he may discover that the problem is one which cannot, after all, be tackled this way, or he may be confident it can and find that, in the event, some necessary support (usually financial) cannot be supplied.

There are some specific aids to this phase of writing which are useful:

The features ideas map. If asked to make brief notes about what we could say about a given learning situation, we would probably list a number of words, phrases, sentences or symbols which, in some way, would all be components of the relevant field of knowledge. We call these components 'elements'. A person who can weave together the elements to give various applications of that knowledge or who can take them apart and put them together at will is a person who understands the situation. Make a list of these elements.

It is very possible that not all of these elements need to be included in the communication event. A comprehensive list of those elements which must enter implicitly or explicitly into the proposed communication should be made. In making this edited list, expressions must be chosen which are the briefest and most direct possible without distorting their meaning. Careful attention must also be given to the scale and detail of the list of elements. It is important to be sure that the theme has been dealt with in sufficient breadth and depth, but that, on the other hand, the scale is not too extensive to be covered with the available resources of time and effort of both writer and learner. The writer must constantly return to his original intention and make sure that he does not include elements which are irrelevant to it.

Such a list of elements will not bring out the allimportant structure of the material that has been gathered. Behind any coherent situation of communication there should be an organized and living body of knowledge. It usually requires quite a lot of work to bring this life into the material. If this is not done effectively, then the communication will not be structured in the dynamic sense in which we use the word.

Elements can be expressed more dynamically in terms of the way they associate together, how they are dependent or independent of each other, and how they combine on one scale to produce more complex notions on a larger scale. Study the possible interactions between the elements that have been assembled and notice the significant patterns which may be formed from them. These patterns may be formulated by using a technique of 'mapping'. Relationships between a set of elements A, B, C, D, E and F can be expressed by various devices some of which are shown in Figure 7.3.

None of these devices can formulate more than a fraction of which can be seen, but the discipline of trying to construct them helps to organize the information and to see new connections.

Using any of the devices illustrated (see above) or any others he comes to use or invent, the writer should construct a draft of a situation or concept map from his elements. The draft should be criticized according to the following criteria:

(a) Is the map very symmetrical? If it is, then the arrangement of the elements is probably too artificial and the writer is probably imposing a limited way of thinking about the situation.

(b) Does the map bring out the tensions or dichotomies in the situation? In other words, do you find elements to have contradictory aspects? (For example, the element of freedom in social life.) If it does not, then the writer probably hasn't gone into the situation deeply enough.

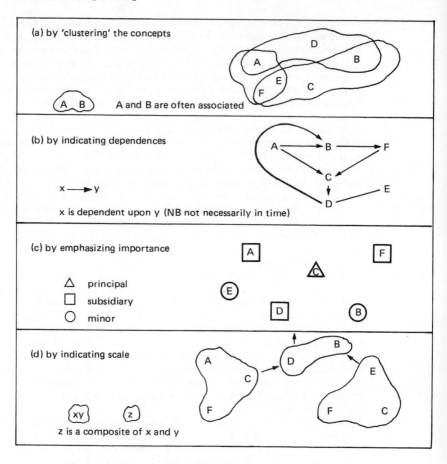

(a) by 'clustering' the concepts

A and B are often associated

(b) by indicating dependences

x ⟶ y

x is dependent upon y (NB not necessarily in time)

(c) by emphasizing importance

△ principal
☐ subsidiary
○ minor

(d) by indicating scale

z is a composite of x and y

Figure 7.3 *Some ways of 'concept mapping'*

The knowledge we build up by attempting to seek out the underlying (or several underlying) organizations of ideas we call 'organic knowledge'. It is a major principle of structural communication that such knowledge cannot be put into the form of a list, a matrix or classification. It is not subject to the simple logic of A and not – A. A key feature of both an organic situation and hence of a structural communication is the presence of more than one viewpoint on a situation, and more than one kind of judgement required to take a decision or to gain an insight. This feature needs to be studied in order to see how to develop an effective communication.

1. Dimensions. An organic situation contains a number of important 'dimensions' which will have some independence from each other. For example, in a management problem, there will be financial, administrative, technological and personnel dimensions which may present conflicting criteria and values to the manager. Such dimensions must be made clear to oneself in any communication involving understanding and decision.

2. Viewpoints. Also, in approaching any situation for understanding, there will be different standpoints or perspectives which may be taken in relation to that situation. For example, in a particular industrial situation there will be the management's viewpoint, the shareholders' viewpoint, the trade union's viewpoint, and so on. All of these viewpoints will be valid and may lead to quite different approaches to the same situation. This multiplicity of viewpoints is found in any real discipline, science, history and art, as well as management. Such differences are rarely found, however, in textbooks, which tend to be written from one viewpoint.

3. *Interaction of dimensions and viewpoints.* The dimensions inherent in a situation, and the variations of viewpoint which may be adopted towards that situation, interact with each other, making each element more complex. This is why there can be no simple right/wrong answers in any real communication problem (except perhaps at a formal level). A simple way of studying this interaction is to construct a dimensions-viewpoints table (see Figure 7.4).

Viewpoints	Dimensions		
	Technical viability	Social necessity	Financial viability
Management	?	?	?
Shareholders	?	?	?
Trade unions	?	?	?

Figure 7.4 *Example of a dimensions-viewpoints table*

Each box of the table will eventually contain a brief indication of that particular dimensions-viewpoints interaction.

4. Such a table should now be constructed on the chosen situation, for use as a reference guide to questioning. The table should include not less than two distinct dimensions and two independent viewpoints.

Finally, the dimensions-viewpoints table and the concept maps should be considered with the aim of trying to see the difficulties which the participants encounter in trying to understand the chosen situation, and assessing whether the proposed learning is within the capabilities of the target population.

7.4.2 The design phase

On completion of the analysis, the phase of analysis changes to a phase of design. During this phase the particular form of communication or learning unit is conceived and developed, based on the strategy emerging from the final stages of analysis.

Good communication can take place only when the writer has a clear objective: he must know *what* he wants to communicate to *whom*. This does not mean that he must know at the outset exactly what he is going to say or write. The objective itself is an invisible guide to the communication: it helps to direct the development of the communication, so that what is eventually said or written really does contribute towards bringing about a recognizable experience for the target population. To put it somewhat tersely: having an objective means knowing *who* we want to get *where*.

Once we have decided who we are going to interact with and where we want to get him, the question arises as to *how* we are going to get him there. This is the problem of communication strategy. Trying to tell someone about something at random as it occurs to us is a limited strategy for communication. It can be counter-productive. A strategy for communication is a plan for building up the participant's experience in such a way that the point of meaningful communication can be reached. Evidently, the strategy is closely dependent on the objective, as the examples below indicate:

Example 1. For one kind of objective, the most appropriate strategy may be to give a full and clear presentation of content and use a guided dialogue to clarify misunderstandings.

Example 2. For another kind of objective, it may be more suitable for the participant to start with a challenge. In this case, the content might be designed as an information pool to which the participants are directed by a diagnosis of their progress in meeting the challenge.

Example 3. For yet another kind of objective, it may be best for the participant to start with stating his viewpoints. In this case, for example, he may be required to select a particular viewpoint from among several, and the guided dialogue to which he then proceeds could be different according to which viewpoint he selects.

Another aspect of strategy which the designer must consider is where he is going to locate the 'focus of communication'. At what stage of the communication event is the critical stage or climax of the communication going to come? When the designer has decided about his objective, and how he is going to go about achieving it (the strategy), then he has to consider what sort of challenge he is going to put in front of the participants.

In an effective communication, the questions which will be put to the participants are not the kind that can be answered by mechanical application of techniques or from already familiar patterns of response. The questions are intended to stretch or 'decondition' the minds of the participants so that it becomes possible for them to make a step in their own understanding. Therefore the questions or tasks must be neither too difficult nor too superficial – they must also correspond to what is possible but demanding for that group, for it is a matter of the development of *their* understanding. The questions will also be designed so that, in trying to bring them into focus and exploring various possible ways of answering them, the participants in fact help each other.

Through the challenge the participant will meet a number of problems which should, so to speak, 'bring his mind alive'. He should not be able to answer the problems by association or by memorized information. In studying the problems the participant has to put things together for himself and he has to invent the organization of ideas and facts. Here, there is something involved which cannot be put in the form of instructions – you cannot really tell somebody to 'think', you can however lead them to the brink of thinking.

The kind of challenge which is appropriate depends upon the strategy and the objective. From the strategy you can see what role the challenge is going to play, and from the objective you can see the outcome which is hoped for. The participant is left by himself to meet the challenge. The problems he has to solve are not insoluble, and help will be available from the feedback which follows – just as in an effective tutorial dialogue. The challenge must bring the participant 'up against it' so he can no longer be passive towards the communication.

Before committing oneself to a particular response matrix or set of problems, it is very important to be clear about the kind of challenge that these are to be designed to induce. The three vital points of a unit – objective, strategy and challenge – all work together in close harmony if the unit is a good communication. At this early stage in designing a unit, it is best to try to think more in terms of *problem areas* which are natural to the situation, rather than focused statements of definite problems, ie to avoid fixing the *problem areas* too quickly into rigid forms. The two steps in problem construction can be indicated figuratively by Figure 7.5 which shows four problems.

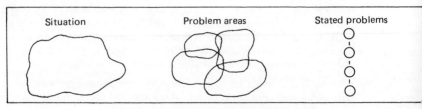

Situation   Problem areas   Stated problems

Figure 7.5 *Resolution into problem areas*

The writer should think about his situation and try to sketch the problem areas which he thinks will provide a challenge which is:

1. 'Natural' to the situation (ie not artificial or contrived).
2. Necessary for his objective (ie not simply for the sake of being challenging).
3. Appropriate to his strategy (ie not putting 'a cart before a horse').

The aim should be to find about three problem areas.

Having decided on a strategy and a challenge he may find that these have 'acted back' on the objective to modify it. The initial objective may have changed quite radically, or it may have developed quite naturally and become clearer. Whatever the case, the proposed strategy, challenge, and specific objective for the unit should now be considered together.

Response, diagnosis, guidance
The next stage of the design task is to translate the problem areas of the situation, which were considered in the previous section, into the form of a set of stated problems and a response matrix. It is neither easy nor useful to try to give generalized instructions for achieving this, partly because the form will vary according to the nature of the actual material which the writer is handling, and partly because it requires an intuitive step. We can, however, offer guiding principles which the writer can use to make sure he is not going astray.

The problems which are presented to the participant constitute the channel through which he comes up against the challenge. Each problem should bring some aspect of the situation into focus. Thus each problem, ideally, should give a different perspective on the situation (see Figure 7.6).

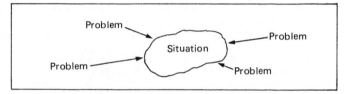

Figure 7.6 *Problems as perspective on a situation*

This should be done in such a way that when all the problems are taken together they bring out the organic structure of the situation.

The problems can be constructed in such a way that they can be attempted either in any order or in one particular order. Whichever the case, they must relate to the situation directly and the writer should make sure that no two problems are merely the converse of each other. Ideally, the problems should be natural ones and as realistic as possible. It is necessary to guard against contrived or highly abstract problems, or any problems which depend on a trick. We have found that problems framed in the form of case studies very often come closest to this ideal. Case studies may take various forms, but essentially they are based on 'real-life' situations.

Each problem should be stated in such a way that it is necessary for the learner to form some mental picture of the problem, before making his response by grouping items from the response matrix. If a problem is stated in such a way that it can be answered by just picking out items from the response matrix, it does not present a challenge. To avoid this, the problems should require the learner to make an act of discrimination. Problems which are simply of the form 'discribe so and so' are to be avoided if possible because they do not require the learner to make decisions based on the interrelationship between statements; at best they require only selection. It is the need for a judgement or interpretation that offers the learner the possibility of understanding something for himself.

The response matrix
The real challenge for the participant is in the combination of the problems with the response matrix. The response matrix plays a dual role: on the one hand it is the means by which the participant can come to an expression of his

understanding of the problem, and hence of the situation; on the other hand it provides a common language for the different parties in the communication, by means of which the participant can 'talk' with the writer, and the latter can see how to guide the learner. The construction of the response matrix requires very careful attention if it is to fulfil both these roles adequately. The following points may be helpful:

1. The response matrix should not consist of a checklist of items, each of which can be accepted or rejected simply by comparing it with the problem. If the items are so obvious that no reflective thought is needed, there will be no challenge. There must be some degree of interaction between items, so that in order to make a response the participant must 'weigh' items with and against each other. Each item may be 'coloured' by the inclusion or omission of other items.
2. Unless the meaning is clear and unambiguous, single word items should be avoided. Such items tend to be open to many interpretations, and when this is the case the challenge 'falls flat'. The writer should remember that the response matrix provides the language with which the learners 'talk-back', and they must be able to do this in a familiar language without feeling baffled and frustrated.
3. Response matrix items do not have to be all of the same type (ie homogeneous), but if they are not (ie heterogeneous), one must be careful that all items in the same class, as names of objects, or names of operations etc, do not all relate to a single problem – providing 'spurious clues'.
4. Each item should have a positive significance for at least one problem, but preferably not for all problems (ie no redundant items).
5. It is sometimes necessary to include some items with the special role of showing which one of various points of view the participant is taking.

The problems and the response matrix emerge together in the design process. For this reason they must be worked on together. The aim should be to find between three and five problems, and between 12 and 25 items for the response matrix. When the problems and response matrix have been 'sketched', it is necessary to word each problem precisely. When each problem has been written down, the designer should read it through carefully and ask himself whether what he has actually written is an adequate expression, as perceived by the learner, of the problem he wishes to pose.

The items on the response matrix must also be carefully written. There must be sufficient items for the participants to be able to give a good coverage of each problem, but not so many that he cannot use the response matrix to consider each problem as a whole. When the items have been written down, it is important to read them carefully to make sure that they are worded in such a way as to be compatible with the wording of the questions in all of the problems (ie the part of the problem where the participant is asked to do something) but without providing simple associative clues. It is important to check that the problems are posed in such a way that they can be answered in the terms of the response matrix.

Criteria for diagnosis   In an effective dialogue, the tutor will try to communicate a situation to the learners by suggesting problems to them or by getting them to raise problems. The problems will be chosen carefully so that, by struggling with them, each participant's experience will be built up in such a way that he becomes capable of entering into the situation for himself. Once the problems have been posed the communication role is to guide the learners to the point of understanding. The tutor does this by observing how each participant approaches the problems, seeing what kind of preliminary responses he makes, and offering comments which he thinks will help him along his way to seeing for himself. The tutor does not tell him solutions directly.

If the problems in the unit are to provide a means to judge how far the learners have gone towards understanding the situation, then the writer must first look at the problems and decide what he thinks the significance of *each* response matrix item is for *each* problem. To do this it is helpful to distinguish different categories of relevance. The way in which these categories are chosen depends upon the nature of each individual unit and upon the writer's discretion. However, we have found that the following set of categories is one of the simplest and most widely used.

1. *Essential (E) items*: these are the items which the writer considers to be the most important for a particular problem; in his view, they are necessary to any response to the problem.
2. *Subsidiary (S) items*: these are 'second order' items, ie they are items which, in the writer's view, contribute to making a complete response to a particular problem, but not so strongly as the essential items. Subsidiary items may depend on essential items, but not vice versa.
3. *Contradictory (C) items*: these are items which the writer considers to contradict his view of the problem, so that they should be firmly rejected in any response to the problem.
4. *Irrelevant (I) items*: these are items which, in the writer's opinion, have no relevance (or at most, very little) to the problem.
5. *Arguable (A) items*: these are items which require further discussion and interpretation.

It is helpful to display the relevance of each item for each problem. This is done most clearly in a relevance grid, an imaginary example of which is given in Figure 7.7. The use of such a relevance grid is twofold:

1. It provides the writer with criteria for diagnosing the learner's response and hence provides the basis upon which he can give guidance.
2. It provides the writer with a picture of the amount of overlap between problems. This enables a check to be made as to whether the problems taken together do give different but overlapping perspectives on the situation, or whether there are 'degenerate cases'. An example of degeneracy which can be detected in this way is the case of a problem which is simply the reverse of another problem, and hence which can be answered (if the reversal is spotted by the learner) without thought, if the first problem has been answered successfully. Another example of degeneracy which can be picked up by the relevance grid is the case where two apparently different problems are shown to have one and the same essential response in terms of the response matrix.

Criteria for diagnostic categories   To provide criteria for diagnosis, the writer must first decide which categories of item relevance he will use. He must look at each problem in turn, together with the response matrix, and try to see what category of relevance each item falls under 'most naturally'. He may find that the five widely used categories which we have given are the most useful, or he may want to use these but with a few additions such as 'questionable', 'possible', 'uncertain' and so on. On the other hand, he may wish to use a set of categories which are quite different, for example numerical weightings such as $-2, -1, 0, +1, +2$, which do not by themselves imply any particular interpretation. When the categories of relevance have been decided the writer must go through the response matrix item by item to decide what relevance he thinks each item has for each problem and how different combinations of these items affect the relevance of any one of them. In this way, a relevance grid can be constructed. When the grid is completed, the amount of overlap between problems must be checked. Cases of degeneracy must also be looked for. If any are found, a way must be found of avoiding them, probably by substituting another problem.

Guidance   The basic observable result of the participant's response to a problem is a set of code numbers from the response matrix. If the author is to know what comment it is appropriate to give the participant he must be able to interpret this set of numbers to elicit its significance. Furthermore, the designer must be able to

Item number	Problem				
	1	2	3	4	
1	E	I	S	S	
2	I	I	S	E	
3	I	C	I	C	
4	S	E	I	C	
5	C	I	E	E	
6	I	E	I	C	
7	S	I	E	E	
8	I	S	C	I	

Note: The amount of overlap gives a crude measure of the degree of structuring in a unit. In a well-structured unit there will be some overlap but not too much.

Figure 7.7 *Relevance grid for diagnosis*

interpret *any* set of numbers which is given in response to a problem. It is not easy to do this. It takes 'feel' and experience as well as method.

The writer's control over the learners' path to understanding is introduced and refined in several steps. The first step should be looked upon as an approximation to guidance which is to be refined subsequently. To illustrate the procedure we will consider a writer of a unit who has constructed a relevance grid with the categories: essential, contradictory, subsidiary and irrelevant. Let us imagine that according to the designer's view, the items (assume there are 20) are distributed in these categories for the first problem, as shown in Figure 7.8. The following is one possible procedure by which the designer might try to provide comments which would (to a first approximation) cover any response from a participant which consisted of a combination of some of these numbers.

Essential	3, 5, 13, 17
Contradictory	2, 4, 8, 20
Subsidiary	1, 9, 10, 14, 15
Irrelevant	the rest

Figure 7.8 *Preparing to write comments*

1. The designer assumes that if the participant has included any of the contradictory items in his response then he has a misconception. The designer looks at the contradictory items as a whole to see if they fall into groups in any natural way. For example, he may find that, in his view, items 2 and 8 go together in one group, and items 4 and 20 go together into another. He then makes a 'hypothesis'. He makes an informed guess as to what he thinks would be in the participant's mind (how he would be thinking, etc) if he were to choose any of the items from one of these groups. In other words, the designer tries to imagine what kind of misconception the participant would have if he included any contradictory items. When the writer has done this he can construct a comment to cover the kind of misconception which he thinks is typical of each group, being careful to include *indirectly* in his comment something about each of the

items in the group which it covers. The whole procedure is shown in Figure 7.9.

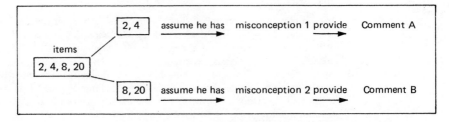

Figure 7.9 *Procedure for comment identification*

2. The writer assumes that if the participant has omitted any of the essential items in his response then he has a misconception. In the example given here, there are only four essential items: 3, 5, 13, 17. They may all fall into one group and hence require only a single comment. Alternatively, the designer may think that one of them is more important than the other three and want to give a comment on that one alone, and another comment on the other three.

3. The subsidiary items are important, but not so central to the problem as the essential items. The designer assumes that the more of the subsidiary items the participant includes in his response, then the richer and fuller is his grasp of the problem, but he does not want to insist that the participant includes all of the subsidiary items in his response. The designer decides how many he will allow him to omit. For example, he may decide that he will give the participant a comment only if he omits more than two of the five from his response.

The writer should go through this type of procedure for all the problems. When he has finished, he has all the control data for guiding the feedback, and the basic meaning to be given to the comment. The writer should also note that:

(a) It is important to give some thought to the *order* in which the comments will be presented to the participant. Which features of his response will he examine first, and which features will be left until later in the process?

(b) For any group of items, it must be considered whether there is a unifying reason for the items being dealt with as a whole. Otherwise, the comment is likely to be either excessively generalized or very fragmented.

(c) There should be some degree of uniformity of significance of the items in a group. For example, if on a grade of importance 1 to 5 the designer puts together items into a comment thus:

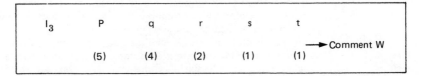

Then if p and q are included, the response may be very different (according to the scale of importance) from a response which omits p and q but includes s and t. However, this difference will not be picked up by the single comment W. This is an example of poor control.

(d) It is important to consider how much 'relaxation' will be allowed in a response before giving a comment. For example, how many items of a certain type will be allowed to be omitted or included.

Writing the dialogue

Now the scenario has been sketched for guidance, but the detail itself must be written. Here are some general points on the way to approach this that the writer might find helpful:

1. It is important to remember that the material is best written in conversational style. It is easy to fall into a 'bookish' way of writing, but difficult to write clearly for the participant. Attention must be given to vocabulary and previous experience.

2. Since the unit endeavours to simulate a real dialogue, care must be taken not to adopt a remote textbook style or use a language which is too technical. A certain amount of 'personalization' gives life to the communication.

3. The different sections interact strongly, both in terms of expression and content. It is important to notice this interaction and to keep cross-checking consistency between sections.

4. If the unit is partly non-verbal and needs illustration or graphic design, it must be remembered that this is not an appendage to the unit but is integral to it. It is often important to get the help and advice of someone experienced in graphics right from the start.

5. A dogmatic style of writing is inconsistent with a unit which is aimed to help the participant to see for himself. Certainly he needs to be given confidence in the information given, but not at the expense of a mode of writing which buries all the assumptions and presuppositions and tends to be conditioning.

6. It is important to make the problems realistic and not simply contrived to fit a response matrix. Such expressions as 'pick out the items which', 'choose those items which' and 'which items exemplify?' must be avoided.

7. It is important to check that the participant has some imaginative step to make, that he has to 'see' something.

8. One can never be too careful about the appropriateness and unambiguous clarity of the problems.

9. Response items may be good and right, but they lose their effectiveness through a choice of words which gives trivial clues to the participant.

10. The designer must try to imagine all the kinds of misinterpretations that the participant might make, and qualify and extend the wording to remove unnecessary ambiguity.

11. It is usually best to have all the items expressed in a consistent way, eg in the same tense, all sentences, or all phrases, or all of the same logical status. However, this is not universally applicable.

12. It is important to make sure that the 'sense' of the problems and the 'sense' of the response items is the same.

13. The writer must be sure that the response items for each diagnostic test are the ones he really intends.

14. If there is room for genuine difference of view or other valid interpretations or if the items are of lesser importance, 'relaxed' functions should be used (eg 'If you have included three or more of', etc).

15. No diagnosis of a response is absolute and applicable to all circumstances. It is better, therefore, to use a conditional tense or its equivalent, eg, 'might', 'probably', 'is likely'. Categorical assertions such as 'you have shown that you understand', or 'you have completely missed the point', or 'you have not understood' should be avoided. It is much better to say 'you seem to be on the right lines', or 'perhaps you did not realize that' and so on. This may seem a little pedestrian, but put yourself in the position of a participant who has a good idea which is valid but different from your own, and which you have not anticipated!

16. The aim of a comment is to help the participant to see for himself. It is better to draw his attention to the area which he has overlooked (or confused) by giving him a clue, than it is to tell him precisely what is

'wrong'. But comments which offer no help, such as simply 'Go back and think again', are best avoided.

17. It is useful to remember that comments can be of various kinds, eg:
    (a) Recollective – 'remember that'.
    (b) Reflective – 'look and consider again'.
    (c) Interrogative – 'did you realize?'
    (d) Imperative – 'refer to this information'.
    (e) Instructional – 'the following information or exercise may help'.
    (f) Paradoxical – 'this seems inconsistent but...'
    (g) Correlative – 'compare this with ...'
    (h) Problematic – 'let's put the problem another way'.

18. In many communication situations it is inappropriate exclusively to use 'corrective' comments – ie comments which suggest to the participant he has 'gone wrong'. Corroboration is equally important. In management training comments often need to be argumentative since the concern is with the development of reasoning powers and independent judgement.

7.5 The potential for structural communication

Structural communication originated in an attempt to make effective use of computers in higher cognitive learning by allocating to the computer the primary task of controlling a process of expressing judgements. At the time of its research phase the microchip revolution was only a dream in silicon valley. With the advent of microcomputers and the development of higher level author languages for computers, the full potential of structural communication is open to exploitation. The logical functions in the method which are complex for the student to administer himself or beyond the scope of early teaching machines can now be done very easily. For example, in the ASET programming language the functions like 'IF included 3 out of (1, 5, 7, 8, 9, 14)' are written 'IF (1/5/7/8/9/14) STRING 3'.

Work is progressing in the UK to develop a computeraided learning library for management education, which incorporates structural communication and develops more advanced ramifications of the method arising from the additional scope computing power provides. This is over a decade since the Harvard Business Review conducted the first computer-managed correspondence case study in 1970 (Hodgson and Dill, 1970). A further advance of the method is a version of structural communication which is open ended in that it is solely a process which the learner uses to shape his own content so that he can gain new perspectives on his knowledge so as to understand it better. In this method the array is articulated as magnetic hexagons (which rest well on a surface to imply clusters of meaning). Procedures are formulated as to how to generate material and how to organize it for provoking insight. This method is helpful for both individual and group problem solving. The device to aid this process is named Communi-Kit.

We can also foresee the possibility of using a computer terminal as a yet more interactive and visual thinking aid. This is already a well-established application in computer-aided design, involving the manipulation, on the terminal screen, of twoand three-dimensional representations of a physical structure being designed. We are all rapidly becoming acquainted with the visual power of this technique through its ever growing application in publicity (for example, car ads on TV). There is no technical barrier to our using a three-dimensional visual display of a group of interrelated ideas as an aid to the formation of more powerful cognitive schemata. If we take structural communication in its literal sense, we are now experiencing the decade in which we can expect the greatest progress yet.

8. Structured Writing or 'Mapping': An Information Processing Approach*

8.1. Origins of structured writing

The structured writing techniques, described here, were developed by Robert E. Horn and his collaborators (Horn *et al*, 1969; Horn, 1973; 1974) under the name of 'Information Mapping' (TM). This term and also the term 'Info-Maps' are now trademarks of Information Resources Inc., of Lexington, Massachusetts, which markets structured writing services under the name of The Information Mapping (TM) Writing Service.

Theoretical basis of 'mapping'

The roots of this technique spread into every psychological 'camp'. It draws its principles from the work of such diverse workers as Gagné, Piaget, Lumsdaine, Skinner, Briggs, Ausubel and Glaser. It knits these principles into a set of rules and procedures for the preparation of written materials which may serve the purposes of instruction, of revision or of reference. It may also serve as the basis of organization of material for a computer-based information system.

If one was to attempt to classify these techniques into one of the major philosophical camps, defined earlier, we feel they would fit best in the 'student-directed learning' category.

What are 'maps'?

A manual of structured pages, or 'maps', somewhat resembles an atlas of geography maps. Each map has a definite purpose, clearly defined. Each map attempts to present the 'shape' or 'structure' of the information it contains, and for this purpose it uses certain standardized conventions. Finally, just like in a geography atlas, one will find maps which give the 'global' view ('overview' maps, 'summary' maps, 'structure' maps) and other maps which give more detailed information on certain aspects of the whole topic (individual 'concept' maps, procedure maps, compare/contrast tables, etc).

Cross-referencing

The whole manual is cross-indexed and cross-referenced to facilitate the student's choice of the material he needs to read. Thus the student can apply his own learning strategy to the material, going from the particular to the general, or vice versa, from the simple to the complex, or vice versa, from rules and principles to examples, or vice versa. The student may follow a holist or a serialist learning procedure. Furthermore, as all exercises (feedback questions) are to be found on separate, labelled pages, he may easily 'skip' them, to use the manual as a reference text, or select those he feels he needs to self-evaluate his learning, or read these pages first, to diagnose which sections he needs to review.

He may use the manual as a quick and general overview of the subject, reading only the review and summary maps, or he may look up detailed definitions and examples of specific concepts, or the steps of specific procedures, as and when he needs the information, thus using the manual as a job-aid. He may find what information he needs and at the level of detail he needs, just when he needs it.

Flexibility in use

This flexibility of use adapts it well to the varied needs of learners, coming from different backgrounds, with different levels of prior knowledge.

The emphasis on the presentation of the structure of the topic, and of each

---

*This chapter has been written in collaboration with Robert Horn of Information Mapping Inc., Lexington, USA.

concept within the topic, should make it particularly suitable for the presentation of highly structured subjects, such as mathematics. Indeed, the technique was first experimentally applied to the teaching of mathematical concepts (Introduction to Probability, Horn *et al*, 1971).

A section of this text on probability has been used at the Middlesex Polytechnic, in London, in an experimental comparison of alternative programming methods for the teaching of mathematics to nonmathematical undergraduates (mainly social science students). This and other 'structured writing' materials proved themselves more effective, quicker and better liked than other forms of programmed selfinstruction that were utilized. Also, potential authors learned to produce materials in this style of selfinstructional programming much more rapidly than in linear or branching programmed text materials.

The techniques were also applied in some research on the teaching of vectors and matrices, performed as part of a project for the Council of Europe (Ellis and Romiszowski, 1972). This project indicated that the use of structured writing, as opposed to linear programmed instruction, may improve learning significantly and cut down the learning time. This experiment was replicated, with similar results, in Brazil, teaching a variety of subject matter content to adult 'second chance' students attempting to make up what they had missed at school (Romiszowski, 1976).

Perhaps the most significant results, and the greatest acceptance, have been obtained in the industrial and commercial context, where manuals of structured maps are now extensively used for initial training, for recycling or review purposes, as overviews for managers and supervisors and as job-aids for reference or guidance.

Many other kinds of documents, not necessarily of a training nature, are now being written, or rewritten more clearly, by means of structured writing techniques. Figures 8.1 and 8.2 show two versions of a communication, before and after a little 'structuring' had been applied.

The changes made to the document illustrated in Figures 8.1 and 8.2 show the power of structured writing and indicate also that in large part the techniques are plain common sense, applied to the layout of the communication that the writer wishes to transmit. There are, however, many rules and general principles that the writer may learn and apply, in order to create the best possible 'structure' for his message. When we are dealing with plain factual communication, as in the case of the example presented above, the rules are mainly concerned with visual layout, labelling the content of each separate paragraph, or 'block', identifying clearly the topic to which the communication refers, using tables, boxes and other aids to clear communication, etc.

In the case of instructional materials writing, many other rules and techniques may be used in addition to the above. The Information Mapping (TM) Writing Service runs seminars and workshops to teach writers to apply these rules and writing standards developed by Information Resources Inc.

One important aspect of structured writing, for instructional purposes, is the initial analysis of the subject matter to be communicated. This analysis classifies the content into categories such as factual information, concepts, procedures, structures, processes, etc (a classification scheme not unlike the one we have been using since Chapter 2). Special rules and standards can be applied for each type of content. For teaching a concept, for example, special attention is paid to the selection and presentation of a sufficiently representative range of examples, nonexamples when appropriate, a clear and concise definition, any diagrammatic or notational representation, synonyms, and so on. For a procedure care is taken to specify clearly and illustrate all the steps, in the correct sequence, use often being made of flow charts and other visual communication aids.

As a rule, one item of content, for example a key concept, would have a page, or 'map', devoted to it. This eases reference and aids with the clear titling of the pages. There may, of course, be 'summary' or 'overview' or 'structure' maps, which illustrate the interrelationship between several concepts and how the

---

MEMORANDUM

Date: 23 January

From: John Rootes, Managing Director

To: All Departments

As all are well aware, our company has recently taken the decision to diversify from our traditional motor car and engine spares manufacturing, to the manufacture of a range of car accessories. This decision, quite naturally, will necessitate a series of changes in the structure of the company and a number of personnel changes at management, supervision and other levels will take place. This memorandum announces the changes already defined and later communications will define other changes, as soon as the new department managers have had a chance to specify their requirements.

We wish to stress that in planning these changes, we have tried to cause the least possible disruption to existing work groups and have made every effort to relocate existing personnel in preference to the creation of new appointments.

The following changes at department manager level will become effective as from the first of February. Bill White, currently Sales Manager, will effectively continue, but will now be called Motor Spares Sales Manager. He will report to Alan Jones, who leaves the post of Chief of Marketing Policy, to become General Sales Director. John Brown, currently Production Manager, will take over one of our new departments, as New Products Development Manager. His position will be filled by Joe Cole, who will be promoted from Production Supervisor to Production Manager.

We take the opportunity to welcome Clarence Black, who is joining the company to head the new accessory sales department, as Accessory Sales Manager.

The vacancies created — Production Supervisor and Chief of Marketing Policy — will be filled by internal promotions, to be announced by the end of the month, together with a full list of other changes at supervision level. These changes will become effective on February 15th, the day on which all other changes at lower levels will be announced.

We hope, therefore, to be fully operational, with the two new departments — New Products Development and Accessory Sales — fully staffed by the first day of March.

Figure 8.1 *Example of a typical (unstructured) communication*

manual's contents 'hang together'. This arrangement enables the reader to get a broad view of the subject, if he so wishes, or delve deeper into the parts of the subject which appear new, or particularly interesting, to him. The use of a manual which has been structured in this manner, is thus highly individualized, each reader selecting the content that meets his needs, at the level of detail that he requires in order to understand it. He may act as a first-time learner, or as a 'browser' wishing to gain a general idea.

In order to give a fuller account and a clearer idea of the techniques of structured writing, applied in practice, we shall continue this chapter using the very techniques we are describing. The 'maps' which follow are adapted from an early report (Horn, 1973), and have been adapted and updated by Robert Horn, especially for inclusion in this chapter.

MEMORANDUM

Date: 23 January
From: John Rootes, Managing Director
To: All Departments

Topic: Impending changes in staffing due to diversification

Motive: The decision to diversify into car accessory production and sales requires
        a re-structuring of the company which will create two new departments:
        — New Products Development
        — Car Accessory Sales
        This in turn will require certain staffing changes at management,
        supervision and other levels.

Staffing changes at management level — effective on February 1st		
Name of employee	From (present post)	To (new post)
Alan Jones	Chief — Mrktng. Policy	General Sales Director
Bill White	Sales Manager	Motor Spares Sales Manager
Clarence Black	(new appointment)	Car Accessory Sales Manager
John Brown	Production Manager	New Products Develop. Manager
Joe Cole	Product. Supervisor	Production Manager

Changes at other levels — to be defined as follows		
Level	Date of announcement	Effective as from
Supervision	January 31st	February 15th
Lower levels	February 15th	March 1st

Welcome: We take this opportunity to welcome Clarence Black, who is joining the
        company to head the new Accessory Sales Department, in the post of
        Accessory Sales Manager.

Please   We wish to stress, however, that our policy in planning these changes, has
note: been aimed at causing the least possible disruption to existing working
        groups and every effort is being made to relocate existing personnel in
        preference to the creation of new appointments. This applies to the two
        vacancies created by the announced moves (Production Supervisor and
        Chief of Marketing Policy) which will be filled by internal promotions.

Figure 8.2 *A 'structured' version of the document
shown in its original 'unstructured' form in Figure 8.1*

8.2. Introduction to structured writing

Map 8.1 *Objectives of structured writing*

Introduction	In the past 20 years, we have seen a significant increase in research projects concerned with the man-information interface. The reasons for this scarcely need repeating. We have more information to handle in almost every job and discipline. This information is increasingly complex. People switch jobs more often, thus requiring more and speedier retraining. Technology changes; people must learn to use the new. The information-generating capabilities of the computer have surpassed all predictions. Researchers are following many lines of inquiry in an attempt to augment the ability of human beings to interact with their new information environment. Hardware and software extend in many new and more flexible directions. Retrieval specialists are seeking new ways of indexing, abstracting, sorting, storing, and retrieving information. Computer-driven display units are becoming widely available. Time-sharing is enabling communities of workers to share the same database. Psychologists and training specialists have given much more attention in recent years to the practical problems of how human beings learn. Enormous efforts are under way to refine programmed instruction and computer-aided instruction in a larger attempt to produce an 'instructional technology'.
Basic aims	As one response to the burgeoning educational demands, structured writing has emerged as a system of organizing databases for self-instructional and reference purposes. Research and development have been concerned with these objectives: ☐ To make learning and reference work easier and quicker. ☐ To make the preparation of learning and reference materials easier and quicker. ☐ To develop economical procedures for designing and maintaining (eg updating) training and reference materials.

Map 8.2 *Structured writing: its scope*

Basic concepts	Structured writing is a system of principles for identifying, categorizing and interrelating the information required for learning-reference purposes. The system can be applied to production of books for self-instruction or to the specification of databases for computer-aided instruction. Most of the research and development work described in this report was concerned with information-mapped books.
Books	Structured books are learning and reference materials in which categories of information are consistently ordered on the page and are clearly identified by marginal labels. The arrangement of information blocks is dictated not only by logical analysis and classification of subject-matter content but also by analysis of the contingencies required for successful learning and reference use. Therefore, in addition to basic content material, structural writing books also have: ☐ Introductory, overview and summary sequences ☐ Diagrams, charts, trees ☐ Feedback questions and answers in close proximity to material to be learned ☐ Self-tests and review questions ☐ Tables of contents, alphabetic indexes and local indexes with connections to related topics.

Computer database	Through our studies with structured books on various subjects, it has become clear that similar techniques could effectively organize a database for computer-assisted instruction. The database would be composed of separable labelled blocks of information together with their interconnections. This would afford a flexibility in using only those parts of the system that are required for a particular purpose. The flexible block-identified database could be rearranged for:
	
	 ☐ Initial learning
	    – for the naive student
	    – for the sopisticated student
	 ☐ Relearning or review
	 ☐ Reference use.

In a book form, maps are . . .

Stored this way . . .	. . . and displayed this way . . .
. . . on printed pages	. . . on the same pages

name	
definition	
example 1	
example 2	
example 3	
connections	

name	
definition	
example 1	
example 2	
example 3	
connections	

Map 8.2 *(continued)*

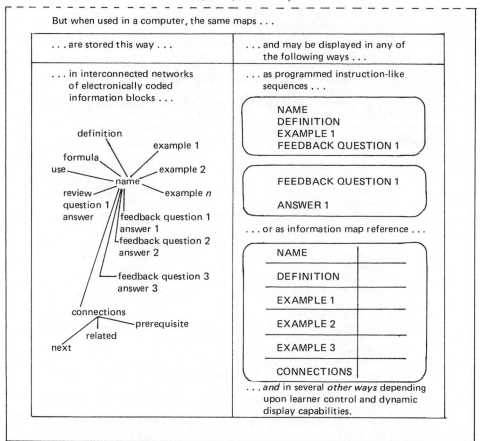

Map 8.3 *Visible and invisible features of structured writing*

Introduction	Maps for self-instructional books are conspicuous for their physical features, the format in which they present information. An equally important aspect of such maps, however, is that the content itself is selected and organized according to a set of underlying principles. The *method of presentation* and the *organization of content* may be thought of as the visible and invisible features of a mapped page.
Visible features	The more obvious visible characteristics are these: ☐ Information is presented in blocks. ☐ Marginal labels identify the kind of information in each block. ☐ A consistent format is used for each kind of information: procedures follow one format, concept maps follow another distinct format, and so on. ☐ Functional and uniform headings and subheadings are used to make scanning easy and to speed up reference work. ☐ Each map begins on a new page and, in programs for initial learning, most maps occupy single pages. ☐ Feedback questions and answers are located in close proximity to the relevant information maps. ☐ A local index at the bottom of maps provides page numbers for quick location of prerequisite topics. (The last two features are not used in technical reports.)

Map 8.3 *(continued)*

Invisible features	The arrangement and sequencing of materials presented in information map formats are the result of: □ Detailed specification of learning and reference objectives in behavioural terms. □ Specification of prerequisites for the subject-matter area. □ Classification of the subject matter into component types (concepts, procedures, etc). □ Definition of the contingencies required for successful learning and reference.

Map 8.4 *Organization and integration of information important for learning*

Introduction	Some important features of structured writing methods owe their origins to a topic of current theoretical interest among learning psychologists, namely, the logical and psychological structures of knowledge and their impact on learning and retention.
Theoretical discussions	Piaget had long ago speculated that 'learning . . . is facilitated by presenting materials in a fashion amenable to organization' (Flavell, 1963), but it is only in recent years that psychologists have actively taken up the problems of how cognitive structures develop and of the role of organization in learning and retention. The 'atomistic' approach of most programmed instruction materials has been criticized and a firm case made for the advantages of 'meaningful organization and holistic presentation of materials'. In a symposium on 'Education and the Structure of Knowledge', P. H. Phenix remarked: 'It is difficult to imagine how any effective learning could take place without regard for the inherent patterns of what is to be learned.' David Ausubel (1963; 1968) has developed a logical and psychological case for believing that learning and long-term retention are facilitated by 'organizers' which provide an 'ideational scaffolding'. He has now amassed considerable experimental support for his hypotheses. Studies with college students pointed up the importance of organization for learning and for retention. The relation of organization of materials to ease of learning also finds support in the area of verbal learning research (Underwood and Schulz, 1960).
Implications for structured books	Although many issues remain to be settled by research, a strong case can be supported both logically and empirically for the advantages of organizing and integrating features in materials for learning. Both verbal and graphical means can be used to inject a sense of organization and direction into a subject-matter presentation. In the practical effort to design effective learning materials, we have incorporated a number of features intended to help the learner integrate and organize the ideas for more efficient storage in memory. These are listed *on the next page*.
List of features	□ Reviews and previews: to take stock of the ideas developed up to that point and to prepare the ground for relating them to new concepts about to be encountered. □ Introductions to each map: to relate new ideas to previous concepts or to familiarize with nature and importance of new ideas. □ Recaps or capsules: to summarize succinctly the essential ideas of rules or principles in nutshell form. □ Tree diagrams: to sketch the ideas and procedures of a topic so as to show the role of each and its links to others. □ Compare-and-contrast tables: to point up the similarities and differences between two concepts that are sometimes confused. □ Summary tables: to chart in easy reference form the main concepts of an area. □ Review tests after short sets of maps and at the end of units: to promote the integration of several concepts and to practise using them in problem solving. □ Prerequisite charts: to show schematically the paths the learner can take through a subject matter in order to reach the learning objectives.

Map 8.5 *Map features for ease of reference*

Introduction	In designing book-type materials for initial learning, we added features to facilitate the return to ideas previously encountered, an activity that is often frustrating with conventional texts where the contents of the paragraphs are unlabelled. Common sense, human factors research, and graphic technology were used in formulating aids for easy access to the learning materials. A list of these aids appears below.
	It is clear also that these same features would be important for reference manuals or job aids. If map materials were designed for those purposes alone, some of the introductions, explanations, and examples needed for initial learning would be omitted.
	Again we note that some of the features needed for easy reference purposes have already been mentioned as desirable on other grounds. For example, labels on information blocks aid in quick retrieval of ideas but they also serve to alert the learner to the nature of his learning task and prepare him to take in a specific kind of information.
List of features	☐ Tables of contents for learning books are organized and formatted to speed location of topics and special features.
	☐ A predictable format for each type of map (concept, procedure, etc) facilitates location of needed information.
	☐ Map headings in consistent typography help in scanning for page topic.
	☐ Marginal labels help not only in locating the kinds of information sought but also in skipping those not required.
	☐ Local indexes at foot of each map permit quick location of concepts relevant to the given map.
	☐ Decision tables display the choices appropriate for each possible situation.
	☐ Summary tables assemble main facts and relations for easy review and reference.
	☐ Capsules provide 'kernel' statements of key rules or concepts.
	☐ Flow charts show graphically the sequences of events in a process.
	☐ Indexes aid information retrieval.

Map 8.6 *Other parts of the structured writing system*

Introduction	So far we have been concerned with what maps look like, how they got that way, and how they are written.
	But the process of writing cannot begin until fundamental curriculum plans are worked out. Furthermore, the end of the writing task is by no means the end of the production process: a crucial part of that process is the series of try-and-revise cycles through which the product is refined and the learning outcomes are brought closer to the programme objectives.
	Our system, then, includes guidelines for curriculum planning and for developmental testing.
Curriculum planning	Once the subject-matter area of the project has been agreed upon, a series of interrelated decisions must be settled, including the type of audience for which the programme is intended, the conditions under which it is to be used and so forth. When the scope of the programme has thus been defined, charts showing the nature of the writing task are evolved through the following steps:
	☐ The nature of the subject matter is explored and the potential topics are listed.
	☐ The learning objectives for the specific programme are determined and are stated in behavioural terms.
	☐ The topics that are required to meet the specified learning objectives are organized into a schematic display called the 'preliminary prerequisite chart' which works backwards from the objectives to the topics that are required to meet those objectives.
	☐ Analyse the nature of the learning tasks and plan the teaching strategies for achieving them.
	☐ Revise the prerequisite chart to show the assembling of concepts into the networks of associations building toward the final instructional goals.

Map 8.6 *(continued)*

The prerequisite chart	This chart of the topics and their sequencing plus special learning materials serve as a guide to the writer in his task.
Successive approximations	Because teaching and writing are both arts, we do not expect the first draft of a learning programme to be totally successful. We rely heavily on the iterative process — cycles of tryouts with students and revisions of the materials in response to their reactions. The most important aspect of these tryouts is that the feedback questions and sets of review questions spaced throughout the programme give us immediate evidence of the topics that need amendment or expansion. Developmental tryouts and revisions are key tools in the production of effective materials.

Map 8.7 *Applications of structured writing*

Introduction	Structured writing was developed for use in that part of science, technology, and business which represents the largest amount of human information processing. It is aimed at the very centre of the paperwork mountain in business and government.
Classification list	The technique can be used in the following types of situation: ☐ For *documentation* of projects 　☐ computer program documentation 　☐ early specification of equipment design 　☐ project records ☐ For *reference* material 　☐ company procedures books 　☐ technical handbooks 　☐ sales reference books ☐ For initial *training* materials 　☐ technical training books 　☐ operator training books 　☐ maintenance training, 　　etc.
Examples	Actual applications in government and industry include: company policy books, typists' training manuals, computer program documentation, accounting procedures books, sales handbooks, teachers' manuals, programmer training in computer industry, management planning training, and many others.
Comment	Can be used in the academic world as well with obvious applications in written material in the sciences, mathematics, social sciences, engineering, etc.

8.3 The process of structured writing

Map 8.8 *The main visible features of maps*

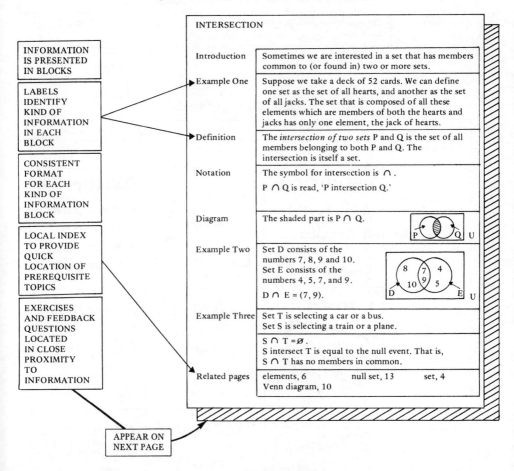

	INTERSECTION	
INFORMATION IS PRESENTED IN BLOCKS	Introduction	Sometimes we are interested in a set that has members common to (or found in) two or more sets.
LABELS IDENTIFY KIND OF INFORMATION IN EACH BLOCK	Example One	Suppose we take a deck of 52 cards. We can define one set as the set of all hearts, and another as the set of all jacks. The set that is composed of all these elements which are members of both the hearts and jacks has only one element, the jack of hearts.
	Definition	The *intersection of two sets* P and Q is the set of all members belonging to both P and Q. The intersection is itself a set.
CONSISTENT FORMAT FOR EACH KIND OF INFORMATION BLOCK	Notation	The symbol for intersection is $\cap$. $P \cap Q$ is read, 'P intersection Q.'
	Diagram	The shaded part is $P \cap Q$.
LOCAL INDEX TO PROVIDE QUICK LOCATION OF PREREQUISITE TOPICS	Example Two	Set D consists of the numbers 7, 8, 9 and 10. Set E consists of the numbers 4, 5, 7, and 9. $D \cap E = (7, 9)$.
EXERCISES AND FEEDBACK QUESTIONS LOCATED IN CLOSE PROXIMITY TO INFORMATION	Example Three	Set T is selecting a car or a bus. Set S is selecting a train or a plane. $S \cap T = \varnothing$. S intersect T is equal to the null event. That is, $S \cap T$ has no members in common.
	Related pages	elements, 6 null set, 13 set, 4 Venn diagram, 10

APPEAR ON NEXT PAGE

Map 8.9 *Blocks*

Introduction	One key innovation of structured writing is the classification of all sentences and diagrams in a subject matter into a new unit of writing called the Block.
Definition	A Block consists of: ☐ One or more sentences (and/or diagrams) about a logically coherent fragment of subject matter, and ☐ A label (which describes the function or contents of the Block, eg 'definition', 'example', etc). A Block is always part of a Map. Blocks are easy to identify on already constructed Maps because they are separated by horizontal lines and have their labels prominently displayed in the margin (or otherwise graphically prominent place).
Comment	There are about 40 basic types of Blocks and about a dozen Map types. We will describe the most frequently used Blocks and Maps later in this section.

Map 8.9 *(continued)*

Exercise	Look at the example of a Map on the previous page.
	How many Blocks does the Map contain? _____
	How many kinds of Block labels are there? _____
Answers	*The map contains eight Blocks and has six types: Introduction, Definition, Notation, Diagram, Example, Related Pages.*

Map 8.10 *Types of Blocks*

Introduction	All sentences and diagrams in a structured book are organized in Blocks.
Examples	Here is a partial list of the types of Blocks:

Comment	As you notice, an attempt has been made to give the Block labels that are 'natural' to the reader and writer.
	There are approximately 40 types of standard Blocks, and a few have sub-types as well; but for now you do not need to know them. You will find a reference to a complete list of the types of Blocks below.
Related pages	Complete List of Blocks, Map 8.20

Map 8.11 *Some properties of Blocks*

Introduction	Unlike paragraphs in ordinary books, technical manuals, etc, which are only loosely defined, and are woven together with frequently elaborate 'transition' phrases and sentences, Information Blocks are standardized and modular.
Properties	1. *Information Blocks of a particular type are all similar.* For example, in a 'Definition' Block you will find *only* sentences which define a specified term. No sentences or words unnecessary to the given definition appear in the Block. 2. *Information Blocks are modular.* Any Block or Map can be taken out, improved and replaced at any given time, with minimum effect on the rest of the Map or Book of which they may be a part.
Example	Here is an example of an Information Block with information which belongs in other Blocks edited out.

Example   The individual numbers in an array are called the elements of the array.
In this array

$$\begin{bmatrix} 4 & 2 \\ 3 & 1 \end{bmatrix}$$

each of the numbers 4, 2, 3, and 1 is called an element.

Map 8.12 *Feedback questions (types of Blocks)*

For each of the following Blocks, choose which Block label is most appropriate.

1.

The *hypnagogic state* is that state of consciousness which occurs between the waking state and the start of the sleeping state of consciousness. Sometimes it is characterized by vivid imagery (visual or auditory).

_____ Example   _____ Fact

_____ Definition   _____ Procedure

2.

Sleeping states of consciousness can be divided into two types, the *dreaming state* characterized by periods of rapid eye movement on the electroencephalograph (EEG) as well as the absence of relatively slow brain waves, and the *sleeping state* which can be identified by absence of rapid eye movements and relatively slow brain waves.

_____ Example   _____ Use

_____ Classification   _____ Analogy

3.

A normal person goes through a cycle of dreaming lasting 85 to 110 minutes approximately 4 to 5 times a night

_____ Definition   _____ Fact

_____ Procedure   _____ Parts

ANSWERS: The first Block is a Definition Block; the second, Classification; the third, Fact.

Map 8.13 *The six types of Basic Maps*

Definition	A *Basic Map* is one of the six types listed below which presents all of the new information presented to a student in a structured book.	
Classification	Map class	Definition
	Structure	A *structure* is a physical thing, or something which can be divided into parts (such as a diagramming system) which have physical or identifiable boundaries.
	Procedure	A *procedure* is a set of steps that a person performs to obtain a specified outcome. This includes the decisions that need to be made.
	Process	A *process* is some structure changing through time for some purpose. The description of a process involves writing about what happens during successive stages of time. (Most of the sciences are descriptions of processes.)
	Classification	*Classification* is used to designate the result of the procedures where a group of specimens are sorted into classes or categories by the use of one or more sorting factors.
	Concept	A *concept* is some sequence of words and sentences in a subject matter for which we can reasonably answer the following questions: ☐ What is its name? ☐ What is its definition (description)? ☐ What are examples of the concept? ☐ What are non-instances (or non-examples) of the concept?
	Fact	*Facts* are sentences containing associations of such things as symbols, measurements, dates associated with events, experimental results, presented without supporting evidence.
Comment	You will note that some of these types of Maps are similar or identical to Blocks by those names (eg Fact Maps always contain Fact Blocks; Procedure Maps always contain a Procedure Block, although it may be of several types).	

Map 8.14 *Writing Structure Maps*

Introduction	In structured writing, 'structure' has a meaning somewhat more restricted than in general usage, referring only to physical objects and things which can visibly be divided into parts and boundaries. It is thus sharply distinguished from classifications and concepts.
Definition	*Structure Maps* contain the following Blocks: ☐ *Necessary* Blocks ☐ Diagram ☐ Name of Map ☐ Related Pages ☐ *Optionally* Used Blocks ☐ Parts (Sub-parts) ☐ Example ☐ Non-example ☐ Description (What does it look like?) ☐ Synonym ☐ Analogy ☐ Use ☐ Introduction ☐ Comment
Comment	You have not at this point encountered examples of all of these types, but they are included for later reference purposes.

Map 8.14 *(continued)*

Rule	Structure Maps contain only structural information, and generally are intended to show only where things are and what they look like. The behaviour a learner performs after looking at such a Map can only have to do with learning nomenclature. If the functions of the parts or the process which is involved in the structure are to be discussed, use a Process Map.
Rule	In almost all structures, you will need to use a diagram, illustration, or picture of the structure. (These graphics help show the structure better than most verbal descriptions.)
How to identify structures	Here are some hints for identifying structures: 1. Look for descriptions (rather than definitions). 2. Ask yourself 'Could someone draw a picture of it?' 3. Look for the words 'part of . . .'. 4. Test to see if the boundaries of the structure can be located (where this structure or part stops and another begins).
Example	

Example of a structure map

The Model 700 Terminal

Description   The Model 700 Terminal uses an advanced CRT with 3,500 character display. It has editing and formatting capability.

Diagram

Related pages   CRT, 15   editing, 19

Map 8.15 *Writing Process Maps*

Introduction	A process is some structure changing through time (usually for some identifiable purpose). Process Maps describe such changes.

Definition	*Process Maps* contain the following Blocks:
	
	☐ *Necessary Blocks*
	  ☐ Name of Map
	  ☐ Related Pages
	  ☐ Description, and/or one of the following:
	    ☐ Stage Table ☐ Cause-Effect Table
	    ☐ State Table ☐ Input-Output Table
	    ☐ WHIF Chart ☐ Block Diagram
	    ☐ Parts-Function Table ☐ PERT Chart
	
	☐ *Optionally Used Blocks:*
	  ☐ Introduction ☐ Synonym
	  ☐ Comment ☐ Analogy
	                        ☐ Use

Comment	You will notice that there are some types of blocks mentioned that you have not encountered as yet. They are included for purposes of completeness at this point. They will be taken up later.
Examples	All of these are examples of processes: operation of a machine (but *not* teaching an operator to run it) the water cycle in the atmosphere how the circulatory system in the body operates chemical changes tadpole changing into a frog cake batter rising in an oven.
How to identify processes	Here are some hints for identifying processes: 1. Look for operations which have stages or phases. Descriptive material covers the changes in the process in the time interval covered by each stage or phase. 2. Look alternatively for changes in state where a process is viewed as transformations from relatively permanent condition (state) to another, each state representing a time instant. 3. Look for descriptions of functioning, relationships of cause-effect, attraction, movement, action-consequences. All of these are methods which people use to describe change (or process).
Example	⟶

Example of a process map

  Paper Feeding

  Description Paper Feeding is the term used to describe the process
         of transporting the copy paper from the paper tray to
         the machine transports.

Stage		
A	When released, the sniffers drop on to the top of the paper stack. Their ends are sealed and a vacuum is created in the tubes.	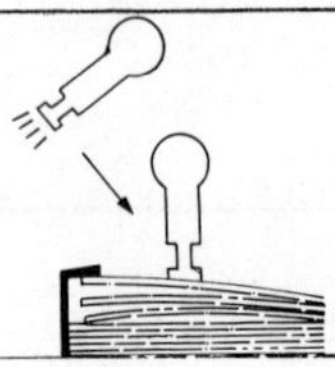

Map 8.15 *(continued)*

B	The vacuum holds the top sheet of paper firmly. As the sniffers rise the top sheet flips out from under the snubbers.	
C	The paper is fed into the paper transports.	

Related pages	Paper transport system, 109

Map 8.16 *Writing Procedure Maps*

Introduction	A procedure is a set of steps a *person* performs to obtain a specified outcome. This includes decisions that need to be made as a part of taking the specified actions.

Definition	*Procedure Maps* contain the following Blocks:
	
	☐ *Necessary Blocks*
	  ☐ Name of Map
	  ☐ Related Pages
	  ☐ One of the following:
	    ☐ Procedure Table
	    ☐ Flow Chart (Algorithm)
	    ☐ Decision Table
	    ☐ Checklist
	    ☐ Worksheet
	
	☐ *Optionally Used Blocks*
	  ☐ Introduction ☐ Example
	  ☐ Use ☐ When to Start
	  ☐ Synonyms ☐ When to Stop
	  ☐ Diagram

Comment	A procedure may be distinguished from a process by examining how you would transform the examples into questions. If you can write a problem which asks the learner to solve a problem, produce a product or take an action/decision, then you have a procedure.

Map 8.16 *(continued)*

Example of a procedure table map

How To Start Up the Terminal

Use	Use this procedure each time you want to begin using the terminal and it has been turned off.	
Procedure	Step	Procedure
	1	Make sure the power cord is connected
	2	Push the POWER ON Button on the left rear of the machine
	3	Push the ACTIVATE Button on the right-hand keyboard
	4	To begin communicating with the main computer, push the ENGAGE Button on the right keyboard

Related pages	Engage, 21	Activate process, 33

Map 8.17 *Writing Concept Maps*

Introduction	Frequently psychologists will refer to persons knowing a concept when they can generalize within a class of things, and discriminate the members of the class from other classes. A person who knows a concept can tell members of a class from non-members. He can also often identify for definitional purposes critical attributes of class membership.
Definition	In writing Concept Maps, we will define a concept as some sequence of words and sentences in a subject matter for which we can reasonably answer the following questions: □ What is its name? □ What is the definition (description) of the concept? □ What are examples of the concept? □ What are non-instances (or non-examples) of the concept? Concept Maps are used to introduce new topics or terms. *Concept Maps* contain the following Blocks: □ *Necessary Blocks:* 　□ Name of Map 　□ Related Pages 　□ Definition (where not stated in the Name of Map) 　□ Example □ *Optionally Used Blocks:* 　□ Formula　　　　　　□ Non-example 　□ Diagram　　　　　　□ Introduction 　□ Use　　　　　　　　□ Comment 　□ Theorem (generalization)　□ Synonym 　□ Rules 　□ Analogy

Map 8.17 *(continued)*

How to identify concepts	Here are some hints for identifying concepts:
	1. Look for:
	  ☐ New terms that are not a part of the user's regular vocabulary, whose usage is peculiar to that field or situation (technical terms).
	  ☐ Important relationships between other concepts.
	  ☐ Important principles or rules or generalizations.
	2. Look for ideas that seem to 'need' an example for clarification. These may often be treated as concepts.

Example	⟶

Example of a concept map

Intersection

Introduction	Sometimes we are interested in a set that has members common to (or found in) two or more sets.
Example One	Suppose we take a deck of 52 cards. We can define one set as the set of all hearts, and another as the set of all jacks. The set that is composed of all these elements which are members of both the hearts and jacks has only one element, the jack of hearts.
Definition	The *intersection of two sets* P and Q is the set of all members belonging to both P and Q. The intersection is itself a set.
Notation	The symbol for intersection is P ∩ Q is read, 'P intersection Q.'
Diagram	The shaded part is P ∩ Q
Example Two	Set D consists of the numbers 7, 8, 9, and 10. Set E consists of the numbers 4, 5, 7, and 9. D ∩ E = (7, 9).
Example Three	Set T is selecting a car or a bus. Set S is selecting a train or a plane. S ∩ T = P. S intersect T is equal to the null event. That is, S ∩ T has no members in common.
Related pages	elements, 6 null set, 13 set, 4

Map 8.18 *Writing Classification Maps*

Introduction	We have already defined classification as the result of the procedure where a group of specimens are sorted into classes or categories by the use of one or more sorting factors. Classification Maps present these results in the form of Blocks with lists (or other more graphic representations) of the items sorted.

Definition	*Classification Maps* contain the following Blocks:
	
	☐ *Necessary Blocks*
	    ☐ Name of Map
	    ☐ Related Pages
	    ☐ One of the following four types of Blocks:
	        ☐ Classification List
	        ☐ Classification Tree
	        ☐ Classification Table
	        ☐ Outline
	
	☐ *Optionally Used Blocks:*
	    ☐ Introduction ☐ Synonym ☐ Use
	    ☐ Comment ☐ Analogy

How to identify classifications	Here is a hint for identifying classifications:
	
	Look for words such as:
	
	    ☐ kinds of
	    ☐ types of
	    ☐ is divided into
	    ☐ may be sorted into
	    ☐ is a sort of
	    ☐ is a part of ('part' is ambiguous and may refer to classifications or structures)

Comment	The specimens which are classified and displayed on Classification Maps are themselves concepts, structures, processes, procedures, and other classifications.
Example	

Example of a classification map

Rectangular Coordinate System synonym
Cartesian Coordinate System

Introduction	The next element of geometry we will take up is the Cartesian (or Rectangular) Coordinate System.

Classification list	Here is a partial (and simplified) list of the topics we will cover:
	
	☐ Uses
	    ☐ Locating points unambiguously
	    ☐ Representing vectors in graphic form
	☐ Visible parts
	    ☐ X-axis and scale
	    ☐ Y-axis and scale
	    ☐ Origin
	    ☐ Points of intersection
	☐ Procedures
	    ☐ How to lay out scales and axes
	    ☐ How to find the values of points
	    ☐ How to plot a point

Related pages	Vectors, 88

Map 8.19 *Writing Fact Maps*

Introduction	We have already defined facts as sentences containing associations of things such as symbols, measurements, dates associated with events, experimental results, and observations presented without supporting evidence.

Definition	*Fact Maps* are collections of Fact Blocks. They may contain:
	
	☐ *Necessary Blocks*
	    ☐ Name of Map
	    ☐ Related Pages
	    ☐ Fact Blocks
	
	☐ *Optionally Used Blocks*
	    ☐ Introduction
	    ☐ Synonym
	    ☐ Diagram
	    ☐ Use
	    ☐ Comment

Example	⟶

Example of a fact map

Specifications of the Model 350 Computer

Speed	200-nsec
Maximum memory	256 K of 16 bit words
Instruction set	164 basic instructions
Channels	8 direct-memory access channels
Registers	32 general purpose registers
Virtual memory	50 million words
Page size	512 words
	etc.

Related pages	Memory, 33	Registers, 18
	Virtual memory, 35	Instruction set, 14
	Channels, 182	

Map 8.20|*Summary of Basic Maps and their Blocks*

BLOCKS COMMON TO *ALL* TYPES OF BASIC MAPS

☐ Name of Map ☐ Analogy
☐ Introduction ☐ Related pages
☐ Comment ☐ Use
☐ Synonym
☐ Diagram

BLOCKS FOR *PROCEDURE MAPS*

☐ When to start ☐ Decision table
☐ When to stop ☐ Checklist
☐ Procedure table ☐ Worksheet
☐ Flow chart (algorithm)

BLOCKS FOR *PROCESS MAPS*

☐ State table ☐ Block diagram
☐ Stage table ☐ PERT chart
☐ WHIF chart ☐ Cycle chart
☐ Parts-function table

BLOCKS COMMON TO BOTH STRUCTURE AND PROCESS MAPS

☐ Description
☐ For structure maps: How does it look?
☐ For process maps: How does it work?

BLOCKS FOR *STRUCTURE MAPS*

☐ Parts (sub-parts)

BLOCKS COMMON TO STRUCTURE AND CONCEPT MAPS

☐ Example ☐ Non-example

BLOCKS FOR *CONCEPT MAPS*

☐ Definition ☐ Rules
☐ Formula ☐ Theorem (generalization)
☐ Properties

BLOCKS FOR *CLASSIFICATION MAPS*

☐ Classification list ☐ Classification table
☐ Classification tree ☐ Outline

BLOCK FOR *FACT MAPS*

☐ Fact

8.4 Supplementary maps for aiding reference and initial learning

Map 8.21 *Summary Maps*

Introduction	One of the things we know about retention of learning is that summarizing and drawing together specific points of a piece of text can help fix the ideas in the reader's memory.
Definition	Summaries contain the important points of the unit in brief form and come *after* a unit.
Classification list	Summaries are of several kinds: ☐ Reference lists (see Map 8.23) ☐ Review (see Map 8.22) ☐ Condensed summaries (see below).

Example (Condensed summary)

CONDENSED SUMMARY OF SET THEORY

Symbol(s)	Name	Meaning	Venn diagram	See page
U	Universal set	The set of all elements in a given study	Shaded part is U	14
0	Null set	The set containing no elements		13
⊂	Subset symbol	'. . . is contained in . . .'	Shaded part is subset	11
P Q	Names of sets	Any capital letters may be used to name sets, eg A, B . . .	P Q	4
P̄	Complement	Set of all elements in U *not* in P	Shaded part is P̄	27
C	Element symbol	'. . . is an element of . . .'		6
∉		'. . . is not an element of . . .'		6
P ∪ Q	Union	Set of all elements in P or Q or both	Shaded part is P ∪ Q	30
P ∩ Q	Inter-section	Set of all elements common to P and Q	Shaded part is P ∩ Q	35
P − Q	Difference	Set of all elements in P but not in Q	Shaded part is P − Q	44
P∩Q=	Disjoint sets	Sets which have no members in common	Shaded part is P∩Q, which is empty (therefore no shading)	42

Map 8.22 *Review Maps*

Definition	*Review Maps* are maps added specially to aid learning and reference. They summarize a unit or section of a course. They may have a non-standard set of block labels; therefore they are treated as a one-block map.
Status remark	Review maps at present are treated loosely in these writing rules in order to give the writers maximum flexibility in providing useful material for the reader. Reviews are one of the places where 'transitions' are possible. The writer may wish to take advantage of this opportunity.
When to use	Reviews should be written at least at the end of every major unit.
Rule	A good rule of thumb for review is about every 10 - 20 maps, depending upon the natural clustering of concepts on the prerequisite chart.
Example	The following is the review part of a review-and-preview map on probability theory. Review. We first learned about the basic building blocks: ☐ The *outcomes* of a *simple experiment* are represented by *sample points* in a *sample space*. ☐ Each sample point is the sole member of an *elementary event*. These elementary events can be grouped in different ways to form *events* representing the classes of experimental outcomes that interest us. Next we saw how operations with events produce new events: ☐ The *union* of events is a new event. ☐ The *intersection* of events is a new event. ☐ The *complement* of an event is a new event.

Map 8.23 *Reference lists*

(A) *List of notation*	
Definition	The *list of notation* is a special kind of summary table which gives symbols (or notation) their meaning and sometimes their pronunciation. The list can encompass symbols from a section of a course or from the entire course.
Rule	The symbols should be ordered in some manner. For example, they may be ordered: ☐ alphabetically ☐ in the order in which they were presented.
Example	Here is an example of the headings of a list of notation: FORMULAE USED IN THIS COURSE <table><tr><td>SYMBOL</td><td>PRONUNCIATION</td><td>MEANING</td><td>SEE PAGE</td></tr><tr><td></td><td></td><td></td><td></td></tr></table>

(B) *List of formulae*	
Definition	The *list of formulae* is a special kind of summary table which gives the name and formula for a section or an entire book.
Rule	The formulae should be ordered in some manner. Typically they are given in alphabetical order by the name of the formula or in sequential order as they appear in the book.
Example	Here is an example of the headings of a list of formulae: FORMULAE USED IN THIS BOOK <table><tr><td>NAME</td><td>FORMULA</td><td>SEE PAGE</td></tr><tr><td></td><td></td><td></td></tr></table>

Map 8.24 *Previews or overviews*

Introduction	Some learning research suggests that learning is easier if the reader is prepared in advance for the kind of materials he is to meet. Such introductory material is sometimes placed in Introduction Blocks. This type of 'advance organizer' can also be placed in a separate map at the beginning of a short group of maps.
Definition	A *Preview Map* is a map added specifically to aid learning. It organizes in advance the material for a unit or chapter. Other information in the various parts of the preview may include 'transition' or 'tying together' information.
When to use	Previews should be written for the beginning of each section and unit.
Example	The following is part of a preview map on probability theory: Present — We have now reached the point where we need to attach probabilities to those events that are formed from other events — by event union, event intersection, and event complementation. The task is complicated by the fact that some events may *overlap* — as happens when any sample point belongs to more than one event. (A point representing a student in a college survey might be a member of all these events: male, senior, French major, Democrat, Baptist, tennis team, glee club, class officer, and red-haired.) Future — The concept of conditional probability will lead us eventually to several convenient tools for evaluating probabilities in more complex situations than we have yet encountered.
Comment	The example shown here in fact appeared in the original book, together with the example of a *review* (shown in Map 8.22), as a composite 'review-and-preview map on probability theory'.

Map 8.25 *Compare and contrast Maps*

Introduction	Sometimes, we find from empirical tryouts of instructional material that two concepts are very similar to each other and are frequently confused. It is important to aid the learner in making a discrimination between the concepts. This can be done in two ways: ☐ with a compare and contrast table ☐ with feedback questions that accompany a compare and contrast table.
Definition	A *compare and contrast Map* is a special Map which contains no new information, but which juxtaposes similar types of information about the concepts.
When to use	Use a compare and contrast table whenever there is evidence that a difficulty of learning may be occurring because of a confusion between two similar things.
Example One	In the study of set theory, the student frequently confuses union and intersection, so a compare and contrast table was made, which compared definition, diagram, and examples.

Map 8.25 *(continued)*

Example Two	Here is an example of a compare and contrast table with three concepts:

COMPARING . . .	YAW . . .	PITCH . . . AND	. . . ROLL
Definition	The rotation of an aircraft or space vehicle about its vertical axis. (Turning right or left.)	The rotation of an aircraft or space vehicle about its transverse axis. (Nosing up or down.)	The rotation of an aircraft or space vehicle about its longitudinal axis. (Side to side rocking.)
Diagram			

Comment	Multiple choice questions are usually the most appropriate methods of evaluating whether the discrimination learning of a compare and contrast Map has taken place.

Map 8.26 *List of course objectives*

Definition	The list of course objectives is a list of objectives which answer the questions: ☐ What will I be expected to do at the end of this course? ☐ Under what conditions? ☐ With what help? ☐ To meet what kind of criteria?
Rule	☐ A list of objectives must be written for each course. ☐ Criteria for writing objectives such as those in Robert Mager's *Preparing Instructional Objectives* apply (Mager, 1962). ☐ The objectives will be organized by unit and will be related to final exam questions.
Example	The following list is an example of a Course Objectives Map:

LIST OF COURSE OBJECTIVES

This unit introduces the student to the topic of set theory. It takes up only those topics which are useful in understanding probability and statistics. At the completion of this unit, the student will be able to:

☐ Identify sets and their elements, and ways of specifying them.
☐ Identify the following:
  ☐ universal sets
  ☐ subsets
  ☐ null sets
  ☐ disjoint sets
  ☐ complement of a set.
☐ Compute the number of subsets in any set.
☐ Identify the elements in sets formed by the operations
  ☐ union
  ☐ intersection
  ☐ difference.
☐ Use Venn diagrams to show the relationships among all of the above concepts.
☐ Use appropriate notation for all of the above concepts.

Map 8.27 *Learning Advice Map (or Block)*

Description	The Learning Advice Map contains information that the course designer believes may help the learner. It might include suggestions about: ☐ How to start the course. ☐ What to pay particular attention to in the course. ☐ The kinds of mistakes or oversights others have made in taking the course. ☐ The optimal amount of time to spend with the course in a session. ☐ The optimal time between study sessions.
Example	In a course on typing in the basic map format, this Learning Advice Block was given: Learning advice ☐ Read the Checklist of Important Things to Remember When Typing. Read it again before you start typing. ☐ Read through the rest of the material. Ask questions if you do not understand what is being said. ☐ Answer the post-test questions. Give your response to whoever is supervising you. ☐ *If you have any questions write them down on the sheet provided* and get answers from whoever is supervising you.
Comment	Learning Advice Maps are most frequently put at the beginning of a course, but they may be added at any point. Frequently in overviews at the beginning of sections, a Learning Advice Map or a Block called 'learning advice' is useful.

Map 8.28 *Prerequisites Maps*

(A) *Prerequisites to course*	
Definition	The *Prerequisites to Course Map* provides an answer to the following questions: ☐ What do I have to know in order to learn this course? ☐ What do I have to be able to do to learn the course? ☐ What are the prerequisites for this course? The answers to these questions will be provided in either list or chart format.
(B) *Prerequisite test*	
Introduction	To determine whether or not a student is qualified to take a course, he may be given a prerequisite test.
Definition	A prerequisite test will cover all of the significant prerequisite information needed by learners for successful study of a course.
Comment	Note the *pretest* refers to material *within* the course and prerequisite test refers to material learners should know *before* taking the course.

Example	In a course on sets, a prerequisite test was given on the learner's 'numerical ability'. The following are the first few questions on a time limited test:
	
	1. $6\frac{2}{3} + 4\frac{1}{6} =$ a. 5
	                b. $6\frac{5}{12}$
	                c. $10\frac{1}{3}$
	                d. $10\frac{5}{6}$
	
	2. $\frac{1}{3} - \frac{1}{5} =$ a. $\frac{1}{4}$
	             b. $\frac{1}{8}$
	             c. $\frac{1}{15}$
	             d. $\frac{2}{15}$

Map 8.29 *Types of maps (a summary and classification)*

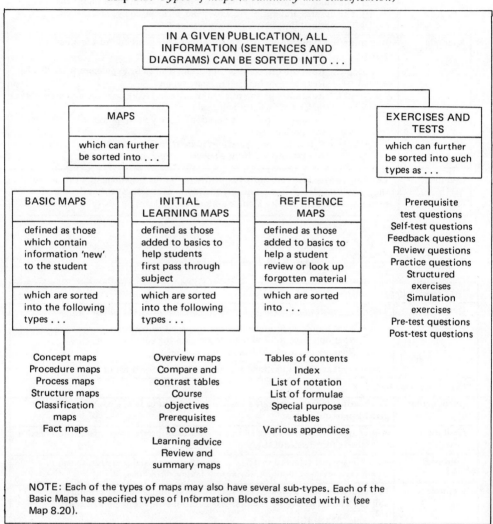

IN A GIVEN PUBLICATION, ALL INFORMATION (SENTENCES AND DIAGRAMS) CAN BE SORTED INTO . . .

MAPS

which can further be sorted into . . .

EXERCISES AND TESTS

which can further be sorted into such types as . . .

BASIC MAPS

defined as those which contain information 'new' to the student

which are sorted into the following types . . .

Concept maps
Procedure maps
Process maps
Structure maps
Classification maps
Fact maps

INITIAL LEARNING MAPS

defined as those added to basics to help students first pass through subject

which are sorted into the following types . . .

Overview maps
Compare and contrast tables
Course objectives
Prerequisites to course
Learning advice
Review and summary maps

REFERENCE MAPS

defined as those added to basics to help a student review or look up forgotten material

which are sorted into . . .

Tables of contents
Index
List of notation
List of formulae
Special purpose tables
Various appendices

Prerequisite test questions
Self-test questions
Feedback questions
Review questions
Practice questions
Structured exercises
Simulation exercises
Pre-test questions
Post-test questions

NOTE: Each of the types of maps may also have several sub-types. Each of the Basic Maps has specified types of Information Blocks associated with it (see Map 8.20).

Map 8.30 *The hierarchical nature of structured writing*

Description	Structured instructional materials are hierarchically arranged into: ☐ Courses ☐ Units (sections, chapters, modules) ☐ Maps ☐ Blocks.
Diagram	STRUCTURED LEARNING AND REFERENCE MATERIALS ARE DIVIDED INTO: COURSES (OR BOOKS OR DATABASES) ⟶ which have the following associated with them: ☐ Previews ☐ Prerequisite tests ☐ Self-tests ☐ Course objectives ☐ List of prerequisites Which are divided into . . . ☐ Time it takes map ☐ Learning advice map ☐ Various aids to reference, etc. UNITS (SECTIONS, CHAPTERS, MODULES) ⟶ which have the following associated with them: ☐ Preview maps ☐ Self-tests ☐ Review questions Which are divided into . . . ☐ Practice questions ☐ Feedback questions ☐ Review maps MAPS ⟶ which have associated with them: Which are divided into . . . ☐ Feedback questions BLOCKS ⟶ which (in some applications for some purposes) may have associated with them: ☐ Feedback questions Which contain sentences and graphic material

9. 'Experience Programming': Writing Structured Case Studies and Role-Plays

9.1 Introduction: the scope of case study and role-playing

In the first volume of this series, we have already discussed the overall planning of case study and role-play exercises as components in many lessons, either in their own right or as part of a more complex simulation-game design. We presented a schema of the progressively greater use that the instructional designer makes of the real system under study, as he progresses from simple case studies, through role-play exercises, to full simulations and simulation-games. The schema is reproduced here for reference (see Figure 9.1).

Figure 9.1 *Relationships between games and simulations used in education and training*

Some examples for analysis

We also discussed several examples of exercises which involve the use of case study materials and the adoption of predetermined roles. These examples included:

* An 'interactive' case study used to develop the skills of front-end analysis in training and development professionals.
* The use of 'microteaching' in teacher training, in which the trainees act out carefully defined roles, to practise specific teaching skills.
* The simulation-game of IDIOTS, in which training personnel take on the roles of instructional designer and subject matter expert, locked in a potential conflict situation and in which a 'hidden agenda' is embedded which only reveals the true objectives of the game at its conclusion.
* The game of STARPOWER, in which the players assume the roles of traders but which also has a hidden plot, concerned with the forces that cause strife between the social classes in society.

All four of these examples can be said to have *some* element of case study, based as they are on data extracted from the analysis of real situations. All four can be said to involve the learners in the adoption of roles other than their own. Should we, however, include such disparate bedfellows in our concept of case study and role-play techniques? Could we perhaps go even further, to include the solution of any problem from a field other than one's own? Can we not argue that the student confronted with a physics problem that involves the description of a spaceship in orbit that must be accelerated to reach another planet, is taking on the role of a NASA scientist and is studying a case based on data extracted from a real system?

Such a broad definition of our concept would, however, be rather useless, as it would include almost all the teaching-learning situations imaginable that involve the development of a new skill or the transfer of knowledge to a new situation. The accepted usage of these terms 'case study' and 'role-play' is somewhat more restricted, although by no means standardized. At least some authors would include all four of the examples cited above in their concept. Others might restrict the terms to 'pure' case studies or role-play exercises, not embedded in more complex, interactive simulations. Yet others might reserve the category for experiential, loosely structured, exercises of the 'sensitivity training' variety. One might include all, some, or none of the examples mentioned in one's concept. It therefore becomes of paramount importance to define what *we* plan to include in this chapter's definition of 'case studies' and 'role-play exercises'!

An analysis of the four examples reveals that the first two are methods of implementing an *expositive* instructional strategy, whereas the last two implement *experiential* strategies. In the *front-end analysis* exercise, as described in Chapter 12 of the previous volume, the trainees first receive information concerning the distinctions made between 'knowledge deficiency' and 'execution deficiency' by Thomas Gilbert (1967) in his article entitled *PRAXEONOMY*. The case study acts as an opportunity to *practise* the application of these concepts in real-lifelike situations. When the trainee errs, the points system of evaluation, built into the exercise, provides corrective feedback, eventually revealing the 'correct' solution of the case. In classical microteaching applications, the trainee teachers are first given 'models' of the expected skilled behaviour, either by 'live' enactment by an expert teacher, or more often by means of a videotape or film of the expected mastery performance. They then proceed to practise the demonstrated skill, comparing their own performance against the model and repeating the exercise to perfection. In both these cases, we see the clear application of our typical 'demonstrate-practise-feedback' model of expositive instruction. We may even discern a reasonably close affinity between the strategies employed in these examples and those of programmed instruction. In the latter two examples, however, the structure of the exercise is largely a 'vehicle', designed to create an interpersonal situation similar to that encountered in certain real-life human-activity systems. The principal objectives are achieved through a process of analysis of the situation presented and synthesis, through 'discovery', of the general principles involved.

The scope of this chapter   The use of case material as practice exercises in expositive instruction has largely been dealt with by our earlier chapters. We shall here limit ourselves to the examination of techniques for the preparation of materials for *experiential* learning. However, we shall not deal with all the domains of experiential learning. The writing of experiential exercises for the cognitive domain was covered in our earlier chapters on programming. We saw that it is possible, though rare, to encounter instructional programmes that lead the student to discover a new relationship between a set of concepts or principles. Attempts to simulate a 'socratic' dialogue in some early branching programmed texts went some way in this direction. In Chapter 7, we saw that the appropriate use of the structural communication technique achieves this ideal to a high degree. In the next section, on computer-assisted learning, we shall again address this area of learning. In the psychomotor domain, it is less common to encounter the systematic application of experiential learning strategies, although some feedback-enhancing simulators could be considered as promoting the 'discovery' of the cause-effect relationships necessary to control skilled activity (though we are really back in the cognitive domain when we talk of cause and effect).

We shall concentrate here on the reactive and interactive domains: feelings and values; social habits and interpersonal skills. It is in this area that case studies and role-play are of especial importance. It is also this area that has so far received less attention from instructional technologists. We shall not, of course, ignore the cognitive domain entirely. As our basic model of the learning process (see Chapter 2) illustrates, knowledge acquisition and cognitive skills development is involved in all but the simplest of learning tasks, whatever the basic 'domain' into which the terminal objective is most conveniently classified. Perhaps the best way to delineate the scope of this chapter is to say that we shall concern ourselves with the development of materials necessary to promote experiential learning about human-activity systems. We should add, however, that our chief preoccupation is with fairly structured learning exercises, for only these require any great efforts expended on the development and validation of special materials. We are now beginning to see the scope of this chapter more clearly. The schema presented in Figure 9.2 is an attempt to define this scope more formally. The figure also serves to establish a general context for this chapter and how it relates to other chapters in these two volumes.

Starting with 'reality', or rather, the real-life systems that we wish our students to study, we may make a convenient distinction between *human-activity* systems and other systems that do not involve 'us' as sub-systems. These 'other' systems include *mechanical and astronomical systems of matter and forces, etc*. The key-word in the 'human-activity' system group is 'activity', so we do not need to get worried about such 'dubious' candidates for classification as, for instance, the biological system of taxonomic classification of animals (which of course includes man as one of the classifications). Excluding all of these, we exclude the consideration of most general problem-solving learning that is involved in the exact sciences, where the use of cases as examples is part of the general process of instructional design, already amply dealt with. The interaction with such cases, in simulations of various types, has been mentioned in the previous volume and will be treated further in the specific context of the design of computer-based simulations.

When we focus on the area of human *activity*, we nevertheless encounter two approaches used in its study, which we shall denominate the *structured* and the *unstructured* approach. This distinction is used by many writers on this subject, including Shaw, Corsini, Blake and Mouton (1980), who tend to equate the unstructured approach with the 'developmental' or 'sensitivity training' movement. A similar distinction is made by Wohlking (1980), who refers to the structured approach as 'method-centred' and the unstructured approach as 'developmental'. We may also see this distinction in the rules for the planning of role-play exercises suggested by McPhail (1974) and quoted in Chapter 12 of the previous volume. McPhail gives ten rules for the planning and use of what he calls 'specific skills' role-plays and a further six rules for what he refers to as 'interpersonal sensitivity' role-plays. In the rules, however, he makes a clear distinction in both of these categories, between tightly structured, materials-based exercises and loosely structured, situation-based exercises. We shall not be dealing here with the loosely structured, developmental, or 'situation-based' type of exercise. The reason for this is largely that such exercises are not within the normal realm of the specialist materials designer. Very little in the way of special material development is required. The design problem is really a Level 3 consideration, requiring a 'bright idea' for an exercise on the part of the course leader. The one exception to this is the general need for some form of previously prepared evaluation instrument, often in the form of an 'observer's checklist'. Such instruments do require special skills in their preparation. However, they differ little from the skills needed to develop similar instruments for evaluating structured, or 'methods-centred', exercises.

Finally, our schema distinguishes between those uses of cases and role enaction used as part of a basically expositive instructional strategy and those that use these methods in the context of an experiential strategy. The examples quoted earlier in this chapter serve to discriminate clearly between these categories.

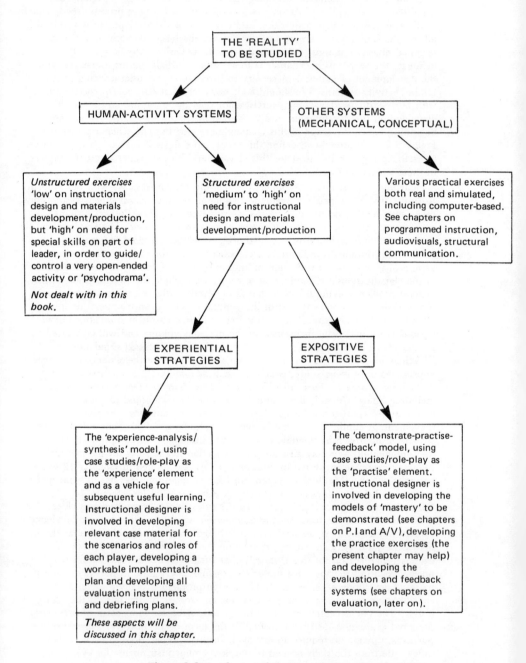

Figure 9.2 *A schema of the case study/role-play field*

In Figure 9.3 we have used the three-column model for lesson planning to illustrate yet another way of distinguishing between these various categories of case study or role-play exercises. The 'cones' superimposed on the grids serve to illustrate where the bulk of the pre-planning and development effort is required. At the right-hand side, in the expositive 'camp', the main work of development is concerned with message design (or 'stimulus design'), with somewhat less effort required to develop appropriate practice exercises. The design of appropriate feedback messages is the least of the three chores. Often, when the instructional materials are developed for very clear and specific objectives, there is no need to design special feedback messages or evaluation instruments, as the very execution of the learning tasks provides all the 'natural' feedback that the learner requires to monitor and evaluate his/her progress. For example, a well-designed assembly procedure in a motor car maintenance manual is often so clear that the learner can judge, unaided, whether the parts have indeed been assembled correctly.

In the centre-space (the experiential 'camp') we see exactly the opposite situation. The bulk of the development effort is devoted to the evaluation system and less to message design. Most briefing documents for a case study or role-play are quite short in relation to the time that the exercise takes to complete. The major skills are concerned with information gathering and structuring it for clear, unambiguous communication. Much less expertise in instructional technology is required by the author. The evaluation of an experiential learning exercise is, however, quite a complex task at times and may require a considerable amount of effort and expertise in order to design a satisfactory evaluation system and develop valid instruments. The matter is complicated as often there are several types of evaluators involved (instructor, peer-observers and the participants themselves), each using a somewhat different evaluation checklist. Often the final de-briefing must be planned to reveal the conceptual structures that the learners have formed – not always an easy aspect of evaluation to plan systematically.

Finally, at the left (the 'unstructured' camp) still less materials design effort and expertise is required. The learning exercise is based on a loosely defined situation, with little or no documentary support. The learning activities may be fairly clearly defined, but generally do not require the preparation of texts or other media prior to the exercise. The evaluation and feedback given to the participants is the main area for prior planning. However, as the activities in such an open-ended, unstructured exercise are somewhat unpredictable, the approach to control and evaluation must be very flexible, depending more on the skill and experience of the group leader/instructor, than on the use of highly elaborate evaluation instruments. In the extreme case, such instructional exercises almost become a form of therapy session, with the group leader acting as the group's 'psychiatrist'. In such exercises, the terminal objectives are rather woolly and often, even if they are stated, little formal effort is made to evaluate to what extent they are being achieved. It is little wonder that George Odiorne (1970) remarked that very little evidence may be found of any real transfer from such learning experiences (when used in management development) to real-life management performance. So, apart from the reason already quoted (that such exercises do not make much call on the skills of materials developers), there may be yet another reason for the little attention that such unstructured exercises have received from instructional technologists in general.

Coming back to 'centre-stage' then, we shall now deal with some techniques for the development of materials for *structured* case study and role-play exercises. On the case study side, we shall start with the techniques for writing case material of the 'Harvard' type, used widely in business management development and other studies of organizational and human-activity systems. From there we shall move on to the somewhat more tightly structured ' interactive' case studies, that border on simulations.

On the role-play side, we shall examine the techniques for the writing of effective roles, the development of role-play exercises on the basis of case materials

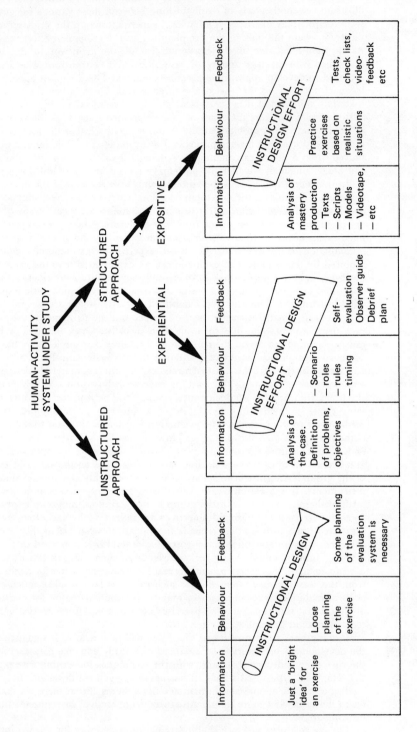

Figure 9.3 *The task of the instructional materials designer related to the three categories of case study and role-play exercises mentioned in the text*

and, finally, consider some techniques for the preparation of more structured role-play exercises, in particular the use of the 'Rolemap' technique (Dormant, 1980).

9.2 The writing of case materials

The development of an effective case study may be seen to follow three distinct stages:

1. the definition, or identification, of a phenomenon or problem-area in a real system, that is deemed worthy of study;
2. the analysis of this problem and of the system in which it 'resides', in order to collect all the data that may throw light on the causes and on its probable consequences, etc;
3. the writing of the case.

Define the problem to be studied

The selection of a problem-area for study is usually motivated by the weight of importance attached to the problems encountered in reality and to the open-endedness of the problem-solving process required. Problems with unique, well-known, or relatively obvious solutions are not a good choice for a case study. The best choices of topic are problem areas where many different types of solution suggest themselves and opinions differ as to which might be the best (if indeed there is any one best solution possible). The benefits to be gained from engaging in the study are as much to do with the processes of analysis, synthesis and evaluation of alternative ideas, as with the actual discovery of a solution. Such problem areas are not hard to come by in the area of business management, interpersonal relations, or indeed in any human-activity system.

Analyse the problem and assemble the data

The analysis and data collection stage relies on the researching of the various sources that may be available. These are in general documentary and/or personal contacts with persons involved with the problem in the system under analysis. One may classify the data sources as follows:

1. available outside the organization (system) in published form or in unpublished form (anecdotes, rumours, news stories, articles);
2. available inside the organization in written form;
3. available inside the organization in unwritten form;
4. not available in any form.

In relation to the second and third categories, it is possible that the information may be readily accessible or very difficult to access, either because of its confidentiality or because of difficulties in locating the appropriate source. All this emphasizes the need for specialist researching and interviewing skills on the part of the case study author. These may not differ substantially from the Level 1 skills of systems analysis/job analysis, etc.

As the data gathered may be confidential, or at any rate not very flattering to the real system under study, the case writer usually takes steps to disguise the real sources of information and, indeed, the very identity of the organization in which the data was gathered. Usually this involves the changing of names, dates and other suchlike 'traceable' data. Very often, in order to obtain agreement for data collection, the researcher must sign a special undertaking to maintain secrecy and to disguise the sources of the data used. The Harvard Business School, when researching the many cases that have been used there and are now widely published, completes a form of legal undertaking to protect the interests of the organization that has agreed to cooperate in furnishing data for the case.

Write the case document

Once the data has been collected, the writing stage may appear relatively straightforward, as compared to the development of special types of instructional text. There are, however, some generally accepted procedures as regards style and layout, which have been found to enhance the quality of the case materials. Some of the most generally quoted heuristics for authoring case studies are listed in Figure 9.4.

Use of appropriate tense

☐ In general, the past tense is preferred for the body of the case description.
☐ However, when the learner is to take decisions concerning a certain problem, the immediate events that lead to that decision should be put into the present.

Use of appropriate style

☐ The case is a report and should be written in a factual, unbiased style.
☐ Avoid statements that imply comment or criticism, unless this is an essential aspect of the presentation of the case.

Use of appropriate language and structure

☐ The case document should appear realistic and 'alive'.
☐ Use the common technical jargon and slang expressions encountered in the real-life situation under study.
☐ Avoid the use of terms that are strange or unknown to the target population.
☐ Avoid the inclusion of conceptual content (definitions of terms, statements of rules, principles or theories) in the body of the case document — these should either be prior knowledge, or are to be 'discovered' by the participants.
☐ Avoid the passive voice, which leads to monotony and clumsy sentences.

Use of communication aids

☐ Tabulate all data in some form of 'easy-reference' format.
☐ Use graphs, diagrams and illustrations, when appropriate.
☐ Print essential data on separate pages, as 'exhibits', to facilitate reference and use during the case study exercise.

Figure 9.4 *Some heuristic rules for the authoring
and organization of case study materials*

Another important aspect of style in the authoring of case studies is the maintenance of an impression of authenticity and reality. Students do not participate so well in exercises which they consider to be totally contrived and quite divorced from reality. Some tactics which tend to promote realism are:

* The use of verbatim quotes from other documents or the quoting of conversations and interviews as a dialogue.
* The identification (even by a fictitious name when necessary) of the source of any information quoted.
* The avoidance of any sympathies, prejudices or bias, expressed in the text.
* Keeping to real, verifiable, or at least plausible 'facts' as much as possible.
* Keep the case consistent; do not contradict yourself or step 'out of time'.
* Explain the background to the events described and the effects on the people involved.

Yet more important is the *structure* of the case. Several aspects of structure should be considered. Among them it is worth including the following:

* Time scale – the case should give the correct impression of the passage of time.
* Narrative – the events which occur in the case should be linked together in a clear narrative style that 'tells it as it was'.
* Exposition – the situations and events which are described should be explained with clarity and conviction.
* Interest – there should be an interesting plot, or some other factor in the style built in, in order to 'grab the reader's interest'.
* Selectivity of viewpoints – whereas one particular viewpoint on the problem

may be favoured, or may have to be pursued by the author, other viewpoints should be mentioned and worked into the plot, so as to avoid the impression of a one-sided argument.

Validate, evaluate and revise

Like any other systematically designed resource material, the case should go through a set of rigorous revisions. First, the factual content should be checked for authenticity (if a real case is being replicated) or plausibility (when the data is being somewhat adapted to the purposes of the study). Then the style should undergo revision, both by the original author and by other specialists in the written word. Some authors suggest up to four or five revisions of style may be necessary. Finally, the case material should undergo a process of developmental and field testing, not unlike that recommended for other self-study printed materials. The developmental testing of the case, for clarity of style and clarity of purpose, is done by giving the material to a series of individual readers, for comments, after each revision. Then, of course, the first application or two of the case material with groups of the target population should be treated as a field-test. The procedures for developmental and field testing need not differ too much from those recommended for other instructional materials (see Chapters 18 and 19).

All in all, the development of effective case study materials may depend less on the specialist knowledge and skills of the instructional technologist, but requires, in compensation, a high level of authorship skills to guarantee clarity, consistency, realism, and all the other attributes of effective case materials that we have mentioned above.

Example of a simple case document

Exhibit 9.1, which follows, is not necessarily a world masterpiece of powerful and persuasive case study writing, but it does seek to follow the heuristic rules suggested in Figure 9.4, especially as relates to:

* Use of appropriate tense – the reader is involved in a decision-making process in the present, based on data presented in the past tense.
* Use of appropriate style – light and factual.
* Language and structure – the technical terminology is self-defining and the context of the case is sufficiently well known to most people, so that few details need to be given in writing.
* Use of communication aids – by boxing in the relevant data (a case study is not merely an exercise in careful reading of a trickily worded puzzle).

This case study has been used by the author on courses dealing with the analysis and solution of performance problems. Sometimes it has been used as a practice exercise, after the principles of performance problem analysis had been taught and exemplified. When used in this way, the case becomes a practice problem, not too difficult in essence as the readers are then already attuned to looking for indications that training may not be the most appropriate solution. It may nevertheless serve as a lively exercise, in which the group leader may be the 'client organization', supplying yet further data on typist performance, on the organization of their work and on the levels of supervision and control exercised.

However, the use which is more interesting in the context of this chapter, is to start the session with the case, and others like it, thus gradually leading the participants to formulate for themselves a strategy for the analysis and solution of performance problems in general. Usually, several such short cases must be studied before the participants are able to form their own model for the analysis of further cases. However, experience shows that the careful design of the case situations can lead to the inevitable discovery of the chief principles and strategies involved.

9.3 The writing of interactive case studies

In the last example (Exhibit 9.1) the document supplied at the beginning of the session and reproduced here is the only case material prepared. All the data deemed necessary in order to engage in the problem-solving process is presented at

Exhibit 9.1

The Civil Service Typists

You are an instructional designer employed by a large Civil Service organization. It has come to the notice of top management that the general performance of typists working in the organization leaves a lot to be desired. Output is low and quality is not as good as it should be. Also, the internal standards for official documentation layout, means of addressing senior officers, etc, are not always followed. You have therefore been called, together with several other experts, to plan an intensive recycling course for the typists. It is suggested that this course should, if possible, be individualized, so as to deal better with the weaknesses of each individual and also to be available on a small-group or one-at-a-time basis, as required.

Before commencing the job, you have the opportunity to work with the other team members on the preparation of a detailed project proposal, to be presented later to top management for approval. You have therefore made some preliminary enquiries and interviewed some typists and their supervisors. When you asked about the standards expected, you found a lot of haziness on the part of the staff. No one interviewed could precisely define what satisfactory typing performance really meant. They talked of improving productivity and quality, but did not quantify the standards required. To help clarify the situation, you checked on what standards are required by local typing schools and found that the norm was '*at least 40 words per minute of continuous production, with not more than five errors per page of typing*'. Many schools would expect much higher standards from 'advanced' typists, but considered the above mentioned standards as the 'minimum required from newly trained typists'.

Armed with this information, you returned to the supervisors and asked to test some of the typists considered to be under-performers and also some of those considered to be adequate performers. Setting some standard typing tasks and timing the typists (telling them not to correct errors in order not to lose speed on the test), you managed to collect the following data.

* The range of performance of the sample tested was from 15 words per minute (typical of the under-performers) to 30 words per minute (typical of the typists considered adequate performers).
* The error rates varied from 10 to over 20 errors per page, under the test conditions. Obviously, most of these errors could be easily corrected by correcting fluid or other such means, but this would bring the productivity down to much less than the already low standards achieved.

In addition to these test results, you found that:

* some of the typists (including some of the slowest and with highest error rates) had been employed as typists for anything up to seven years;
* all the typists had, at some time early on in their career, taken a typing course and obtained a certificate;
* attempts had been made in the past to improve typing productivity, by encouraging typists to take refresher courses at local typing schools, for which they were released during office hours and the organization paid the fees. It is the inability of these local schools to improve the typists' performance that has led to the launching of the project in which you are about to work.

Armed with this information, you meet with your colleagues to plan the detailed proposals to top management. What would you suggest?

Using a simple case study in an interactive manner

once to all the participants. We did, however, mention that the session leader may assume the role of the 'client organization' and supply any further data that the participants request. When this occurs, we are not necessarily transforming the case study session into a role-playing session, as the leader does not 'act out' the role, but is merely a store of further data (which must be asked for). The session does, however, in that case, take on a certain degree of interactivity. Many case study sessions are written as a single-document case, but are then implemented in this interactive manner. In the typist example, the leader has quite an important role to play in keeping the problem-solvers within the bounds of reality. If, for example, participants come to the conclusion that the real causes of poor performance are lack of incentive and lack of disciplinary measures, suggesting productivity bonuses and the threat of sacking, the leader must remind them that in the context of most Civil Service organizations both these measures are outside the realm of possibility (unless the whole country gets reorganized in order to improve typing performance).

The leader, if new to the context of the case, may therefore need to be briefed on how to supply extra information and what possible solutions to the problem are to be considered legitimate. This does not, in fact, limit the creativity of the participants, for they only get informed that sacking or bonuses are 'not on' once they come to suggest them, and the skilful leader will let the implications of these possible solutions be analysed and criticized by the participants themselves, with the minimum of prompting.

Writing interactive case documents

The writer of the case may therefore write a leader's guide suggesting how the session might be run and coordinated. However, this overall guide would not make the material a full 'interactive case study' in the sense that we use the term here. We defined what we mean by this term in Volume 1, when discussing simulations and games. The true interactive case study has some elements of simulation planned into it. This is done by the preparation of new data or new scenarios that only become available to the participants if certain intermediate decisions are taken. Thus the interactive case study requires much fuller 'scripting' of a series of stages in the problem-solving process. The example presented in Exhibit 9.2. is in three stages and could indeed have even more subsequent stages by extending the 'story' from cause-analysis and definition of the *type* of probable solution, to the design and implementation of a solution together with all the unpredicted side effects that often occur at those stages.

An example

Exhibit 9.2 is adapted from an interactive case study used by the author in printed correspondence course materials originally produced in Brazil. The overall subject matter content of this case is similar to the previous exhibit, both dealing with the analysis of performance problems. Indeed, the typist case was also used in the above-mentioned correspondence course. However, the materials of this second case, as presented here, are in a version suitable for group use in a typical case study environment. The data is, however, only supplied to the participants as and when they complete the previous stage of the problem. The data received by the participants is, to some extent at least, dependent on the decisions they take earlier on in the problem analysis. The exhibit, as presented here, does not illustrate the full interactive potential of the technique. No data documents have been prepared to allow the participants to follow up all three of the alternatives set up in document number one, separately. Instead, the questions in document number two allow the participants to follow their predispositions expressed in the first stage of the exercise. There is a possibility, however, that a person who felt, at stage one, that training was quite clearly the answer, might be prompted to change his mind when confronted with the full set of eight specific questions presented in document number two. Document number three deals only with the three questions considered to be most important by the author of the exercise. One could have written alternatives to this document, presenting the results of interviews based on other questions. This would make the exercise more interactive and indeed more open-ended. It would, indeed, become a fully-fledged simulation (and could even be computer-based, though this is not strictly necessary).

In practice, however, a stage-by-stage case study as shown in Exhibit 9.2 generally does not follow up totally blind alleys, as the group interactions are sufficient to analyse the implications of most fruitless directions of analysis. We use this case study by asking each individual to make a decision (in writing) on each question posed, before sharing his views with the others. The ensuing group discussion then leads (with the minimum of guidance by the leader) to a consensus opinion which almost invariably coincides with the author's. The interesting aspect is that the questions finally selected are the least likely to be suggested by the participants at the beginning of the exercise (unless, of course, the exercise is being used to practise a previously taught strategy for performance problem analysis).

We view the interactive case study as a most useful technique, even when the full range of alternative scenarios is not written out, but group discussion and guidance is used to limit the range of alternative paths that may be followed. As a technique of case writing, it does not matter too much whether all or only some of the alternative scenarios are written – the skills are the same. One should note as well that these skills are the same as those required to develop the full print-based simulation, as in the majority of management games – and this includes the more elaborate computer-based management games. Although the media aspects are quite different, there is also much in common between the development of such interactive case studies and the preparation of worthwhile interactive video programmes, as we shall see in later chapters. After all, the key is to break the case into meaningful and useful stages.

9.4. The use of structural communication for the writing of case studies

The example of an interactive (or semi-interactive) case study, presented in the last section (Exhibit 9.2), could be rendered yet more interactive by means of the structural communication technique of writing, that we described in Chapter 7. This would be especially useful if the case study material were to be used on an individual, self-study basis as opposed to the more normal group-study mode. Of course, one of the benefits of the case study method is the sharing of viewpoints between the members of a group, but this is not to be taken as reason for abandoning the use of case study when, for reasons beyond our control, learners have to study alone and, perhaps, at a distance.

A rigidly programmed case study exercise

We mentioned that the case material presented in Exhibit 9.2 is based on a case originally developed for use in a correspondence course. In order to be able to supply adequate feedback, the case study had to be 'programmed' so that learners could assess their own progress through each of the stages (the case study was but one exercise in a much bigger study module and not the basis for three modules, with personalized feedback from the tutor at each stage). In order to supply feedback, the learner filled in a special response sheet as he or she came to a decision at each stage of the case (this sheet is shown in Figure 9.5). After each decision-making stage, the learner would read a comments sheet which, apart from giving the author's opinions on the validity or otherwise of each choice, would award a number of points which the learner could transfer to the response sheet and thus 'score' his progress. Although the use of such a scoring system makes the case study much more directed, it does not fully close the open-ended nature of the exercise, as the scores for a given question (in document number two for example) vary from zero to ten points, allowing for 'shades' of approval or disapproval.

Without a doubt, however, our case study is moving ever closer to a 'programmed' case study, as was quite commonly encountered in the 1960s Crowderian-style multiple-choice branching programmes (eg, Patrick Thornhill, Waterloo Campaign, 1965).

Exhibit 9.2

The under-performing salesmen – Document No 1
Background

You are a training specialist working for a Brazilian-based training consultancy company, named Training Technology for Human Competence, or TTHC for short. This company has a reputation for effective work in the training of salespersons. One of the current contracts of TTHC involves the general reorganization of the administrative and human resources functions of Roberto Carlos Ltd, a company that imports and sells all manner of specialist scientific equipment to research centres, university departments, etc. Most of the importation and sales are by special order, once the needs of the clients have been analysed and compared to what is available on the world market.

TTHC has performed other training jobs for Roberto Carlos Ltd in the past and a good relationship exists between the directors of the two companies. Recently, over dinner, the sales director of Roberto Carlos Ltd expressed dissatisfaction with the performance of his sales force. He said to the managing director of TTHC, 'I don't know exactly what's wrong, but I feel that our salesmen could produce much better monthly figures than they are achieving at present.' By the end of dinner, TTHC had committed itself to look into the problem and you are one of the team charged with this project.

The problem in greater detail

Roberto Carlos Ltd is not a very large organization. At the moment it employs about 70 people, of which 35 are salesmen, all men with a scientific or technical background, though not as highly educated as most of their client contacts. The sales force is divided into two teams of roughly equal numbers, headed up by two area sales managers, who report to Joe Soares, the Sales Director, who was the instigator of this new project for TTHC.

Roberto Carlos Ltd does not manufacture its own products, but acts as representative for a wide range of manufacturers, both Brazilian and foreign. The company prides itself on being able to supply any of the hard-to-find items of scientific equipment that other similar companies fail to deliver. The company's clients include just about all the nation's 'high tech' industries, as well as most government research centres and both public and privately owned universities. Relations are generally good with the client organizations and the turnover of Roberto Carlos Ltd has been slowly rising in real terms (after allowing for the country's high rate of inflation) at about 3 per cent per annum over the last four years. However, given Brazil's very rapid rate of industrialization during these same years, coupled to the very massive investment in new research and development, the sales director feels that Roberto Carlos Ltd is not maintaining its share of the market. He estimates that company turnover should be at least half as much again in order to represent the same percentage of the total market for scientific equipment that the company was supplying five years ago. Furthermore, Joe Soares is convinced that the performance of the sales force is the key to achieving this greater share of the market.

Your first step was to seek more information about the salesmen and their work. Each salesman is assigned to a particular area, being responsible for visiting all potential clients in this area, whatever their type of research activity. He generally has to make several visits to a client before making a sale, as first he has to find out their needs, then make specific proposals and then follow these up whilst internal decisions are taken and finance is liberated. Often there are changes to be made in the original proposal at various stages of this process, as the technical staff of the client organization analyse the proposal and research projects take new directions. Much depends on the patience and good relations of the salesman with the technical staff working in the research departments. Often, however, the final purchase is put out to public tender.

Due to this rather slow sales process, the sales staff is remunerated mainly by a fixed salary, receiving a 5 per cent commission on sales made during the year.

Exhibit 9.2 *(continued)*

This is paid at the end of the financial year on the basis of sales actually completed within that year. The overall salaries and commissions received by the sales staff place them in average salary brackets, as compared to sales staff of this type in other companies. They are neither exceptionally paid nor poorly paid as compared to other technical sales staff in similar jobs. They seem to be relatively happy in the organization, as turnover is well below the norm encountered in other technical sales forces (say, computer salesmen).

Given this information, you are asked by your project chief to state your opinion as regards the causes of the problem. Which of the following three statements is closest to your own opinion?

(a) The salesmen need training in product-knowledge, so as to be better able to judge the needs of the client organization and thus reduce the time between initial contact and final sale. This reduction in time will allow them to make more visits to new potential clients, which should further increase sales. I need further information in order to plan the training programme in detail. What are the main types of product? What type of technical person is most commonly encountered by the salesman?

(b) It seems to be more a motivation problem than a training problem. Clients don't seem to complain about the knowledgeability of the salesmen. The variety of technical areas is so large that no one can be expected to be expert in all of them. Rather, the salesman must be prepared to learn from the customer on the job, and this is a problem of attitude and motivation. I need more information to be able to design a solution, but it's clear that the salesmen are not sufficiently motivated to increase sales.

(c) There is probably some problem associated with the salesmen's performance, but I don't think I've received enough information to be able to judge the causes. I need more information before coming to an opinion.

Document No 2

(Second handout: distributed after a decision has been argued out in relation to the problem question posed)

You have decided that you need more information about the case. You therefore set up a set of interviews with the sales director, Joe Soares, and the two regional sales managers, for tomorrow. In order to prepare yourself for these interviews, you wish to consider the types of questions you will ask. Look at the list below. If you had only time to investigate three of the questions listed, which would you concentrate on. Select the three questions in the list which you think are the most important.

(a) What type of training do *you* think that your salesmen need?

(b) Do the salesmen have sufficient knowledge of all the products you distribute and do they know how to explain them and their special advantages to the client?

(c) How many of the salesmen need to be trained? Are they all equally poor performers?

(d) What happens when the salesmen have success in their visits? And what happens if they don't manage to close a sale, even after several visits?

(e) What is the present level of motivation of the sales force? Do you think that in general the salesmen enjoy their work?

(f) What is your opinion of the salaries paid? Do you think that the commissions are adequate compensation?

(g) Do the salesmen have to do other things with their time, other than visiting prospective clients?

(h) What is the overall budget that can be spent in the training of the salesmen?

(i) Do the salesmen have clear performance standards by which to measure how well they are doing? Are sales targets specified? Do salesmen know how close they are to these targets?

Exhibit 9.2 *(continued)*

Document No 3
(Third handout, after the questions selected individually have been compared and analysed and a group consensus has been reached)
Let us assume that you have chosen to pursue the three questions posed below. You carry out the interviews you have planned and, later, write a summary of your findings. These findings are as described below each of the questions.

(d) What happens when salesmen have success during a visit? What happens if they don't manage to close a sale?

Summary – It is not easy to define the 'success' of a visit. Many visits are made, and several contact persons are generally spoken to. Sales are not 'closed' in the normal sense. A list of required equipment and a request to tender are as close as the salesman gets to really feeling success. From then on, the list is handed to the tenders department and the salesman knows no more about the process until, perhaps months later, an order is finally placed. He only knows he made a sale on a subsequent visit to the client, or at the end of the year when commissions are computed.

(g) Do the salesmen have other things to do with their time, other than visiting prospective clients?

Summary – Not really. The visits they make take up a lot of time, due to the large geographical area covered by each salesman. Also, the contacts made are generally with key technical personnel working on 'high tech' projects, who have many potential needs for equipment and generally take their time in explaining these needs. What time is left, the salesman spends on product-knowledge improvement, using the excellent facilities at his disposal at head office.

(h) Do the salesmen have clear performance standards by which to measure how well they are doing?

Summary – We got a variety of responses here. There appear to be no fixed standards set in relation to the number of visits per week, or clients per month. Salesmen are encouraged to seek out new prospects, which they do by asking existing clients on their visits, studying the technical press, etc. It is felt that the salesman is the best person to judge the frequency with which to visit a given client, as this is a factor that varies enormously. The most objective performance standard is sales volume, but this is only computed formally, on a salesman-by-salesman basis, at the end of the financial year. Of course, salesmen themselves follow up their sales record, by asking clients on subsequent visits or by badgering the financial and tendering departments. However, the usual delay between a salesman submitting a list of requirements and a final sale is upwards of three months, when all goes smoothly.

In addition to these replies, you discovered that the two sales managers are very proud of their sales force and think them an excellent working team. They are generally well educated in some technical or scientific subject. More than half of them have a university degree. They seem very interested in the products they sell, taking time (even their own spare time) to 'play' with new items of imported equipment, so as to understand them fully. They make a lot of use of the excellent technical library at head office and any literature that is supplied by manufacturers, in order to be well informed when meeting their clients. In addition, they always take the opportunities of learning from the people they meet and are able to explain much of the research that is being performed. They stay a long time (average of three years) with Roberto Carlos Ltd, but when they do leave, it is generally to a better job, where the knowledge and experience gained on their sales job can be put to greater use.

Given this new information, what is your opinion on the sources of any performance deficiencies and what are your recommendations to the sales director of Roberto Carlos Ltd.

In order to keep closer to the *experiential* model (as we set out to do in this chapter) yet achieve a fully self-study version of our case, we might turn to the 'response matrix' and multiple comment technique used in structural communication. Returning once more to document number two of Exhibit 9.2, we could, instead of limiting our reader to selecting only three of the eight questions suggested, allow a free choice of any combination of questions (and could indeed include a large number of alternative questions). Using the 'inclusion and exclusion analysis' described in Chapter 7, the author would write a large number of specific comments, which could all be kept open-ended, in the sense that they would not openly judge and 'score' the validity of the reader's choice, but would instead explore the implications of the choice.

Such case studies have been written and have proved to be both effective and quite captivating. Hodgson himself has used such a set of structurally written case studies in a project that offered a tutorial service, at-a-distance, in support of case materials distributed by means of articles in the *Harvard Business Review*. The number of managers who participated, quite voluntarily, in the interchange and continued to do so over a series of case studies, bears witness to the practical use and attractiveness of the interactive open-ended case study as a technique of distance education. Furthermore, the experiment was probably the first computer-managed case study session run at a distance (Hodgson and Dill, 1970). See also the excellent book on structural communication by Kiern Egan (Egan, 1976) for further examples of its use as case study.

We see once more that the various techniques brought together in this book have indeed many aspects in common and others which complement one another. Instead of being kept in water-tight compartments which do not allow any intermingling, they should be brought together into a more comprehensive approach to instructional development and materials design.

Stage		Your decision	Score
1. The problem			
2. Questions to follow up in interviews	1		
	2		
	3		
3. Decisions — the components of a probably effective solution*	1		
	2		
	3		
	4		

* In the self-study version, document number three is linked to the selection of up to four comments of an ideal solution, selected from a list of ten alternatives (the list is not shown in Exhibit 9.2)

Figure 9.5 *A scoring sheet for the individual use of the case study shown in Exhibit 9.2*

9.5 The development of role-play materials

As outlined in the opening section of this chapter, role-play may be used as an *unstructured*, totally open-ended experiential activity (as when people are simply instructed to act as if they were spacemen who had just encountered other life forms on a distant planet), or as a *structured* exercise, in which specific roles are carefully defined, a situation or scenario may be described in some detail, objectives may be specified and observer guides may be developed.

Varieties of role-play

In this book we are interested in the *structured* variety, as this demands pre-planning of materials by an instructional designer. However, the structured role-plays may be further sub-divided, into those designed to drill and practise a specific set of previously demonstrated interpersonal skills (as in the case of microteaching or similar exercises to develop specific sales or interview skills), and those designed to create a situation from which the learner is expected to extract or discover a general principle. We would call these two groups 'expositive-strategy-based' and 'experiential-strategy-based', in line with our terminology as used throughout this series. Other writers have similar classifications. Wohlking (1980) uses the term 'method-centred role-play' for the expositive approach and 'developmental role-play' for the experiential. However, we must be careful, for Shaw *et al* (1980) use the term 'developmental role-play' as a synonym for the *unstructured* variety. We shall therefore stick to our terminology and stress that the development of expositive role-play has much in common with the planning of any expositive 'demonstrate-practise-feedback' lesson materials. Such materials must be based on a detailed analysis of the task to be mastered, as was the case with the Stanford microteaching system, based on the identification of over a dozen specific classroom teaching skills (Allen and Ryan, 1969). Such instructional development has been dealt with in other chapters and we shall not repeat our discussion of it here. In this chapter we shall consider mainly the use of role-play in experiential learning situations. However, the observations and examples concerned with the writing of role descriptions may be of equal application to the development of other materials, whatever the basic strategy of implementation. Just as we saw in the previous sections that the same case information could be used as part of an expositive or an experiential case study exercise, so it is possible to build the same role descriptions into expositive and experiential role-play.

We shall thus consider the development of role-play materials, albeit quite briefly, from the vantage point of experiential learning. Maier *et al* (1975) present a list of principal benefits of the role-play technique (summarized in Figure 9.6), which, although not restricted to the experiential aspect, illustrate very clearly that much of the justification for the technique is associated with incidental learning of feelings, attitudes, self-analysis and so on.

Case study-based role-play

Before we go any further, however, we should note that structured role-play exercises (both expositive and experiential) may be classified in yet another way – whether or not based on a detailed case study. The first benefit in Maier's list (Figure 9.6) obviously refers to role-play based on some form of case description. Indeed most of the exercises described by Maier *et al* (1975) are case studies transformed into role-play exercises by the addition of role descriptions, scenarios, observer checklists and the like.

In contrast, most of the exercises described by Shaw *et al* (1980) are structured (in that they carefully spell out the role descriptions and the play procedures) but are not based on a detailed factual description of a specific case. In order to illustrate this difference (which is important to the materials designer), we include in Exhibits 9.3 and 9.4, a sample exercise from each of the two books mentioned.

Examples of role-play not based on case study

Exhibit 9.3 (adapted from Shaw *et al*, 1980) is a condensed version of the instructions for a role-play exercise on 'handling disagreement through effective listening'. We note that there are quite specific roles defined for two players, who are to engage in a discussion. Although the role descriptions do give some hints on how to proceed, the methodology is not expositive as no demonstrations are initially given of the required behaviour. Also a comparison of the observer guide

1. It requires the person to carry out a thought or decision. From a case study, for example, a conferee may conclude that Mr A should apologize to Mr B. In role playing, A would go to B and apologize. Role-playing experience soon demonstrates the difference between *thinking* and *doing*.
2. It permits practice in carrying out an action and makes clear the fact that good human relations require skill in the same sense that playing golf requires skill.
3. It accomplishes attitude changes effectively by placing persons in specified roles. This demonstrates that a person's behaviour is not only a function of his personality, but also of the situation in which he finds himself.
4. It trains a person to be aware of and sensitive to the feelings of others. This awareness functions as feedback on the effect of his behaviour.
5. It develops a fuller appreciation of the important part played by feelings in determining behaviour in social situations.
6. It enables each person to discover his personal faults. For example, the person who enjoys making wisecracks may discover that these often hurt others.
7. It permits training in the control of feelings and emotions. For example, by repeatedly playing the role of a supervisor, a person can practise not becoming irritated by complaints.

Figure 9.6 *The principal benefits of the role-play technique*
(Adapted from The Role-Play Technique, *R F Maier et al, 1975)*

notes with the roles will show that the players should gain a lot more from this experience than is described in print. A discussion topic and position is defined in the role description, but this is so brief as not to qualify as a case. It is no more than a convenient vehicle for the practice of the general listening skills to be learned and may readily be changed for another controversial topic, if the group thinks fit, without requiring the alteration of anything else in the exercise structure.

The exercise materials are composed of a general goal (to be communicated to the group), specific objectives, instructions and a debriefing summary (for the eyes of the facilitator or leader only, as the learning is to be experiential, through 'discovery'), an observer guide (for the non-players) and role instructions (for the players). Note that much of the learning occurs in the summary or debriefing session when observations are generalized and lists of heuristics are drawn up by the learners. Very often, this type of exercise would be repeated several times, observers and players changing places, for different discussion topics, so that sufficient 'play' is enacted to make it possible to formulate all the important heuristics purely by analysis of what occurred in the discussions.

Example of role-play based on a case study

Exhibit 9.4, in contrast to the previous one, is based on a fairly specific case situation. Although the problem under analysis (an unscheduled coffee break) could occur in many work situations, it is put into a highly specific context by the descriptions given to the learners. We are asked to analyse a specific case, as if we were in a case study session and then interpret our reactions to the situation and act them out in a role-play. This type of exercise demands much more thorough documentation, as is illustrated by the sample in Exhibit 9.4 which, by the way, is incomplete. The full role-play materials include, apart from what is reproduced, a lengthy analysis of the principles involved and a four-page guide to conducting the debriefing session (for use by the facilitator), as well as a detailed observer questionnaire.

The case study origins of this exercise are very clear in the suggested debriefing (not reproduced), which stresses both the delicate aspects of interpersonal relations involved in the simulated meeting and the organization and management aspects of the specific case. Members not only discuss how the meeting was handled and conducted, but what was decided, why the decisions were taken and what are the general principles involved in the planning of coffee breaks. We see a clear contrast between the types of objective closest to the heart of the two authors concerned. One is concerned with general interpersonal skills, that may be practised in any situation (hence the 'situational' approach), whilst the other is concerned with a specific problem and its solution, in the best 'case study' tradition.

Some differences between the two approaches

There are yet other differences between the two exhibits which are of interest to the materials designer. The first is a 'two-player, many-observer' model, which may readily be repeated by other pairs of players – what is called a 'role-rotation' technique. The second is a 'many-player, few-observer' model, which cannot be repeated with the same group, once the specific problem of the case has been fully aired in the debriefing session. There are many other models that may be used for the implementation of role-play. Morry van Ments lists many of these in his excellent book on the technique (van Ments, 1983), including 'role-reversal' in which two players (usually antagonist and protagonist in a dispute) repeat the play after exchanging roles, 'multiple role-play', in which a large group splits into several smaller groups who enact the same role-play in parallel sessions, the 'fish-bowl' technique and many others. Whereas the variety of implementation techniques and points to watch in the classroom are outside our scope here, interested readers are recommended to read van Ments' book in order to get the broader picture of role-play in all its varieties. It is important, however, for the materials designer to have an idea of how the role-play will be enacted, in order to design the materials in a suitable manner.

Some hints on the writing of role-play materials

There is little new in the way of writing skills, required to produce role-play materials. The initial design stages, of 'inventing' a situation or a case (which are Level 2 or 3 considerations that precede materials development) are where the major creative effort occurs. The comments and heuristics listed earlier for the writing of case study materials apply equally to the writing of the basic materials for role-play. The new element is the role description. This should be clear and unambiguous, written so as to attract and involve the player. There are several points to watch in the *design* of the roles in order to avoid the inadvertent introduction of unwanted learning – what van Ments calls the 'hidden agenda'. These include care in avoiding stereotyped roles (macho males, helpless females, etc). Otherwise, basic clear writing skills are the main requirement. Like most other writing tasks, the study of examples, both good and bad, may help to form a personal style which nevertheless applies the main heuristics specified earlier. The reader is directed to study the many collections of role-play exercises available (for example Maier *et al*, 1975; Shaw *et al*, 1980; Pfeiffer and Jones, 1979; Horn and Cleaves, 1980). Two aspects which do require special attention are the observer guides and the debriefing guides, however. The observer guides are often 'instrumented', meaning that they include checklists, questionnaires or other instruments by which the players' performance may be evaluated. As these instruments are similar to those which may be used for the evaluation of other forms of group-activity exercises, we shall leave them to be discussed in later chapters on evaluation.

The debriefing design is, however, of special importance in the case of role-play exercises and we shall devote some time to it now. As Exhibit 9.4 illustrated, the role-play exercise may have multiple objectives and so the debriefing must be performed on various levels of discussion. We also saw in Exhibit 9.3 that much of the real learning in an experiential role-play exercise is during the final analytic/synthetic discussions. Morry van Ments (1983) sees three principal purposes in the role-play debriefing session:

* To 'clear up' the session, by establishing the facts of what happened, clear up misunderstandings, find out how players felt about the roles, how they saw themselves and others.
* To 'draw conclusions' about the session, by deciding why things happened as they did, identify causes and effects, generalize and formulate some principles which can be applied to the real world.

Exhibit **9.3** *A condensed set of instructions for a structured role-play not based on a detailed case study (Adapted from* Role-Playing: A practical manual for group facilitators, *M E Shaw, R J Corvin, R R Blake and J S Mouton, 1980)*

HANDLING DISAGREEMENT THROUGH EFFECTIVE LISTENING

Goal (stated to the group)

To increase the ability of supervisors and managers to listen effectively when confronted with disagreement or resistance.

Specific Behavioural Objectives (for the facilitator)

1. Participants will increase their skill in using questions, general statements, paraphrasing, and other nondirective or active-listening responses, specifically when confronted with feelings or ideas with which they may disagree.
2. Supervisors will increase their capacity to give and receive feedback without defensiveness.

Instructions (for the facilitator)

1. After a brief discussion of the purpose of the session, each participant is given a brief role and background information sheet. The group is then divided into teams of three members each.
2. One member of each subgroup is designated as member A, another as member B, and the third as the observer. Each member is given a sheet of instructions pertinent to his or her role.

Role Instructions (Member A)

Your role during the first phase of this activity is to listen and ask questions without pursuing your own opinion or attempting to force the other person into a conclusion that agrees with your own. You may, from time to time, state ideas; however, you are not to attempt to change the other person's point of view. Your goal is to listen and understand. The following information is provided to guide you in this effort.

Appropriate Listening Responses:

1. General questions (What do you think? What happened?) are useful in finding out more about what another person thinks. Loaded questions or questions that lead to a predetermined answer (Don't you think you should be more reasonable?) are inappropriate.
2. Statements that encourage the person to talk (Tell me more about it; I'm interested in some of your other feelings) tend to increase communication.
3. Paraphrasing or nondirective responses (sometimes called active-listening responses) that pick up what the other person has said and feed it back may be useful. For example, if a person says, 'I think women are becoming too aggressive and belligerent in trying to get ahead,' the interviewer might say, 'I see, you think that women are abrasive in the way in which they are pursuing their goals.' This kind of response encourages the other person to talk.

Your partner (role player B) will initiate the discussion.

Role Instructions (Member B)

During the role-play situation you are to express your opinions freely. If you encounter resistance or disagreement you may deal with it any way you see fit. You are asked not to play a part but to express your genuine feelings about the issue and pursue your point of view in order to clarify the issue, influence the other person's opinion, or simply get your own position stated. You are to initiate the discussion by introducing your point of view about this statement: *Women are treated unfairly in most organizations. Males dominate the scene and often are unaware of their chauvinistic behaviour.*

Exhibit 9.3 *(continued)*

Observer Guide

During the course of the role-play discussion, observe Member A. Did he or she (check statements that apply):

— Quickly reveal his or her own point of view and push for a favourite point?
— Ask questions that suggest a desired answer (leading or loaded questions)?
— Seem interested in drawing out and understanding the viewpoint of the other?
— Occasionally paraphrase what the other person said or use other nondirective techniques to draw out additional information?

Comment on ways in which the interviewer kept the discussion going:

Summary (for the facilitator)

Conduct a brief discussion of the role-play activity, drawing out observers and participants in order to clarify the nature of the listening process and ways in which diverse viewpoints can be pursued. Discussion items include:

1. Which responses by the supervisor seemed to encourage the interviewee to talk more?
2. What comments seemed to discourage the interviewee or put him or her on the defensive?
3. Were the observers or member B quickly able to discern the interviewer's motives and goals or did the interviewer maintain an open listening posture?
4. Construct a list of the kinds of comments, approaches, and attitudes that seem to improve the quality of listening.

Exhibit 9.4 *Some of the documentation of a role-play based on a case study — the unscheduled coffee break (From R F Maier et al, 1975)*

Preparation

1. Thirteen persons can participate in the role play. (If the class has less than thirteen persons, some participants may combine the individual instructions for two persons.)
2. The instructor selects a class member to play the part of Mr Johnson, the supervisor.
3. The twelve persons seated nearest the centre of the room become the crew members. They count off from one to twelve and remember their numbers, but each retains his own name in the role-playing scene. (In the event that less than twelve persons are available to play the parts of crew members, the counting continues with each person taking a second number until the twelve numbers are used.)
4. Persons who are not role players arrange themselves on the two sides of the crew in order to be able to observe the overall actions. The person assuming the role of Mr Johnson remains on the side-lines until the action begins.
5. When the room arrangements are completed, everyone reads the General Instructions for the Class.
6. The person playing Mr Johnson reads his role sheet, crew members read the role sheet for all crew members, and observers read their instructions.
7. Crew members decide among themselves who are especially good friends and go places together. These subgroups of three to five persons may assume that their work territories lie close together.
8. Crew members read the attitude instructions opposite the number they received when they counted from one to twelve during step 3. They adopt the suggested attitude, but should not feel obligated to reveal it to the supervisor unless

Exhibit 9.4 *(continued)*

they consider it relevant and appropriate to the discussion. They should also feel free to reveal any of their opinions or attitudes that develop in the discussion.

9. Observers take notes but do not converse with role players.
10. The instructor makes sure that everyone reads the proper roles or instructions and is prepared for the action. He observes whether the crew members form subgroups and read their personal attitudes as well as the roles for the crew. He also checks that there is a chair in front of the crew for Mr Johnson and that Mr Johnson has access to a blackboard or newsprint in case he wishes to use it.

General Instructions for the Class

Mr Harold Johnson is the supervisor of one of the repair crews in the American Telephone Company. Repairmen are highly skilled craftsmen and take pride in their ability to diagnose difficulties and do high quality work. They are well paid and most of them regard telephone work as a career.

Mr Johnson's office is located in a large garage, which is in a central location for the western half of the city. Besides Mr Johnson's crew, three other repair crews work out of this garage. The building houses the repair trucks and serves as a storage place for the supplies used in making repairs. There are also several tool shops in the building as well as offices and conference rooms.

Each repairman reports to his supervisor's office in the morning in order to pick up his individual assignment. When Mr Johnson has a general problem to discuss with his crew, he schedules a meeting in the morning before they go out on their calls. These are held in the conference room adjacent to his office. One of these meetings is arranged for this morning and the repairmen are gathered in the conference room awaiting the arrival of Mr Johnson.

Role Sheet: Mr Johnson, Supervisor

You are a supervisor in a large utility, the American Telephone Company, and are in charge of a crew of twelve repairmen who leave the garage and go to work in different sections of the city. Your supervisor has reported to you that there has been too much time wasted by repair crews stopping at restaurants for coffee in the morning. It seems that groups of them meet at certain places right after leaving the garage and have a morning visit over coffee. This condition seems to apply to all groups who work out of garages, so your group has not been selected as a bad example. However, your boss points out that it has gone too far and the abuse must be stopped. The company is very sensitive about public opinion and wonders what people will think if several company trucks are parked in front of a restaurant.

You have called your group together for a meeting and want to go over the issue with them before they go out on their calls. You feel that if an authorized rest period were allowed in the middle of the morning, you would be able to improve the situation. You feel it may be necessary to define rest periods for your group. The production of your crew is slightly, but consistently, above average. You feel that your crew should be given the same consideration as office workers. They are waiting for you now. You have meetings of this kind whenever a general problem has to be discussed and clarified.

Role Sheet: All Crew Members

You are members of a crew in the telephone company. You do repair work on phones that are out of order and make five to eight repairs per day. All of you leave the garage in the morning and return at night. It is your practice to drive off in groups of three or four and stop at favourite spots for a cup of coffee. When the load isn't too heavy, you may spend as much as half an hour talking. Although the company has never stated its policy on the matter, you assume it is all right. Office workers get fifteen-minute relief periods both morning and afternoon, and they

Exhibit 9.4 *(continued)*

can visit the company restaurants. Your boss has never raised the question of coffee stops and you are not sure that he knows about them. Some supervisors don't permit their crews to stop for coffee, but the crews do it anyway. The boss has asked you to meet in his office in the garage this morning to discuss a problem. He has these meetings about twice a month.

Attitudes of Different Members of the Crew

No. 1. You find that if you go directly to your job assignment, you frequently cannot go to work because customers are not ready and you have to wait. If you arrive a bit later, people are more likely to be up and ready for you.

No. 2. You find that coffee in the morning makes you feel more civil towards your customers.

No. 3. You like the coffee at one particular restaurant and do not want to get coffee elsewhere.

No. 4. You find that a visit with the crew keeps you interested in the job.

No. 5. You stop for coffee because everybody else does.

No. 6. You often meet your friends who work for another utility at the restaurant and you like to kid with them.

No. 7. You need a cup of coffee in the morning. Your wife is an invalid and you get the kids off to school in the morning. Then you like to relax with the crew over a cup of coffee.

No. 8. You have stomach ulcers and drink a glass of milk rather than coffee. You carry some milk with you, but if you stop for milk in the morning, your thermos bottle supply holds out longer.

No. 9. Your friend works in the restaurant where you stop and you insist on going to this place, a half mile from the garage.

No. 10. You like the group you stop with and join them at these stops.

No. 11. You believe you can work better if you stop for coffee. It starts the day out right.

No. 12. It is your understanding that coffee privileges are company practice. Office workers have them. Why shouldn't you get your coffee when you want it? You prefer having it early, particularly on cold mornings.

* To 'plan action', both immediate and future, aimed at repeating the exercise if learning proved to be inadequate, develop new alternative exercises, see how what was learned may be applied outside the classroom and plan for such application actually to occur.

Each of these three phases can be visualized as a series of steps, as summarized in the flow chart in Figure 9.7. For each phase, van Ments suggests a series of heuristics for the execution of the debriefing (see Figure 9.8). It is clear that a set of general heuristics may not be enough, however, to ensure that the group leader (or 'facilitator') remembers to carry out all the debriefing functions in ways appropriate to the exercise. For this reason, the materials designer should develop a debriefing guide, which will suggest exactly how in a given exercise, the general heuristics suggested by van Ments may be implemented.

Once again, the best way to learn how to develop a debriefing guide is to analyse a few examples. Most of the sources quoted earlier contain examples of debriefing guides. It is a good idea to read them with Figures 9.7 and 9.8 to hand, in order to analyse exactly how the general suggestions made in these figures may be put into practice.

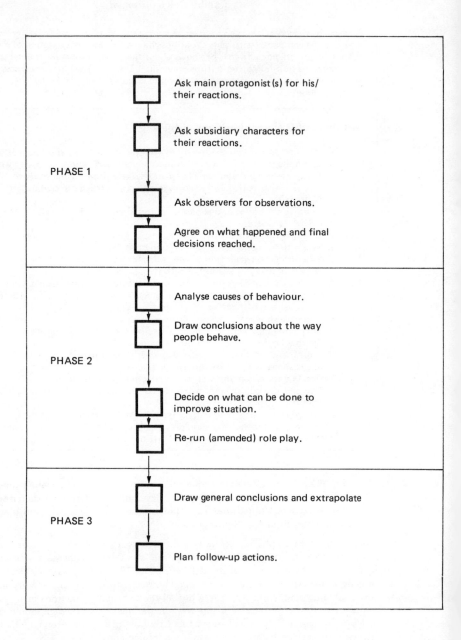

Figure 9.7 *Flow chart for debriefing (from van Ments, 1983)*

Phase 1

Use open-ended questions, How? Why? What?

Concentrate on individual players.

Explore alternative actions.

Reflect feelings.

Insist on descriptive not evaluative comments.

Give feedback in terms of observer's own experience rather than someone else's.

Use group discussion of reaction sheets.

Do not evaluate quality of performance.

Do not argue about misunderstood instructions.

Do not assign motives or make judgements about underlying attitudes.

Emphasize what was done rather than what could have been done.

Use role titles in discussion, not the player's name.

Phase 2

Ask for reasons. Why? How? Who?

Probe answers, Why not? What if?

Seek alternative theories. Is there another possibility?

Collect other examples. Where else has this happened?

Test conclusions against alternatives. Which makes more sense?

Give views of outside experts.

Phase 3

Get students to commit themselves to actions.

Write up actions on wall posters.

Organize students into action groups or pairs.

Put time scale on actions.

Agree criteria for success.

Figure 9.8 *Techniques of debriefing (from van Ments, 1983)*

9.6 The rolemap – a systematic approach to the development of role-play

It is possible that our discussion of the writing of role-play materials in the last section has seemed somewhat less systematically organized, as compared to previous chapters on programmed instruction and the like. It is true that in the experiential learning field the art of clear but imaginative writing may yet count for more than the technology of exercise design. However, the *rolemap* approach has done much to redress the balance and has in some applications shown itself to be especially beneficial.

The name 'rolemap' arose from a combination of two aspects in one exercise design – the use of role-play and the careful analytical mapping of the structure and content of the exercise during the design stages. The technique aims to develop both affective and cognitive domains, using the role-play dynamic to achieve attitudinal changes and the mapped structure to ensure that important knowledge and skills are also mastered. In general, therefore, the rolemap

Principal characteristics of the rolemap technique

technique is case study-based, having a carefully defined scenario (or, as we shall see, various scenarios) in which the players act out their defined roles.

One characteristic aspect of a rolemap session is that it goes through a series of stages. The problem situation under study is analysed into a series of sub-problems, each giving rise to its own scenario and set of defined roles. The learners follow the problem through its stages, often taking on different and conflicting roles at different stages.

Another special characteristic is the use of audiotape to present and control the sessions. Not all instructions are taped (the players still have their role definitions in easy-to-refer-to booklets) but the introduction to each session is tape-recorded and played to the participants at the opportune moment. The tape then runs on in silence, acting as a 'timekeeper' of each stage of the exercise – when time is up a recorded voice announces this fact and defines what should be done next. This attempt to semi-automate the delivery and control of a role-play session may seem artificial, but its protagonists claim that in some cases the recorded commentaries and strict timekeeping overcome the negative attitudes and fears that some students demonstrate in 'normal' role-play situations.

Finally, another special characteristic is the systematic design model used.

The design model and matrix   This involves the execution of a series of pre-design activities, design activities and production activities (see Figure 9.9). The design activities do not differ much from the suggestions made by other authors and already discussed in earlier sections of this chapter. However, the way of going about these activities is very specific, controlled by the plans laid in the *pre-design* activities. These analytical activities break down the problem (or issue) into separate sub-problems (or sub-issues). Significant roles are identified for the people generally involved with each one of these sub-issues (they may not be the same people at each stage as the problem may move from one area to another). Another pre-design activity involves the careful statement of instructional objectives (after taking into account the participants' entry level of skills, knowledge and attitudes). The objectives-based approach to instructional design of this method is most apparent in the suggestion to write the objectives in the form of questions to be used during debriefing (the post-test) *before* proceeding to the design of the rolemaps themselves.

These pre-design activities lay the foundations for a systematic approach to the design stage. The prior definition of a series of exercise stages, or sessions, each

---

Pre-Design Activities

1. Analyse controversial issue into sub-issues.
2. Identify significant role-models.
3. Compare with participants' entry level.
4. Select sub-issues and state instructional objectives as de-briefing questions.

Design Activities

1. Specify sessions.
2. Specify roles.
3. Fill in design matrix.
4. Write script and role descriptions.
5. Rewrite de-briefing questions.
6. Prepare and debug prototype on convenient audience.
7. Try out on representatives of target population.

Production Activities

1. Prepare final audiotape.
2. Prepare final role description booklets or cards and role labels.
3. Prepare leader's guide.

Figure 9.9 *Steps in development of rolemaps (from Dormant, 1980)*

	SESSION I	SESSION II	SESSION III
PLAYER A	O	+	−
PLAYER B	−	O	+
PLAYER C	+	−	O

Observations:

1. Each session will have a different scenario.
2. Each session will involve three players.
3. The roles of each player are different in each session.
4. The roles are so distributed that each player rotates through positive (+), negative (−) and neutral (O) positions in respect of the principal issue or problem being studied.

Figure 9.10 *A design matrix for a three-session rolemap*

SESSION I:

The Coordinator of the Learning Resources Centre at the university has just returned from a workshop on simulations and games and has invited the chairpersons of various academic departments to meet to discuss the strengths and weaknesses of S/G and their possible use in each of the departments.

SESSION II:

After a month of faculty tryouts of S/G (as a result of an administrative memo), the students have been talking so much that the campus newspaper has sent its star reporter to interview some students about their own preferences about the use of S/G in general, and in specific classes.

SESSION III:

After the appearance of a newspaper article on S/G, which quoted some faculty and students out of context, there have been public complaints and the Academic Vice-President has called a meeting of faculty and student representatives to discuss how well S/G are meeting specific instructional objectives. That is, 'Are the students really learning anything using S/G?'

Figure 9.11 *Descriptions of sessions of a rolemap*
(from Playing Your Way Through School, *Dormant, 1980)*

	SESSION I	SESSION II	SESSION III
PLAYER A	○ J Sanford Coordinator, LRC	— A Smith Student, Economics	+ R Reston Representative Student Union
PLAYER B	+ J Rodriguez Chairperson, Language Dept	○ M McCullough Reporter, *Globe U Gazette*	— V Pappas Representative, Student Union
PLAYER C	— G Navarro Chairperson, Mathematics Dept	+ C Mason Student, Business	○ M Tracy Vice-President, Academic Affairs
PLAYER D	+ H Kent Chairperson, Political Science Dept	— D Russell Student, Mathematics	+ J Sanford Coordinator, LRC
PLAYER E	— M Boyd Chairperson, Business Dept	+ F Powers Student, Political Science	— G Navarro Chairperson, Mathematics Dept

Figure 9.12 *Example of the use of the design matrix
(from* Playing Your Way Through School, *Dormant, 1980)*

with a different 'cast', enable one to construct a form of session/player matrix. Figure 9.10 shows the form of such a matrix for a three-session exercise, each session involving three players. As the issue in question generally divides opinions, the plan includes provision for the allocation of positive, negative and neutral roles, as indicated by the symbols in the cells of the matrix. In her book on the rolemap technique, Diane Dormant (1980) shows several examples of 'rolemapped' exercises, including one taken from a course on simulation and gaming, which aims to prepare the participants to implement simulation and game techniques in an institution that has not previously used them. The exercise was initially designed for participants from developing countries who were to act as 'experts' on these methods on their return home. The problem situation was analysed into three sessions (see the descriptions in Figure 9.11). Further analysis of the problem and the three sessions identified a total of five separate roles for each of the three sessions. This gave, therefore, a three-by-five design matrix as shown in Figure 9.12. The next design step is to write a total of 15 role descriptions and other support materials, including the script for the audiotape presentation.

A much fuller description of this interesting technique may be found in Diane Dormant's book mentioned earlier. The technique is full of 'good ideas' which may be used in totality, or 'borrowed' separately. Even if you do not plan to prepare an audiotape, why not use the design matrix?

PART 3.
The Impact of the Computer

Overview

In the preceding chapters, we have mentioned several times the potential impact
that the computer may have on the individualization of instruction. We have also
hinted that some early 'good ideas' seem to be forgotten, or are deliberately being
ignored, by the CAL/CAI movement of the 1980s. There are signs, at the time of
writing, that some practitioners are at last 'seeing the light' and realizing that
much of current CAI materials design practice and the rapid dissemination of
these materials through the microcomputer boom, may indeed 'kill' computer-
based instruction as effectively as the mass of mediocre print-based programmed
instruction of the 1960s contributed to the undeserved 'death' of the programmed
instruction movement as a whole.

In the following chapters, we start by analysing the CAL/CAI field as a whole,
in order to clarify concepts, terminology and current trends. We shall note much
'carry over' of ideas and techniques already mentioned in the earlier chapters on
print-based auto-instructional materials. Some of this transfer is healthy and
should be encouraged. Unfortunately, some weaknesses and 'bad ID habits' are
also being transferred. One function of the present book is to help arrest this
unsatisfactory trend.

Chapter 10 is devoted to the establishment of clear definitions of CAL, CAI,
CMI and the myriad other cryptograms that are used, often indiscriminately, by
authors in this field. Using the specific definition of 'instruction' that forms part of
the general ID model adopted, it is possible to define with precision the difference
between CAL and CAI and to delineate the area of interest of this book.

In Chapter 11, we examine the role of the instructional designer in the process
of courseware development, distinguishing it carefully from the role of the
computer programmer. We highlight some of the dangers of ignoring the generic
ID process and becoming too mesmerized by the 'bells and whistles' of the
computer itself as a presentation medium. The use of a computer does not, in
itself, alter the principles of learning and teaching upon which the generic ID
process is based. Therefore, the methodology suggested in this chapter does not
depart radically from the overall approach previously decribed in Chapter 3 and in
previous books in this series. One key point, seldom practised by courseware
designers, is the development of a lesson plan, prior to the planning of individual
screens, in much the same way as a teacher might plan a conventional lesson.
Such a lesson plan then keeps the designer 'on track' during the more detailed
screen development stages. The chapter presents a detailed case example of the
design and development of a small drill-and-practice exercise in arithmetic. The
example was chosen for a number of reasons. Firstly, the content is known to all
and simple enough not to cause interference between 'understanding the content'
and 'understanding the ID process'. Secondly, despite its simplicity, the topic is
real, has been taught by means of CAI on many occasions and the existing
packages have not all been identical, neither in terms of ID factors, nor in terms
of effectiveness and efficiency. Thirdly, there is a large body of research evidence
on this topic, showing what ID factors are important for effectiveness and
efficiency. Fourthly, the drill-and-practice mode of CAI is one which shows how a
computer contributes to the instructional process in a way that cannot be realized
easily by other means. Fifthly, the drill-and-practice mode is currently much

maligned by some authors and it is our intention to redress the balance somewhat. Finally, the example serves to illustrate that quite worthwhile courseware can be designed to run on the simplest of computers. We have purposely chosen to implement the example package on the simplest of widely available microcomputers: the Sinclair (Sinclair-Timex in the USA) TX81, which may have as little as two kilobytes of usable (RAM) memory. In its original form, this computer was almost a 'toy', at a 'toy' price of about £200. Today the original may be had for as little as £20 ($30 in the USA) and 'clones' abound in many developing countries (there are at least four Sinclair copies on sale in Brazil at the time of writing). The choice of this computer seemed to serve our purposes well in showing the simplicity and insignificance of the hardware, as compared to the skill of the instructional designer in putting together an effective and efficient instructional design.

In Chapter 12, we enlarge our view, to encompass the full range of CAI modes currently serving as models for the design of computer-based courseware. We may identify all the 1960's 'programmed instruction' models as still in use and note close parallels between the ID methodologies used then for the design of paper-based materials and those currently used for computer-based courseware. What is the innovative role of the computer, we might ask. The reply would be that in some 80 per cent of cases, the same instructional designs could be implemented with much the same success on simpler and cheaper, non-computer-based media. There are some exceptions to this trend and the modes that show innovative promise are singled out. We also note how more use could be made of the lessons learned in the 1960s and 1970s in order to produce better quality courseware (from an ID standpoint) and to make better use of the computer's own unique characteristics (as a presentation and control medium).

Finally, in Chapter 13, a brief account is given of the techniques of coding the courseware for the computer. This chapter is quite short and superficial, as it is not our purpose here to go deeply into the production of material in any specific medium. We did not deal with writing style and choice of type fonts in the previous section, and we shall not be dealing with technical aspects of sound recording or of photography in the next. All these do have certain implications for the design and development of instruction, but are dealt with in other sources. We limit ourselves here to the message design and learning activity design aspects of materials development. For this reason we deal, in this chapter, only with the differences between using a general-purpose programming language, a special-purpose courseware authoring language, or a specific authoring system. These differences must be understood by the instructional designer, even if some other team member is to produce the final courseware, as the choice of language or system imposes specific restrictions and offers specific opportunities, as regards the presentation, layout, style and control of the final courseware package.

In all, this part of the book reviews the current status of CAI and the design of computer-based courseware, indicating the, as yet untapped, potential of the computer as a medium of instruction. In later chapters, notably in Chapter 17 on interactive video and in the last chapter (20), we shall be highlighting in more detail and evaluating some of this potential.

10. 'CAL', 'CAI' and 'CMI': Definitions, Scope and Potential

10.1 Computer-assisted instruction: the concept and its variations

In Chapter 2 of *Producing Instructional Systems*, we discussed, in a general way, the use of computers as a means of delivering or controlling instruction. We drew up a list of modes of computer-assisted learning, indicating which of these could in general be considered 'instruction'. We also identified the differences and the areas of overlap between the terms used in this field – computer-managed learning and instruction (CML and CMI), computer-assisted learning and instruction (CAL and CAI), computer-based learning, instruction or training (CBL, CBI, CBT, all meaning much the same as CAL/CAI to most authors, but endowed by others with slight shades of special meaning). It will be useful to review and complete the schema we presented, if only as an aid to our understanding the overall scope of the 'field' of computers in education and defining what this section will and will not deal with.

The 'field' of informatics in education
Figure 10.1 presents our view of the 'field' of discussion in respect of computers and education, or perhaps more appropriately, *informatics and education*, (to include aspects of information processing, use of microprocessors and telematics, which might not be recognized by most people as 'computer' applications). Our field is divided into two main sectors, which we may call 'informatics as content of education' and 'informatics as an instrument of education'. This distinction separates the problems of designing curricula for computer specialists or for general computer literacy, from the problems of utilizing computers in the educational process. Interesting and important as the first of these two sectors is, we shall not be able to devote time and space to it in this book. Our aims here are concerned with the use of computers as an instrument of education. However, this is still an enormous field, as the right-hand half of Figure 10.1 attempts to show. We have divided this sector of the field into three sub-sectors, dealing respectively with the use of computers as an *administrative tool*, as a tool to assist (or substitute) the teacher (a teaching tool) and as a tool to assist the learner – we prefer to call it a *learner's tool* (rather than a learning tool) for, as the diagram illustrates, the learner may use the computer to help him learn or to take over routine tasks and thus reduce the drudgery of the learning process.

CAI or CAL?
Alongside the specific examples of applications of computers in education (which we shall discuss more deeply later on), we have placed some brackets to indicate how we are using the three principal technical terms CAL, CAI and CML. Our usage does not necessarily agree with that of all other authors. Indeed, every author seems to have his own definitions and classifications for the same technical terms. One marked difference is in the use of the terms 'Instruction' and 'Learning' on either side of the Atlantic. In the USA particularly, the term 'Instruction' tends to be used in a more global sense, for any type of teacher/learner interchange, whereas in England the term CAI has of late become restricted to 'programmed instruction' types of exercises, the preferred generic term being 'computer-assisted learning' (CAL). It seems that Canada, caught between the two influences, cannot make up its mind, although recent publications, such as *The Elements of CAL* (Godfrey and Sterling, 1982) suggest that perhaps the British view – that *learning* may be assisted in many diverse ways but

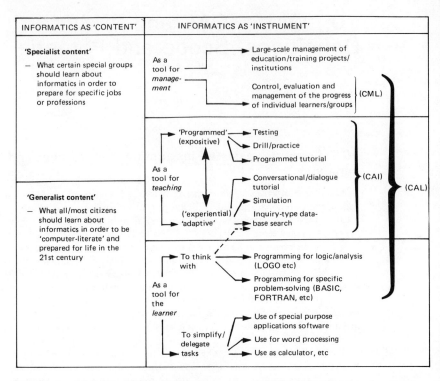

Figure 10.1 *The 'field' of 'computers in education' divided into 'sectors'*
(as an aid to the analysis and discussion of the topic)

only some of them have the characteristics of true *instruction* – seems to be gaining ground.

The position of CML in the overall schema
However, detailed differences exist between the usage of CAL/CML/CAI in the works of British authors, even though they do agree that CAI is a subset of CAL. Some consider CMI (computer-managed instruction) as a variety of CAL. After all, the continuous formative evaluation of a student's progress and the feedback of the results, either to improve course materials, select different options or merely to inform the student of his progress/errors/weaknesses is one very effective way of assisting the learning process. Others exclude CML/CMI as a separate category, not a part of CAL in the strict sense. For Godfrey and Sterling, the development of a CML element is part and parcel of the development of *any* CAL materials (they refer to this as the 'tracking layer' in their model for CAL materials design). Nicholas Rushby, on the other hand, in his book *An Introduction to Educational Computing*, seems to separate these two modes, dealing with them in quite distinct chapters (Rushby, 1979). He also differs from us slightly in other aspects of terminology, using a four-paradigm classification of CAL:

1. The *instructional* paradigm (including our programmed tutorial, drill-and-practice and, *perhaps*, the conversational or diagnostic tutorial modes).
2. The *revelatory* paradigm (including, principally, the simulation mode, but perhaps as well the objectives-directed search of a complex informational database);
3. The *conjectural* paradigm (including model building and problem solving through programming – roughly equivalent to our category of 'learner's tool, to think with');
4. The *emancipatory* paradigm (including principally the use of the computer as a calculator/ problem-solver (to take over the routine tasks) and also its use

for 'serendipity' learning, or the 'browsing' through databases without very clearly defined specific objectives.

A fairly wide concept of CAI

In our schema, we have tried to indicate that some of the 'emancipatory' uses may be assisting learning, whilst others are simply 'labour-saving' applications that do not lead directly to any new learning. We have also indicated, at the opposite (CML) end of the scale, that some applications of the computer as a management tool may assist learning, whilst others, generally at a more large-scale, 'total system management' level do so only very indirectly. We have thus tried to define a fairly wide, but not all-inclusive, concept of CAL. Within this concept, CAI exists as a subset which includes perhaps more examples than some other authors would include. For example, our concept of CAI includes Rushby's 'instructional' and 'revelatory' paradigms.

Matches our fairly wide concept of instruction

We find this concept most in keeping with the way that we defined instruction at the beginning of this book and with the distinction that we have made throughout between *expositive* and *experiential* instructional strategies. So long as specific learning objectives exist, and pre-planned and tested procedures have been developed to achieve these objectives, then we are in the business of 'instruction'. Therefore, open-ended conversational sequences, such as the problems and comments in a structural communication unit, classify as instruction in so far as the unit aims to develop a specific cognitive schema in the mind of the learner (albeit with some allowable variations of viewpoint). Most simulations, *especially* those that are computer-based, have specific objectives concerned with the understanding of the mathematical model that underlies the phenomenon being simulated. They are developed and carefully tested to verify that these objectives are achieved in practice, so they classify as examples of instructional materials, albeit based on *experiential* strategies. The use of a database on some specific topic within the context of a specific exercise, with pre-set objectives and the inclusion, in the system, of sufficient data to make it possible for the learner to solve the problems posed, is also an example of instruction. It is quite different from the use of, perhaps, the same database for unstructured, 'serendipity' learning. It is not so easy to classify the strategy of this example. In so far as the learner reads, or otherwise absorbs, the information stored in the database, the strategy is expositive, but in so far as the learner masters, or develops, strategies for the analysis of the problems posed and the searching of databases in an efficient manner, the learning is experiential.

We have, therefore, chosen to include all six of the examples of use of computers as a 'tool for teaching' within our concept of CAI. In later chapters we shall indicate exactly what is involved in the use and the design of each of these modalities of CAI. In so doing, we are using neither the very broad, all enveloping notion of CAI, as is current in the USA, nor the very narrow, 'expositive' notion that seems to have gained currency in the UK. We have set the bounds of our concept of CAI, and therefore of the scope of this section to be in line with the other concepts and models of the instructional process that we have developed in an earlier book (*Designing Instructional Systems*, Romiszowski, 1981) and have been applying throughout the current work.

10.2 The scope and limitations of this section

In this section it has been possible to give only a very brief outline of the various types of CAI materials that may be readily produced, without great experience in computing. We have stressed that the major part of the courseware design task is in the initial pedagogical planning and organization of the material to be studied. This planning is essentially identical to that required for any other type of self-instructional material. The facilities that the computer offers enable us to implement teaching strategies that would be difficult to realize by other means. However, the planning of these strategies does not require high expertise in computing.

To make effective use of the computer as a tool in education, one must, however, have a basic understanding of the computer and its capabilities. We do not therefore recommend a total divorce between the initial courseware planners and the computer experts who might be employed to encode the lessons. A teamwork approach is extremely effective and makes more efficient use of the special skills of the instructional designers and the programmers, so long as there is a basis of mutual understanding. Despite the apparent 'high-tech' image of informatics and the commonly held beliefs that 'anyone is an expert on education', it is our view, expressed and exemplified throughout this book, that the intellectually more difficult task is the pedagogical design of good instruction. It would therefore seem more logical that the instructional designer, or the design team, should fully master the concepts and possibilities of informatics, in order to make full use of this important new technology in the search for better solutions to the real problems of education.

A look to the future

We stress once more that instruction is only one aspect of education. Therefore CAI is only one aspect of the impact that the computer will have on education. We have limited ourselves to the discussion of the computer as a tool for instruction, as this is the principal area of interest in this book. However, as our schema (Figure 10.1) indicated, there are many other important areas of education which are feeling the impact of informatics. One area of particular promise is the use of the computer, *by the learner*, as a 'tool to think with'. The work of Seymour Papert (1980) with the language LOGO, illustrates the potential of harnessing the computer as a means of developing and extending the reasoning powers of the child. We only regret that space does not permit us to step outside the area of 'instruction' in order to analyse more fully the potential of the computer as an 'extension of the brain'.

11. Initial Design and Development: The Educator's Role

11.1 Materials design for CAI: the initial stages

Depending on the aspects stressed, the development of CAI materials may be presented as an extremely new and specialized type of instructional design, or it may be seen as one branch of the general instructional design process, which has its special techniques and skills at the detailed design and production level (as has, for example the development of instructional audiovisuals) but is in general terms very similar to the design of any other instructional materials. We shall attempt to show that the latter view is perhaps closer to the truth and, certainly, more useful as a starting point into this relatively new instructional technology.

In *Designing Instructional Systems* (Romiszowski, 1981) we developed a four-level model of instructional design. We furthermore based our model on the application of the *systems approach*, conceived as a heuristic problem-solving process composed of five basic stages, or activities. The systems approach can be seen as the basic model at each of the four levels of instructional design.

This led us to present the total instructional design process as a complex *matrix* of questions and decisions. We tried to illustrate this matrix, at various points, both in this and the earlier book, in order to emphasize specific aspects of the total instructional design process. These visual representations were not always identical in the detailed comments put into the cells of the matrix, for in each case, the content of the cells is but a sample of the possible content – the instructional design process is in reality so complex and involves so many interrelated questions and decisions that it defeats any effort to 'map it completely' in one chart, however detailed. Nevertheless, we believe that the 'matrix' concept and attempts to partially represent, on this matrix some important aspects of the instructional design process, are useful aids to communication and help us – author and reader – to compare our cognitive schemata on the subject with a bit more precision than is possible with plain prose.

We therefore present, in Figure 11.1, yet one more version of our matrix, illustrating this time, within the context of the overall decision-making process, the principal special decisions and tasks that have to be performed in order to design, produce and effectively utilize CAI. In order to highlight the special aspects, we have used the expedient of circling, or ringing, those items that refer specifically to CAI design/production/use. As, very often, a CAI implementation also involves some aspect of score-keeping, grading, etc, we have included in the diagram, references to 'CAI/CML' systems.

Decisions to use CAI are taken at Level 2

First of all, we note that the CAI/CML decisions generally are taken well into the instructional design process – at Levels 3 and 4 principally. At Level 2, initial decisions may be taken *as part of the general methods/media selection process*, as to whether CAI/CML has any role to play in the course being planned. These decisions are, of course, taken on grounds similar to those for the selection of any other methods and media – instructional effectiveness/efficiency, cost, production factors and utilization factors. In some projects, overall media decisions are taken at an earlier stage – Level 1 of the design process. This occurs when the project designers decide right at the start to exploit a given medium, as might be the case of a project specifically launched to exploit the potential of Prestel or other videotext systems for education. In such cases, the instructional designer has the

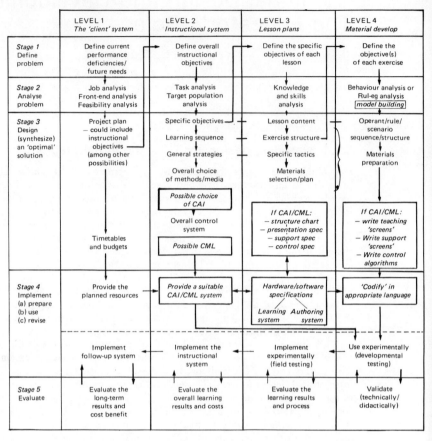

Figure 11.1 *The overall instructional design process, illustrating
the points at which decisions regarding CAI/CML
are taken and a system is designed*

'back-to-front' task of selecting his objectives in relation to the capabilities of the
predetermined instructional media. However, we shall, for the moment, ignore
these 'solutions in search of problems' projects (without necessarily condemning
them as always 'a bad thing') and shall concentrate on the 'instructional design as
a problem-solving process' type of project.

In such a project, CAI/CML might appear as a promising alternative during
the initial selection of methods and media, at Level 2. This decision of course
implies the provision of a suitable system (composed of hardware, software and
courseware). The hardware (computers and special peripherals) and software
(operating systems, course authoring languages and/or general programming
languages) will have to be selected and bought from among the alternatives on the
market (or already available to the organization). The courseware (the CAI
lessons, support materials, documentation, testing and management procedures)
will also have to be selected and bought or, more probably, will have to be
specially developed. This is the principal topic of the present chapter. However, in
passing, it is worth mentioning that the courseware design process defines, to some
extent, the hardware and software needs. It is therefore too early, at Level 2 of
the design process, to define these needs fully. Notwithstanding, it is necessary to
make some estimate at Level 2 of the cost of these needs (as well as the cost of
courseware development) in order to budget for these items correctly in the overall
project.

Planning the CAI lesson at Level 3

Passing now to the lower levels of the instructional design process, we note that the planning of a CAI lesson, at Level 3, is not very different from the planning of any other lesson. A detailed lesson plan of the type we suggested in Chapter 3, presents the lesson's structure in a way that shows quite clearly the *information* to be transmitted by the teacher to the learners (Teaching activity), the *responses* or *tasks* that the learners should perform in order to learn and to *show* the teacher that they are indeed learning (Learning activity) and the actions to be taken by the teacher to reward and reinforce learning, or to correct errors, reorganize conceptual schemata and so on (Feedback activity). The same approach may be used to the initial planning of a CAI unit. The only difference is that other media (mainly the computer) take on the teaching and feedback activities and the learning activities must take a form which results in input by the learner to the computer. We say 'mainly the computer' for there is nothing against our developing a 'mixed' CAI unit, involving the use of printed materials or audiovisual media as parts of the overall sequence. Indeed, research shows that such mixed-media designs are often much superior, in terms of learning results, as well as being more motivational and sometimes cheaper and quicker to produce. In such 'mixed' units, the large units of information presentation are packaged in print or other 'off-line' media, whilst the overall control system and much of the 'Feedback activity' is computer-based.

Methods of presenting the lesson plan

This initial planning can, indeed, be performed using the same three-column planning sheets that we recommended for conventional lesson planning. However, it is probable that some CAI lesson designs will be quite complex, expecially in the number and variety of feedback activities that may be offered, depending on the particular response pattern of an individual learner. In such cases, the three-column format may be a bit clumsy and flow-charting, similar to that used by computer programmers for mapping-out a problem-solving algorithm, may be a clearer method of expressing one's intentions.

Figures 11.2 and 11.3 illustrate examples of a simple design, using the standard planning sheet and flow chart methods. It is worth noting another possible advantage of using flow charts to illustrate the lesson structure of a CAI unit. It may often be the case that the instructional designer acts as the subject matter and teaching method expert in a team that also includes computer programming experts who take over the final development/production stages of transforming the design into a working lesson. In such a case it may be an advantage for the instructional designer to adopt a method of presentation that is familiar to the programmer. However, most programmers could work from either of Figures 11.2 and 11.3.

Other documentation produced at Level 3

In addition to the basic lesson plan, or 'structure chart', the instructional designer will, at Level 3 of the design process, be in a position to specify a number of other important aspects of the system under design. These include the following documentation:

(a) Presentation spec.

* *The 'presentation specifications'* – the output characteristics that the system must offer in order to be able to present the lesson being planned, the size and layout of the output on the screen, whether any special split-screen or 'windowing' techniques are necessary, any other special output characteristics such as graphics capability, colour, sound or timing control (use of a computer-clock), special input characteristics or devices that must be available to the learner, such as joysticks, paddles, special keyboards, graphic pads, light pens and the like. The example in Figure 11.4 is for the simple addition drill-and-practice exercise. Hence the specifications are not very sophisticated.

(b) Support spec.

* *The 'support material specifications'* – any form of *help* that the learner should have available to him in order that he may use the system and the courseware correctly; any form of supplementary and reference materials that should be made available to the learner to enable him to review technical terms (a glossary), read further (a bibliography), get a quick

	Stimulus information	Learner responses	Feedback control
	Introduction Present an explanation of the objectives of the exercise. Demonstrate three examples of sums (types a + b = ?)	Sign-on in standard manner. Read the opening presentation.	— — — — — — —
	Practice cycle Present a randomly generated problem of the type 'a + b = ?' where both a and b are single-digit numbers. (This cycle will be repeated as often as is necessary to reach the criteria specified in the feedback column.)	Key-in the answer to the problem presented. Check and correct before finally entering it into the computer.	*If error?* Present the correct answer and then repeat the cycle, with a new addition problem. *If correct?* Give confirmation of response, then: *if* last five problems were correct *and* at least ten problems have been solved, *then* present the 'sign-off' screen. *Otherwise*, repeat the cycle, presenting a new problem.
	Sign-off Present to the student (a) number of problems attempted; (b) number of problems answered correctly; (c) a suitable remark of congratulation or motivation; (d) sign-off instructions and say 'goodbye'.	Sign-off in the standard manner.	— — — — — — —

Observations:

1. We have used our usual three-column planning sheet, though with slightly modified column titles, as there is no teacher or instructor involved. The left-hand column describes the information that will appear on the screen (as will some of the feedback information, presented in the right-hand (feedback control) column).

2. We have drawn arrows, as an afterthought, to better illustrate the cycle of practice.

Figure 11.2 *A lesson structure for a simple drill and practice exercise on addition of integers*

overview (introduction and summary pages), get a deeper view than is obligatory (a database or 'expert system' available for 'interrogation' by the learner), learn or revise key concepts which are prerequisites (diagnostic testing and remedial sequences not included in the mainstream of the lesson), or even student 'mail' or 'comment' facilities (optional comments that may be made concerning the lesson, by the learners, intended to be read by the author, the instructor or other learners, in order to comment on personal difficulties, disagree with viewpoints in the material or simply open up the discussion). See the example in Figure 11.4.

(c) Control spec.
  * *The 'control systems specifications'* – any form of control that may be exercised over the process, or the results, of study; a specification of the way that individual students' responses should be recorded and classified in computer memory; the way that cumulative records should be kept of error rates of

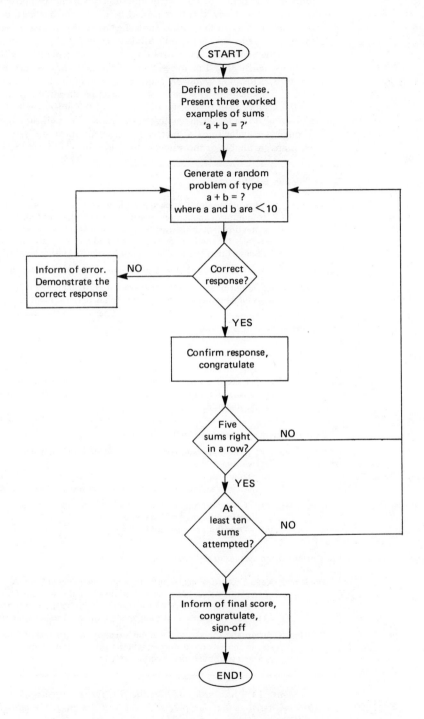

Figure 11.3 *A flow-chart version of the lesson structure shown in tabular three-column form in Figure 11.2*

individual students, or on specific questions; whether response times should be recorded and how this data should be used; whether all this data should be available only to the author/controller or to instructors at key points in the lesson, or to each individual student as and when requested, etc. All this can be considered as control data needed to generate feedback to students, teachers and/or the authors of the materials. Another form of control specification would be required to define the guidance control to be offered – indications of what to study next; of alternative materials available for the achievement of specific objectives; of the objectives to be achieved in a given unit/lesson; of alternative study routes through the objectives or through particular units of material. These specifications may be thought of as 'maps' available to the students to guide them through the study unit.

Presentation specifications

Output: Screen presentation, capacity minimum ten lines of text (for the introductory explanations and demonstrations) and four lines of text for the problem drills (two lines for the instruction and the problem and two lines for the feedback comment and correct answer). Preferably 20 lines of text so as to have more freedom of visual layout.

Input: A ten-key numerical pad is sufficient, as all responses are numbers.

Effects: No colour or special effects are required, though they may be used, if available (especially colour to emphasize the student's response).

Support material specifications

Essential: None.

Optional: If the screen size of the equipment to be used is very small, it may be more convenient to print the introductory explanations and the sample demonstration problems on a card or folder. No responses are expected to this material.

Control system specifications (minimum specs.)

Feedback: (a)   Immediate feedback of correct answer to the student after every problem.
           (b)   Summary of attempts and errors/number correct supplied to student at end of exercise, on screen (no printed hard copy).

Guidance: (a)   'Why and how to study' — on the presentation screen(s), or optionally in a printed guide.
           (b)   After the exercise, none.

(Optional CML system to be considered):

Feedback: Keep a record of each student's score in a 'personal' file. A 'profile' of the student's success record should be available to staff on request. Also a 'mean' profile of all the students who have taken the series of exercises. (Use for identifying and helping the weaker students.)

Guidance: Automatic analysis of a student's record on a given skill. The sign-off screen should advise student whether to continue on current exercise or proceed to some other specific exercise.

Figure 11.4 *Examples of possible further specifications prepared by the instructional designer at Level 3 of design of the addition drills*

We note that so far we have not really specified anything that we could not have specified for a non-computer-based study unit of individualized materials. The lesson structure is like a normal interactive lesson plan, or a plan of a

Note the
analogy to
good
instructional
design in
other media

simulation-game session, or a flow chart of the frames that we plan to write for a programmed instruction chapter. The presentation specification is equivalent to a detailed listing of the materials, presentation media and special equipment and materials that we must assemble before the start of any lesson. The support materials specification is equivalent to the preparation of bibliographies, glossaries, user manuals, job aids, specialist libraries of selected texts, diagnostic tests and remedial self-study units or student-evaluation questionnaires, for any other form of instruction. The control system specification is akin to the design of any other course monitoring and management system, as regards the 'control by feedback' aspects and similar to the design of a study guide for any course, as regards the 'control by guidance' aspects. In order to perform these design tasks, the instructional designer needs only to have a basic knowledge of the capabilities of CAI systems in respect to the aspects discussed, in order to exploit available capabilities to the best advantage and not to design into the system any non-existent capabilities. Other than this, well developed skills in the normal tasks of instructional design, already discussed at length in Parts 2 and 3 of this book (and in *Designing Instructional Systems*) will suffice.

Minimum configuration (for mass manufacture)

A 'Little Professor' special purpose teaching machine, the size of a small calculator, with a liquid crystal screen of four lines of text (could be less with a little re-designing of the lesson), ten numerical keys and an 'equals' key for inputting the result (probably a 'cancel' key for correcting mis-keyed responses before inputting the final answer). This configuration would have the necessary programming resident in the circuitry, as part of the microprocessor 'chip' that would probably be specially made (the machine could easily be made with several different arithmetic drills available on request, by keying in the appropriate drill identification code number. All instructions in a printed leaflet.

Minimum configuration (for use of equipment on the market)

Any small personal computer will do. If programmable in BASIC, then one kilobyte of RAM (random access memory) is sufficient for the drill described. A (Timex) Sinclair ZX81 (without expansion memory) would serve, coupled to any home TV and cassette tape recorder.

(Note: Neither of these configurations would be suitable for the implementation of the optional CML system. The 'Little Professor' lacks the circuitry, power, etc. The Sinclair is powerful enough to do this task for limited numbers of students (say 50) if equipped with a 16K memory expansion, but due to the storage of programmes and data on audio cassette, it is awkward to keep records updated from one study session to the next. (All records would have to be loaded and then re-recorded, at slow speed.)

Ideal configuration (justified if other uses are envisaged)

Any home computer with minimum of 32K (preferably 48K) of RAM, two disc drives and a colour monitor and at least four-colour capability. Will handle all the optional CMI functions for large numbers of students. Can offer more attractive presentation, with colour for emphasis. Can have several alternative drills available, rapidly, on request by student or automatically under CML control, stored on one diskette and will automatically update the student and course files, on the other diskette. The CML data is available at any moment, on the screen and, if a printer is available, a hard copy of the data could be requested, for detailed analysis.

Figure 11.5 *Some possible system specifications, drawn up
on the basis of the documentation so far presented
by the instructional designer, for the proposed addition drill
exercises (could be prepared by a computing expert)*

(d)The
system
specification
One more set of specifications should be developed during the Level 3 design stages. This is the detailed *specification of the system required to present and operate the courseware being designed*. The four types of specifications discussed above are sufficient to define the *hardware* (computer, memory, size, peripherals, speed, etc) and the *software* (programming languages and/or authoring systems and/or operating systems, database systems, etc) which would be ideal for implementation of the courseware being designed.

These specifications, of course, require a good deal of specialist knowledge as regards the systems, costs, reliability, 'user-friendliness' and other factors involved in making a well balanced choice from the systems available on the market. This specialist knowledge can of course be 'bought in' by asking a computer expert (not necessarily a programmer) to examine the instructional design specifications and transform them into appropriate hardware/software specifications. It is a good idea, however, for the instructional designer to gain at least a reasonable grasp of this body of knowledge in order not to fall foul of the 'salesman' type, eager to promote a particular type of equipment or the well-intentioned enthusiast eager to implement the latest 'All-Purpose-Accelerated-Learning Language for Educational Design' (APALLED?).

This aspect of design is one that is in a state of constant development, with new products and software hitting the market on an almost daily basis. The only way to keep abreast is to do a fair amount of regular reading of specialist journals. On the other hand, the problem may be more academic than real in many situations, where the available hardware and software options are known in advance (as in the case of instructional design for a particular make of computer system already installed). The instructional designer's task reverts to producing lesson structures, presentation specifications, support system specifications and control system specifications that do not step outside the bounds of the currrently available computer technology.

11.2 CAI materials development: the ID task at Level 4

We now have an overall lesson structure and a specification of all the components/sub-systems that will make up the lesson. Probably we have planned a series of lessons that make up a CAI course-unit, some of the support and control systems being common to all the lessons in the unit, whilst other design aspects, of presentation, support or control, will be peculiar to a particular lesson. We now face the task of developing the materials. This is a 'Level 4' instructional design task. Once again we may refer to Figure 11.1 to see the general flow of the task and to identify those steps in the task that are 'special' to the development of CAI materials.

Simple
exercise
structures
may be
'coded' from
the Level 3
design
We note that the first stage, of defining the objectives (and the content, probably) of each step, or exercise of the lesson, has already been done in the mapping of the overall lesson structure, as part of the Level 3 of design. We may sometimes require to do a little more analysis of these exercise specifications in order to decide on an appropriate exercise structure.

For our addition drilling example, the lesson plan, or structure chart, specifies the objective 'add any two single-digit numbers correctly', the mode of 'drill-and-practice', the structure as 'a set of randomly generated addition sums', the control 'practised until at least ten have been performed and at least the last five have been answered correctly'. The guidance control will be an introductory display of the 'rules' of the exercise. No support materials or systems are envisaged. In order to write materials for this objective and exercise specification, very little further analysis is required. One could proceed almost directly to the design of the introductory 'frame', or rather 'screen' of general instructions, followed by the general layout of the question 'screen' that would present each of the problems to be solved, the confirmation or correction comments that should appear on the screen, depending on the learner's (correct or incorrect) responses and finally the 'control' screen that would show the learner how he is progressing (how many sums attempted to date, how many were correct, how many error-free attempts were achieved in a row). We shall come back to this example in more detail later, showing that the design of these three or four basic 'screens' is sufficient to define what the learner will see as he works through the drill exercise.

The only remaining materials development task is the clear specification of the 'control algorithms' that serve to guide the programmer, who will 'codify' the exercise by means of an appropriate programming language or authoring system. This 'coding' task may be performed by the instructional designer, or may be passed on to a programmer who should be able to follow the documentation that the designer has produced (on paper) to produce an exercise that 'looks' and 'works' just as the designer had planned. In our simple addition drill example the flow chart, or the lesson plan (Figures 11.2 and 11.3), could act as the control algorithm.

More complex exercise structures may need further analysis

If, however, the exercise structure were a bit more ambitious, then some further analysis of the exercise objective and content may be required. For example, if the learner were to be free to define the range of numbers to be added (he can define the largest number that might appear in a problem and this might be between 0 and 99), or if wrong answers were to be followed by a presentation of a simpler grade of problem, or if the learner were to have access, at will, to support materials that present a suitable level of 'worked example', then it is not yet clear how the exercise should be designed. One problem to clarify is what exactly are 'simple' and 'complex' addition problems. Another is what do learners typically find difficult – what are the typical mistakes that they make? And why – what do they confuse? The first problem might be analysed by some form of topic analysis, aimed at defining the procedure of addition as a sequence of steps (referred to as 'rules' in the Rul-Eg system of topic analysis). The one important rule which complicates the structure of an addition problem is the 'carrying' of tens to the next column. In the case of adding fractions, or subtraction or long division, the number of rules is much greater and so the 'scale of complexity' of all possible problems is more extensive, with various part-way positions between the simplest of problems and the most complex.

The sources of typical errors made by learners are not always directly associated with this scale of 'complexity'. Strange, difficult to explain, psychologically based difficulties sometimes creep in. For example, some research showed that many children have a higher probability of errors in arithmetic problems that involve the digits 4 and 7, than for other digits. Other children have particular difficulty in handling zeros or nines. The reason for this may be the physical shape similarities between 4, 7 and 9 that lead to perception errors when under stress. The difficulty with zero may on the other hand be one of concept formation. An analysis of the frequency and the probable causes of errors of this type is a form of behavioural analysis, which would possibly utilize some of the detailed 'further analyses' illustrated in Chapter 4.

An example of a more complex control algorithm

We may be exaggerating the depths of analysis really worth carrying out in the particular quite simple example that we are considering. However, it was exactly this sort of approach that was used by Patrick Suppes at Stanford University when performing the first large-scale research into computer-assisted drill-and-practice exercises for arithmetic skills (Suppes *et al*, 1968). Suppes analysed the tasks involved in most of the basic arithmetic procedures and classified problems in each task on a scale of 'complexity'. He also analysed typical errors committed by hundreds of school children, being thus able to classify problems of a given complexity according to level of difficulty. In this way a 'matrix' of problems may be constructed, classified in the two dimensions described, as shown in Figure 11.6.

A particular learner receiving a particular problem (say problem 'A') at a given level of complexity and difficulty, makes an error. This error is used to assess what level of problem the child should receive in order to ensure success, but not waste time on problems that are easier than his/her current skills level. Analysis of the child's pattern of errors may be used to select the next problem at a lower level of complexity (say problem 'B') or at a lower level of difficulty, but the same

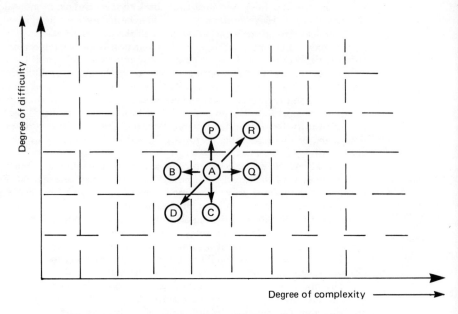

Figure 11.6 *A matrix for the classifying of problems according to complexity (number of different operations to be carried out) and difficulty (probability of error)*

level of complexity (say problem 'C'), or at a lower level on both scales (say problem 'D'). Consistent correct responses at a given level of complexity and difficulty, on the other hand, lead to the advancement of the learner to higher levels – to 'P', 'Q' or 'R', for example. The algorithm for deciding in which direction to advance in the matrix of problems, and how far, may be quite simple and mechanistic (one up for each correct answer, then one right for the next – one down for each error then one left for the next). It may, however, be much more sophisticated, based on a given learner's history, or 'pattern' of errors, which is compared to standard patterns that have been found, in past experience, to indicate particular weaknesses, best remedied by particular sequences of practice problems.

A self-improving control algorithm

This 'pattern recognition' and matching to 'optimal problem sequences' may be researched and improved by the same computer program that is presenting the drills to learner. In this way, the program furnishes the research data needed to further refine the control algorithms built into it. The program is thus, to a certain extent adaptive, or at least, self-improving, in that it can 'learn' to do a better job of selecting appropriate practice problems for a given learner. This type of design, and many similar ways of controlling and improving drill-and-practice exercises in arithmetic skills, were researched by Suppes *et al* (1968).

The paper simulation of a CAI drill

Thus we can observe that even the apparently simple 'drill-and-practice mode' of CAI can often be much more sophisticated than meets the eye. However, it is important to point out that the computer only makes it easier, faster and more efficient to apply a predetermined algorithm of decisions on how to react to particular error patterns. The same algorithm could, in theory, be implemented by a teacher in the classroom, administering problem sheets classified according to the parameters of our grid, marking and checking the errors and error pattern and using this data to select the next problem sheet. This was, in fact done by several researchers in the early days when computers were scarce and very expensive. Dodd and Sime, at the University of Sheffield, carried out much research in the early 1960s on experimental CAI designs, 'simulating' the computer, by means of putting a learner at a terminal, which connected to one in another room, manned by the experimenter who 'played' the role of the computer, behaving just as the computer was supposed to. As far as the learner was concerned, he was learning in exactly the same way as if at a computer terminal. Various types of skill training were researched in this manner. Among other applications, Bernard Dodd developed a set of fractions problems classified on a difficulty/complexity matrix such as the one we have illustrated. The problems were administered in this simulated-computer situation replicating Suppes' expensive research at Stanford (Suppes, 1968). The drill-and-practice exercises were most successful. Much was learnt about the construction of such exercises. The materials were later used (without the pretence of a 'computer out there in the other room') in the form of problem-sheets and self-marking feedback sheets, administered by a tutor or a secretary/clerk/office boy.

We mention this research here partly for its historic interest, but principally to illustrate that the development of what the learner sees and studies (we are calling this the 'screens' of information and instructions) and the algorithm of how the material should be presented and how different learner responses to the material should be catered for, may be prepared without access to a computer and without the use of any programming or authoring language.

CAI materials are developed first on paper

Not only *can* CAI material be designed and developed in this 'off-line', paper-and-pencil manner, but it *should* generally be developed in this way. Even when a fairly easy-to-use authoring language is being utilized, one needs a paper 'script' of the courseware to be produced before sitting down at a terminal or a microcomputer to transform one's plans into computer code. There are many advantages to this way of working:

(a) the courseware can be analysed and technically validated in its paper version, as a series of pages, or 'screens' that will be studied by the learner;

(b) it can often be validated for learner acceptability and teaching effectiveness in its paper version, by 'simulating' the computer presentation, just as Dodd and Sime did in the above mentioned experiments;

(c) up to this point in the development process, the instructional designer needs no programming skills or experience in the use of any authoring system – he does, however, need to have some knowledge of the capabilities of the programming language or authoring system to be used, in order to plan sequences, branches, random-access facilities to reference materials and so on, which are within the capabilities of the system to be used for implementing the courseware;

(d) if he so wishes, the designer may at this point delegate the task of transforming the paper plan into a computer programme to a skilled programmer, in the knowledge that what appears finally on the computer screen will turn out just as specified;

(e) to a large degree, the courseware plan at this point of development is independent of specific hardware/software. It has not yet been programmed in one highly specific dialect of BASIC, or put together into screens by means of an authoring system that only works on a particular manufacturer's equipment. It is hardware-dependent to the extent that it uses facilities (say, high resolution graphics, or sound effects, or colours) which are available only on some computer systems. But any system that offers the planned facilities might be used to implement the courseware. The degree of software 'compatibility' is not always great, in reality, when some of the more specialist programming languages (such as LISP, or SMALLTALK or PASCAL) are initially considered and then the decision is taken to implement the courseware in, say, BASIC. The structure and programming possibilities of the currently available languages differ greatly from one another. But when a courseware package is to be implemented in, say, BASIC (of which there are no end of dialects), the one paper 'script' would serve to produce courseware in any one of the dialects and would, indeed facilitate even the adaptation of a programme from one language or dialect to another.

For all these reasons, it is worth spending time on the initial paper-and-pencil design of the intended package.

The similarity to other instructional materials development procedures

Thus we see that in yet one more way, the development of a CAI courseware package is not that different from the development of any other package of individualized instruction materials. We have seen that the same procedures for specifying objectives and deriving useful content may be used. The same techniques of 'deep' (Levels 3 and 4) analysis may be used to clarify what is important to teach, why and how it should be taught. We face the same choices between expositive and experiential strategies and the existing modalities of CAI offer us alternatives that apply each of these strategies. At a more detailed planning level, we face the same decision problems of choosing our words, our examples and non-examples, our analogies, our visual illustrations and other aids to clear and complete communication. We face the same decisions regarding the best methods to control and evaluate the course as a whole and the progress of individual learners through the course. We look at the content and the learners carefully to decide who should direct the learning – should it be 'autocratically' controlled, or is a 'cybernetic', adaptive approach, which takes individualization decisions on the basis of individual sucess, more indicated? Or should the learning be entirely student-directed? Should we apply norm-referenced or criterion-referenced methods to the measurement and evaluation of learning results?

All these questions are part of any serious instructional design project. They occur in just the same way in a CAI materials development project. The only difference is that the computer, especially a mid-range mini or larger, with 'bags' of memory, opens up so many new ways of implementing the decisions we take in reply to our questions. The computer is an all-pervasive medium – it may present information, may provide the 'scenario' for interactive learner activity, it may provide feedback in so many different ways and it can implement the most sophisticated control/evaluation systems.

An example of the design of an addition drill

As a simple example, we may continue with the addition drill-and-practice exercise already discussed. Let us assume that our intention is to prepare a courseware package to be used on simple, inexpensive microcomputers. In order to plan the layout of the information we would like to present to our learners, we must make certain assumptions regarding the screen size and the facilities that it offers to the programmer. Some systems have a screen size that can be defined as '40 column – 24 line' meaning that a total of 24 lines of up to 40 characters each, may be presented on the screen at one time. Other systems offer larger screen sizes – a common standard on business computer systems is '80 column – 24 line'. Yet others have even more restricted screen characteristics – the original 'Little Professor' of Texas Instruments, presented its arithmetic drills in one line only, similar to the display window of a hand-held calculator. As we have already noted, such a restricted screen size would not be appropriate for our proposed courseware, which plans to present messages of motivation and congratulation and some worked examples, as well as the exercise problems. If we have done our Level 3 design as suggested earlier on, we should already have defined the screen size requirements in our presentation specifications.

The question of screen size

In the sample specifications presented in Figure 11.4, we suggested a minimum of ten lines, but a preference for 20 lines of text, so as to have ample freedom of visual layout of our information.

Let us assume that, as we wish to make our package widely available, we shall opt for the smallest normally available screen size on the most widely used home computers. Our systems specifications, presented in Figure 11.4, suggest the use of a Timex-Sinclair ZX81. This is thought of as a hobbyist's, or even 'toy', computer and is not among those officially recommended for school purchase. However, it is not as much a 'toy' as may at first appear. We have selected the Timex-Sinclair as our example, exactly because of its simplicity. Although there are, of course, very strict limitations to what can be done, it is interesting to see that some CAI can be implemented. In the UK, the Sinclair is probably the most widely used of the small home computers and is certainly the cheapest. Its screen size is '32 column – 24 line', which is just a little more than the desired specifications. If we plan our information layout for this size of screen, then it will look alright on screens of that size and slightly bigger. For example, in the case of a '40 column – 24 line' screen, our material will still be well planned, provided we centre it (print everything four columns to the right to allow for the eight extra columns). Our material will be somewhat under the maximum size for the screen, but the extra margin space round the edges is small and may even improve the appearance and readability of the message. It would be a different matter, of course, if we resolved to use a system that had 80 columns by 24 lines, as the layout we had planned for the Sinclair would be using less than half of the screen width. In such a case, it would be better to replan the layout of the information to take advantage of the space available. To this extent, therefore, we are now entering a phase of materials design which is not totally independent of the hardware to be used for presentation.

The layout of the screen

In Exhibit 11.1 we present the 'key screens' designed for our hypothetical addition drill-and-practice exercise. The instructional designer has decided to plan, in principle, for the Sinclair, or other equipment of a similar screen size. Note that the plans have been drawn on squared paper, which has 32 columns and 22 lines. In so doing, the designer has simplified the task of planning the layout of the screen. Each square represents a legitimate position for a letter or other character. It is just a question of planning the information to fit the space (22 lines because the bottom two lines are unavailable for printing messages).

Exhibit 11.1 *The 'key screens' for the drill-and-practice exercises on simple addition*

(A) The introduction – objectives and how to study

```
        ADDITION   PRACTICE

   I  SHALL  SET  YOU  A  NUMBER
  OF  ADDITION  SUMS  LIKE  THIS

             4 + 3 = ?

             YOU  SHOULD -
 *1*  TYPE  THE  ANSWER  TO  THE  SUM

 *2*  CHECK  WHAT  YOU  HAVE  TYPED:
      [LOOK  AT  WHAT  APPEARS  HERE]

        *3*  USE  THE  DELETE  KEY  TO
             RUB  OUT  ANY  ERROR  AND
             THEN  RETYPE  CORRECTLY

        *4*  THEN  PRESS  NEWLINE  TO
             TELL  ME  YOUR  ANSWER

   7     OK?  PRESS  **S**  TO  START
```

(B) The introduction — three worked examples

```
JUST    TO   HELP  YOU    REMEMBER    WHAT
YOU   HAVE   LEARNT   ABOUT    ADDING,
HERE   ARE  3  SUMS   OF   THE   TYPE  THAT
I  SHALL   ASK   YOU   TO   DO.  LOOK   AT
THEM   CAREFULLY   AND   REMEMBER   HOW
THEY   SHOULD   BE   ADDED.

    1 + 7 = 8          7 + 0 = 7          7 + 5 = 1 2
    ▬▬▬▬▬              ▬▬▬▬▬              ▬▬▬▬▬

        1                  7                  7
       + 7                + 0                + 5
       ▬▬▬               ▬▬▬                ▬▬▬
       = 8                = 7                = 1 2
       ▬▬▬               ▬▬▬                ▬▬▬
THIS   ONE        ADDING   A         SOMETIMES
IS   QUITE        ZERO  HAS          YOU   GET   2
EASY              NO   EFFECT        DIGITS   IN
                  ON  THE  SUM       THE   SUM

     PRESS   **C**   TO   CONTINUE.
```

(C) The stimulus information. Each problem is presented to the learner in the same format; only the numbers in the circles drawn with dotted lines change in each successive problem. Note that the learner has typed in a tentative answer, which appears at the bottom left and which happens to be wrong.

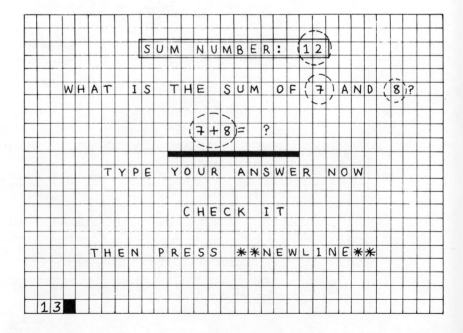

(D) The response and the feedback information. This is the case of a wrong response. Note that the only change in the upper part of the screen is that the answer '13' has appeared in the sum. The bottom half of the screen supplies corrective feedback and instructions on what to do next.

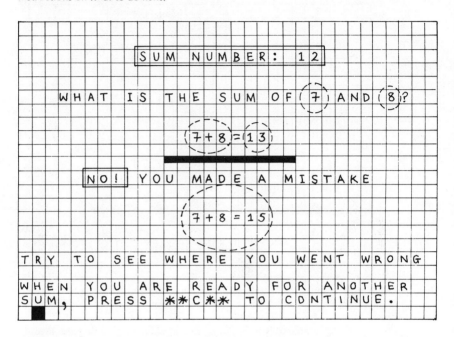

(E) The complete screen, after a correct response. As in the previous case, this screen appears in stages — first the stimulus information, above the horizontal bar, then the student's response (first at the bottom left, as it is typed, and then in the 'sum' when entered by pressing 'NEWLINE') and finally the feedback, which this time is confirmatory.

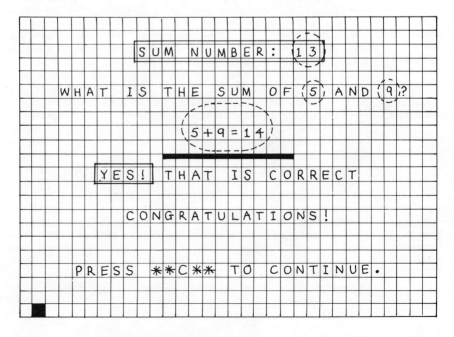

(F) The final 'sign-off' screen. Note the parts of this message marked by dotted circles — these parts change depending on the student's performance on the exercise.

```
                    | VERY  GOOD! |

 YOU   GOT   THE   LAST  5  SUMS   RIGHT!


 IN  THIS  EXERCISE  YOU   HAVE:

 ATTEMPTED  IN  ALL  ( 17 ) SUMS,
 ADDED  CORRECTLY,  ( 12 ) SUMS  AND
 MADE  MISTAKES  IN  ( 5 ) SUMS.

 I  THINK  YOUR  GENERAL  SCORE  IS
(FAIRLY  GOOD.
 YOU  COULD  IMPROVE  WITH  PRACTICE.)

 DO  YOU  WISH  TO  PRACTISE  NOW?
 TO  DO  ANOTHER  EXERCISE, KEY  **Y**

 TO  SAY  GOODBYE  FOR  NOW, KEY  **N**
```

Well, of course it is not quite that simple. In planning the layout and the other aspects of the presentation, the designer should take into consideration several other presentation factors which may, or may not, be offered by the hardware to be used. And, of course, the pedagogical aspects of the visual layout and structure of the messages should also be taken into account. We shall consider some of the hardware-related factors first and then shall come back to the pedagogical considerations, making reference to Exhibit 11.1 to illustrate our points.

Some hardware-related aspects of screen design

Our hypothetical instructional designer has set out to produce a package principally with the Sinclair computer in mind. He or she has shown knowledge of many of the limitations and some of the benefits of this piece of equipment, in the layout plans suggested in Exhibit 11.1. Some aspects have also been misunderstood, or incorrectly used. We shall discuss these aspects, in turn.

-upper or lower case

Firstly, as the screen plans show, only upper-case letters will be used. This is a limitation of the Sinclair and of many other simple home microcomputers. However, the trend now is towards the availability of both upper and lower-case letters on most 'serious' computers. In the case of planning for various types of equipment, or when the ultimate hardware to be used is not yet known, the designer could plan the messages in upper and lower-case, to take full advantage of the possibilities for emphasis and greater clarity that this offers. In the case of implementation on a system that only presents upper-case, the programmer would merely ignore this aspect of the plan, treating all the message as if it were in upper-case.

-methods used to emphasize words or phrases

Secondly, the designer has shown knowledge of the Sinclair computer's capability of emphasizing words or characters by printing them in 'inverse', or white-on-black. This gives the effect of a black oblong with the appropriate message appearing in white letters. This has been suggested for most titles (eg 'ADDITION PRACTICE' on screen A) and has been indicated by putting the appropriate characters in an oblong box. The technique has also been suggested to emphasize some key-words in the text (eg DELETE; NEW LINE; YES; VERY GOOD).

Our designer has, however, shown an incomplete knowledge of how the computer actually produces the inverse characters. An inverse 3 is not the same character, for the computer, as a normal three – they are two separate characters with two different codes in the machine's language. Thus when, as on screen C, the designer has specified that the whole title 'SUM NUMBER: 12' should be printed in 'inverse' characters, the programmer will encounter a problem. It is easy to instruct the computer to print 'SUM NUMBER: 12' in inverse characters, so long as the message always is to be that. But, as the dotted circle indicates on the plan, the '12' changes every time that the student gets a new problem (keeping count of how many problems have been attempted). So really, the programmer must instruct the computer to print 'SUM NUMBER: X', where X is the number of the problem (a variable that is being calculated by the computer, as it gives new problems to the student). So the computer must have a register reserved somewhere in its memory, that keeps tally of the number of problems given. At some point, this is '12' – but if asked, the computer will say 'twelve' the only way it knows how, by printing a normal '12'. All this is to explain that it is very easy to programme the Sinclair to print

> SUM NUMBER: | 12

where the words are in 'inverse' white-on-black, for emphasis, but it is more difficult to programme it to present the whole title in 'inverse', like this

> SUM NUMBER: 12

which is what the instructional designer's plan is asking for at the moment. This is not, of course, a great problem. The designer will probably accept the easy solution as just as good as the original idea. If so desired, some other way of emphasizing the number could be used – for example, building a 'box' around it, or underlining. However, the importance of this number is insufficient to warrant the effort and too much emphasis may indeed introduce 'noise' into the communication, by distracting the learner's attention from what is really important in the screen.

-the use of graphics capability

This last point brings us to the use of graphics. Our designer has shown some knowledge of the graphics capabilities of the Sinclair computer, in the use made of underlining and the attempt to draw an arrow down to the bottom left-hand corner, in screen A. Obviously, the Sinclair computer's graphics are of low resolution, all pictures being made up of 'quarters' of a square that is usually occupied by one character. Thus each square in the planning sheet can be divided, like a Battenberg cake, into four 'pictorial elements' or 'PIXELS', which may be either left white or filled in black (it is also possible to get some of them to go 'grey'). You can produce some very clever effects with these 'pixels', but being of low resolution, the designer could not see how to draw a decent arrow head on the arrow in screen A.

We present, in Figure 11.7, the complete set of standard graphics characters available on the Sinclair computers. Comparing these characters with the grids used to plan the screens in Exhibit 11.1, or the more informative grid presented in Figure 11.8, one can visualize the sorts of shapes that can be plotted. Horizontal and vertical lines are quite easy to produce, though they have a minimum width of half a character space (one pixel-width). Diagonal lines, on the other hand, are step-like sets of squares, rather than lines. The instructional designer in our hypothetical planning example has shown knowledge of this, in the way that the arrows and lines have been kept horizontal and vertical.

-parts of the screen which are 'out-of-bounds'

One point of misunderstanding, or possibly a slip of the moment, is shown in how the designer has planned the message presented in the introductory screen A. In trying to prompt the student to check his answer, the designer has asked for the figure '7' to appear alongside the cursor. However, a study of Figure 11.8 will show that the cursor actually is located on the bottom two lines, which are *not*

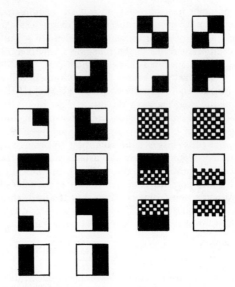

Compare this figure with the screen format shown in Figure 11.8, to get an idea of the degree of resolution possible when 'drawing' pictures or graphs.

Figure 11.7 *The graphic characters normally available on the SINCLAIR computers*

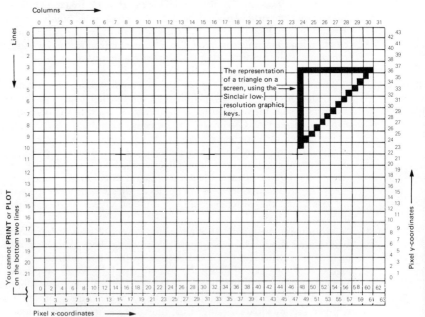

Note that the screen is really '32 column × 24 line', but the bottom two lines are not available for printing or plotting information. Normally, when information is typed in, it first appears in these lines (as does the cursor) and is only placed in other positions on the screen when 'NEWLINE' is pressed.

Figure 11.8 *The screen format of the SINCLAIR computers*

available for printing messages to the student. Indeed, in Exhibit 11.1, which is prepared on grids of 32 x 22, the cursor should not appear at all (as it does in all the screens planned in this exhibit). It may be impossible to do exactly what the designer has specified, but it is easy to think of other equally effective prompts, like saying 'look at what appears in the bottom left corner' or drawing a picture of what should appear, immediately above where it should appear, or changing the prompt to 'in this example you should have typed a 7 right here' with the arrow pointing out where to look.

The hardware-related aspects are not critical

We have left in these small errors to illustrate that the designer should have a reasonable grasp of the capabilities of the hardware to be used and of its limitations, but that an incomplete grasp of this technical aspect does not generally lead to serious problems. All the errors and weaknesses shown here can easily be corrected by the programmer, in discussion with the designer, during the codifying of the plan into an appropriate computer program.

The role of a pro-grammer as a 'media expert'

We also note that up to now, the instructional designer need not be an expert in any specific programming language. He may, indeed, stop at this point of the design process and pass the design documents to a competent programmer, in order to get the package coded and working as desired. It is for this reason that we have placed the codifying task as part of the implementation and preparation phase, in our matrix of the design process, shown in Figure 11.1, earlier in this chapter. To draw an analogy, the designer who takes CAI development as far as we have done in our example, is akin to the designer who produces the sketches for a set of overhead projector transparencies and then contracts a graphic artist to produce the final artwork, or the designer who writes a script for an audiovisual and plans all the slides and then passes the work of production to specialist photographers and recording studio staff. In all these cases, the original design, if detailed enough, should orient the media specialists exactly how the final product should appear and work. It is possible that these specialists may come up with their own alternative ideas on how the product should be made. These alternative approaches should not be implemented, however, without the original designer's approval. They may be good ideas, aimed at better exploiting the characteristics of the medium. Or they may be attempts at gimmickry or over-use of the special effects possible with the medium, to the detriment of communication effectiveness and learning efficiency. In the TV studios we meet over-use of camera movement and zoom effects, in audiovisual production the over-use of background music and in CAI the over-use of flashing words, colour and sound effects, just because 'we have them available'.

The role of the instructional designer as a 'methods expert'

All of these tendencies can be controlled and evaluated by the instructional designer, from a pedagogical viewpoint, without the need for the designer to learn how to produce the effects. It is necessary to know what effects are possible and to decide what effects are desirable in a particular case. 'What is possible' implies a knowledge of the medium to be used, as we have already indicated above in relation to the possibilities and limitations of the Sinclair computer.

Deciding 'what is desirable' depends on the application of the instructional design principles that we have been outlining throughout this book and its predecessors. We have, indeed, been leading up to this chapter, through our treatment of the design and development of other forms of individualized instruction materials, in the preceding chapters. We gave so much space to the analysis of alternative techniques of preparing programmed instruction, not only because they are useful for the preparation of programmed texts, but also because they are the basis for the design of CAI sequences.

Analysis of the screen design process – the introductory examples screen

Referring, once more, to the sample 'screens' developed in Exhibit 11.1, it may be interesting to comment how the designer set about developing the layout and the content of each screen. The process is very similar to that used in the design of mathetical exercises, in that one starts with a sheet of paper and 'draws' the message, starting with the most important parts. Thus, in Screen B, the first items of information to be put down on the paper were the three examples. These were chosen to illustrate the variety of problems that are to be encountered in the exercise. There are three examples not because there is just space for three (we could have used two or more screens) or because it 'seems' an adequate number, but because the designer judged that it was the minimum of examples necessary to illustrate fully the variety of problems to be included. Note that the numbers vary little (the '7' occurs in all three examples) in order to focus the student's attention on the important differences between the three examples.

The next part of the screen to be designed was the prompt following each example, which aims to focus, yet more exactly, the student's attention on the differences between the problems. Finally, the upper part of the screen was written, to make the best possible use of the remaining space in orienting the student and introducing the exercise. This part could, indeed, be cut to a fraction of its length, without weakening the exercise. It may be argued that the exercise would benefit by cutting some of this non-essential introduction. Alternatively, the programmer may comment that the memory of the computer to be used is inadequate for the storage of so much textual information (text uses up an inordinate amount of computer memory). If this is the case, the instructional designer can see at once what parts of the text could be 'pruned'.

The practice exercise

A similar approach was used to develop the screens of the main 'practice cycle' – screens C, D and E. In this case we can observe a yet closer analogy to the development of a mathetics exercise:

1. The first item to be planned and 'positioned on the page', is the *stimulus information* – the 'sums' (eg 7 1 8) which are printed as near to the centre of the screen as is convenient. Note that the examples will be generated by the computer – the designer need simply plan the space necessary to display them. (The general form of the examples has already been planned at an earlier stage and is contained in the lesson 'structure' chart or plan.)

2. Then the designer plans the space for the *response* that the student should make. This is indicated by the question mark in screen C.

3. The next part of the message is an *instruction* to the student, telling him what he should do. This is an important part of any frame of a programme that ensures that the learner is left in no doubt as regards what is expected of him. It could be argued that in a drill-and-practice exercise of this sort (that presents a series of problems of the same form) the instructions soon become superfluous. The instructions in screen C are made up of two parts – the question 'what is the sum of ...?' and the directions on when and what to type by way of a response. The first part of the instruction could be considered a legitimate part of every problem presented, being a verbal statement of the question that is also stated symbolically immediately below. As each form of statement may in future be encountered, it may be argued that both forms should be presented in every problem. (One could indeed argue that the verbal question is a part of the stimulus information.) The second part of the instruction, however, will soon be learned, as the student proceeds through the exercise and will thus become redundant. It could be argued that it should not be repeated in all the screens, on the grounds that it may slow down the students who insist on reading it, or that learning to 'skip' the bottom half of the screen may transfer to 'skipping' the study of the feedback information also presented there. The visual effect of a (feedback) message suddenly appearing in what was previously a blank space, is probably of more impact than the changing of a message. A student with a low level of attention may not notice that the content of three lines of text has changed, but he must really be asleep not to notice the 'flash' or three new lines appearing in an empty half-screen. For all these reasons, it may be a good idea to 'fade' the keying instructions after a few problems. This can be easily done in the program, by including a command that makes the printing of these instructions conditional on the number of problems so far presented – say, 'if the number of problems presented is less than five, then print ... (the instructions)'.

4. The final part of the screen to be planned is the *general context* or *title*. In this case, the designer has used this to keep the student informed of how many 'sums' he has so far attempted. This part is, of course, quite optional and the exercise would not suffer much if it were to be eliminated. However, our screen size permits one to put in a title and it probably adds some interest to the layout, if nothing else.

A model identical to a test frame in a mathetics programme

We note, therefore, that the components of the basic information screen of our exercise are identical, and were developed in the same design sequence, as was suggested for a test frame of a mathetically designed programmed sequence, by Gilbert way back in 1961 and refined by Imbucon Learning Systems in the UK during the late 1960s (see our discussion in Chapter 5). It should come as no surprise that our drill-and-practice exercise should follow the model for a test frame, as, after all, it is a sequence of test items, followed by immediate feedback (the feedback is the only aspect that differentiates our exercise from a test). Being a drill on something that was previously taught (by some other means), the exercise does not contain any teaching frames (especially full-blown demonstration frames). Screen B of the introduction could, however, be considered a form of 'prompt' frame, in that it recalls the layout of an addition sum and the main aspects of this level of addition sums that most commonly lead to errors. However, screen B is not a 'classic' example of a prompt frame (as discussed in Chapters 5 and 6) as it does not involve the learner in performing the behaviour that is to be mastered. This is reasonable, as we are assuming that the learner has mastered addition in some form of prior learning and now only requires further practice. There would be nothing against the development of some demonstration and prompt frames, as a teaching sequence, to be included on other screens at the beginning of our exercise, thus transforming it into a teaching *and* practice package. This would however complicate the programming somewhat, as we shall see later on.

The design of the feedback information

Before we leave the example presented in Exhibit 11.1, we should comment on the feedback information. Our screen size is adequate to devote half the screen to this information. This has enabled the instructional designer to include quite a bit more feedback than may be considered the minimum. The minimum, in the case of a correct response, our could be the one word 'CORRECT' and the instruction shortened to 'C – CONTINUE'. In the case of a wrong response, our designer wishes to furnish the correct response (considering that just saying WRONG is pedagogically inadequate). However, a message like

'NO! 7 + 8 = 15'

'C – CONTINUE'

would be quite adequate. The remainder of the feedback message is perhaps desirable, to stimulate the checking of one's errors or to motivate one to try harder. As this part of the message will be the same for every error committed, the designer must be careful to word it in a manner which does not become tedious to read. In the early days of programmed instruction, especially of 'branching' programmes, much was written on the alternative ways of telling a learner that he is wrong. We learnt to vary the wording so as to reduce the chances of monotony. In a fully programmed instructional sequence, one can relate the feedback directly to the type of error that was made in a specific problem. In a cyclic, drill-and-practice exercise of the type that we are analysing, the feedback message generally does not vary. It could do – it is very easy to programme the computer to print one of a selection of alternative error messages, picking them randomly for each wrong answer registered by the student. However, there would be no special relation between the message printed and the error made, just a variation of language to avoid tedious repetition. Many designers would question if it was 'worth the programming effort' and would aim to cut down the feedback to short and general statements, even simpler than the ones included in screens D and E, that will not pall with repetition.

Finally, we come to the 'sign-off' screen, that presents the student's score on the exercise and offers him the option to have another go, if he so wishes. In the example presented as screen F, the instructional designer has included a lot of text, using the available space to the limit. Once again, another designer might choose to keep the text down to a bare minimum. This will save valuable computer memory space (a consideration that is critical only when working with very small computers). It will also save reading time, however, and will make smaller demands on the student's reading capabilities (an important factor for a simple arithmetic drill, which may be studied in the very early days of schooling). It may also improve the layout of the information, using space to emphasize what is of the most importance. However, the alternative, of providing a fairly full evaluation of the student's progress, has its attractions. We could do without any comment that evaluates the student's score, leaving this for the student himself, or his teacher, to judge. (It could even be the basis of a competition between the students in a group.) However, the assessment and the suggestion of further practice shown in screen F, may be important elements for the orientation of individual study, outside the classroom hours (say in the library, the resource centre or at home). If the package were to be implemented on a larger system, with full CML record keeping (suggested in the Level 3 design documents as a desirable option) the final feedback to the student could be even richer. The computer could compute the error rates on this and past exercises and tell the student whether he is improving or not.

The 'drill-and-practice' example analysed in this chapter is a very simple, one-exercise package which aims to develop a specific sub-set of arithmetic skills. It does not teach the requisite knowledge, as this is assumed to have been learned previously. This previous learning may have occurred in a variety of ways, including conventional classroom instruction. It may also have been acquired by the prior study of another CAI package (or, rather, an earlier 'chapter' of our package).

Similarly, the learning of the basic arithmetic skills may be embedded in a more complex instructional design, involving the application of the skills to real-life problems, the building of new, higher order skills, the discovery of further mathematical laws, etc. However, each exercise in such a package is, as it were, a 'module', which interlinks with the other modules in a specific way. The overall plan that specifies the 'modules' and their interrelationship, is developed as part of our Level 3 instructional design. At Level 4, we turn our attention to the detailed design of each individual 'module' or 'exercise'.

Our intention, in this chapter, was to illustrate the design of one such module, showing that there is much in common between the design process of a CAI exercise and one which is to be presented by some other means. In our specific case, we made use of some of the principles of *mathetics*, in order to design the most effective and economical screens. In another case, we might have used the *Rul-Eg* approach to design screens that communicate and teach new concepts or principles. In yet another case, when we wish the learner to use a screen as a reference (as in the case of the 'help' facilities usually built into CAI courseware) we might use the principles of *mapping*. In short, the techniques we use to design the screens of a CAI package are similar to those we use to design printed or audiovisual packages. There are some differences of detail (which spring from the special characteristics of the medium being used) but the general approach is the same.

In the next chapter, we examine in more detail some of the more common forms of CAI packages and attempt to identify the differences – the special techniques which may be applied when using the computer as an instructional medium.

12. The Range of CAI Designs: Some Examples of Courseware

At the beginning of Chapter 10, we defined the part of the total field of computers in education that we consider to be within the scope of this book. This part, which we choose to call Computer Assisted Instruction (CAI), was sub-divided into six different modes (see Figure 10.1):

1. The TESTING mode.
2. The DRILL-AND-PRACTICE mode.
3. The PROGRAMMED TUTORIAL mode.
4. The CONVERSATIONAL or DIALOGUE TUTORIAL mode.
5. The SIMULATION mode.
6. The INQUIRY or DATABASE SEARCH mode.

We shall deal with each of these modes in turn, explaining their principal characteristics and giving one or two examples of the application of each.

12.1 The test mode

The main difference between the test and the drill-and-practice modes

We have already mentioned that the principal difference between the TEST mode and the DRILL-AND-PRACTICE mode is in the *feedback* provided to the student. In drill-and-practice, the student generally is told immediately whether his response is correct or not. He may be allowed two or more attempts before receiving the correct response, if he fails to furnish it. Thus, he is learning, as well as practising. In some systems, the items answered incorrectly are recycled, so that eventually the student must answer all the items correctly at the first attempt. Still other tactics may be incorporated to ensure that the student develops the knowledge or the skill being drilled.

In the test mode, there is less emphasis on drilling to perfection and more on the assessment of current levels of knowledge or skill. Thus the student may be given a set of test items to respond to, and only at the end of the test be informed that he got 'X' items right and made 'Y' errors. Alternatively, he may be told, item by item, whether he was right or wrong, but he is not given a chance to attempt again. In yet other systems, he may attempt the item several times, but be awarded full points for a first-time correct response, half points for a second-attempt correct response and no points thereafter.

The design problems of a computer-based test are thus very similar to a drill-and-practice exercise, in relation to the *stimulus information*, *response* and *instruction* components of a typical screen. The differences lie principally in the design of *feedback information* received by the student and of the *control algorithm* that defines the sequence of presentation and/or re-presentation of the items, the computation of the results and the keeping of records.

Use of computer-assisted testing

Computer-assisted testing may be used for several reasons:

* To automate the application and the marking of routine, objective tests, that are checked by reference to a standard set of responses – this reduces some of the drudgery of the setting and marking of such tests and speeds up enormously the availability of the results;
* To make it possible to collect and process the kind of detailed data on the progress of individual students, or groups, that is needed in order to take

reliable decisions as regards individualization, common-interest grouping, mixed-ability grouping, careers guidance, etc;

* To inject a greater measure of interest and motivation into the testing procedure, by clothing it in the form of a computer-based game.

There is something of interest, therefore, in the computerization of testing, to all the groups concerned in the instructional process – administrators and planners, teachers and instructors, and the learners themselves.

The first two benefits outlined above may be gained, at least in part, by the partial computerization of the process – eg by the application of written tests, which are responded to on special cards and the results rapidly processed by the computer. The third benefit generally requires the direct student-computer interaction, only possible when both the questions and the assessment system are computer based.

Macro-applications   An example of the first type of application is the extensive use of computer-marked assignments in the Open University's correspondence courses. Given the number of students enrolled, the use of traditional methods of marking and assessing the work of students would place an impossible burden on the lecturers and tutors involved. Of course, some of the assignments are open-ended and creative, necessitating personal assessment and open comment. The bulk of the intermediate assignments, however, are objective in nature and therefore capable of being marked automatically. This provides rapid and objective feedback of results to the students, whilst not precluding individual comments and assessments of particular students' work, selected by the course tutors as worthy of special attention.

A large-scale example of the second type of application is PROJECT PLAN, which was developed in the late 1960s in the USA. This project used computer-marked objective tests, to accompany all the primary and some of the secondary school curriculum in American schools. The schools subscribing to the system were equipped with special 'post office' terminals, one to a classroom. After covering a particular unit, of, say, three or four objectives, either individually or as groups, would take a pre-printed objective test, in multiple-choice format. They would respond on special cards, marking the choices by scoring the appropriate letters (A, B, C, or D) alongside the question's number, already pre-printed on the cards. They would also mark their individual code numbers. Then they would 'post' the cards into a slot of the terminal, to update their individual file, kept in a computer at a distance. The individual student received feedback of results and suggestions for further study. The teacher could compare the individual student's progress with that of his group and of other students and groups in other schools, right across the country. The computer was programmed to follow a highly complex control algorithm, that could use this extremely large bank of data to generate individual student profiles, individual guidance, suggestions of the best methods and materials of study for particular students or particular course objectives, predictions of success in particular types of career and so on. The system involved several million American schoolchildren and used an even larger database on long-term success in life, related to school performance, which had been collected in an earlier project – PROJECT TALENT – that had involved five million children.

Micro-applications   Coming down from these two examples of 'macro' applications of computer-assisted testing, we now have hundreds of 'micro' applications of tests-as-games, being used in home and school, as part of the video-game and home-computer revolution. Many of the simple computer games are in fact just tests of skill, wrapped up as a game. Others, like the well-known HANGMAN game, are computer-based versions of quiz-games of knowledge that have a long history. The game of HANGMAN has been used as a pencil-and-paper game for word-guessing or spelling practice. The computer version is identical: the screen displays a series of dashes or other symbols that indicate the number of letters in a given word and a graphic image of a scaffold. The player, or players, have to guess the letters, which appear in the correct positions in the word-outline if guessed correctly. Every incorrect guess results in the printing, beneath the scaffold, of one part of a human body – first the head, then the trunk, arms and so on. The aim is to guess the word before the body is complete. Generally, this implies that all the letters must be correctly guessed before about six incorrect guesses are made. The game can be adapted for the testing or drilling of arithmetic skills or for other types of questions. Each type of question, however, requires the program to be modified. Primitive versions of this game have been programmed with a fixed set of questions – a single purpose test. More sophisticated versions allow the teacher, or even the children, to invent and to type in their own test data, for example ten new words of up to eight letters each. The game interrogates the user, presenting a 'menu' of choices such as:

1. See the words.
2. Play the game.
3. Change the words.

By selecting option 1, the user can check whether the test's content is suitable for the students. Students may also use this option in order to study the words before being tested. Option 2 allows one to do the test. By selecting option 3, the user enters into a sub-routine that asks him to type in the words that are to be used in the game, thus allowing a suitable vocabulary to be injected for each week of the school year, if the teacher so desires.

A whole range of special test-games has been developed and marketed by Wida Software. These include a straightforward multiple-choice test format, a short-answer test, crossword puzzle formats, and several other varieties of entertaining test formats. In each case, the test is a mere format, into which the teacher may put his/her own content, without needing to know anything about computer programming. Thus with a battery of half a dozen different testing formats, the teacher may create hundreds of varied tests, accompanying a whole school curriculum in a given subject, without needing to know all that much about computers or computing.

Many other similar test-games are available off-the-shelf. One, used for arithmetic drills and similar simple test situations, is called CRASH. It has a basic structure very similar to HANGMAN, but the scaffold is replaced by the graphic images of two cars on opposite sides of the screen. For every error committed, the cars come one step closer. After about six errors, the cars collide, with all the visual or sound effects of which the particular computer system is capable.

A note of caution is necessary here. Computer games like CRASH and HANGMAN are sometimes criticized for their 'violence' or 'morbidity' and the potential negative impact that this can have on children. There is not much concrete evidence either for or against this hypothesis, but one should most certainly avoid excesses of 'blood and guts'. Whatever the evidence for or against this aspect, there is another one which all too often shows its negative effects. In all too many games, the 'bells and whistles' are all associated with the incorrect responses – you only get the cars to crash if you make errors. This leads many children to make errors deliberately, just to 'see what will happen'. One should keep a lookout for such examples of inadvertent reinforcement of the wrong behaviour. If possible the consequences of success should include most of the special effects.

12.2 The drill-and-practice mode

We have already studied, in some depth, a hypothetical example of a drill-and-practice exercise on addition. We also discussed some more sophisticated variations of this model that are possible and, indeed, were the basis of thorough research studies of the effectiveness of drill-and-practice CAI. These are the studies performed by Patrick Suppes and his collaborators at Stanford University, California, in the 1960s (Suppes *et al*, 1968). The outcome of these studies showed that, at least in the area of mathematical skills, well designed drill-and-practice routines could develop higher levels of competence, in less time than was normally the case when other means of organizing the practice sessions were tried.

The growth and importance of drill-and-practice

Since those early days, the number of different drill applications has increased and, of course, the costs of computers and software development have fallen, turning Suppes' effective, but exceedingly expensive, techniques into everyday reality. The type of sophisticated, matrix-based, model of complexity and difficulty in a given set of problems, may be quite easily accommodated in even quite a modest computer. As a guide, the memory needed to run the type of complex drill-and-practice routines designed by Suppes is much less than that required to run a reasonably sophisticated version of VISICALC, SUPERCALC or one of the other spreadsheet programs now available for most home computers in the 32K upwards range.

We chose this mode of CAI for our basic example in the early part of this chapter, mainly because it is much simpler to illustrate in the first instance. However, we also feel that this mode is unduly criticized by some people, as being a 'lesser form' of CAI. We would not agree. There are, of course, whole areas of the curriculum and whole classes of objectives, for which this mode of practice is not appropriate. However, there are also whole areas where some form of drilling is necessary in order to develop basic skills or to reinforce essential knowledge. No one learns a foreign vocabulary without the need to practise and repeat the words. This may occur in the natural context of using the language. But if one is in a hurry, some artificial drilling exercise is needed. If the computer can offer entertaining and efficient drill exercises at an economic cost, there is nothing 'second class' about it. If a clerk has to develop a high level of skill in the performance of a routine task, such as the error-free calculation of pay-as-you-earn tax deductions from employees' wage packets and a computer-based drill proves to be the most effective and efficient means of developing this skill, why should we frown on this form of application?

We are rather of the opinion that drill-and-practice, though not very spectacular, is destined to be one of the main uses of CAI, as long as people have to learn routine and repetitive tasks of any nature.

Teacher-programmed drill exercises

A number of versatile packages, rather like the test packages described earlier, are now available, that allow the teacher to change the content of the exercise while maintaining the general format of the exercise. The difference between these drill exercises and the test packages mentioned earlier, is in the form of feedback supplied to the student and in the conscious effort to develop the mastery of specific objectives. Once again, many of these exercises are disguised as games and are often, indeed, based on well known games, adapted to the computer-based format. Some of these, marketed by Wida Software, are specifically designed to develop associations, as in the learning of a new vocabulary. One version is an adaptation of the card game of SNAP. Two words or phrases appear on the screen at any one time, one in each language. The words change in a random fashion, first one then the other. At times, the two words which are visible are equivalent – in that case, the player must register that he/she observed the association, by depressing a key. If the key is not depressed within a certain time limit, the computer gains a point. If the player responds within the time limit, he/she gains a point. There are nine different speeds of play that can be pre-set at the beginning of the exercise. The game can be used for any association-learning, where rapid identification and response is an asset; eg in understanding a foreign spoken conversation, in learning map codes and conventions, in identifying traffic signs and in many other association and discrimination learning situations.

Another exercise, named ODD MAN OUT, presents four words or short phrases on the screen, three of which belong to one class and one (the 'odd man out') belongs to another. The player must identify the odd man out, by keying in the appropriate code number. If correct, the player obtains confirmative feedback, together with an explanation of the classifying concept used. If incorrect, the player receives corrective feedback and, once again, an explanation of the basis for the classification. In this way, a whole range of concepts can be drilled to perfection. One application would be in the recognition and classification of parts of speech. The game would be 'loaded' with four adjectives, four verbs, four adverbs, four prepositions and so on. The computer then selects, randomly, three examples from one group and one from another (see the sample screens in Figure 19.10). Other applications for which we have used this format of drill include the training of supermarket checkout staff in the classification of products sold according to stock-control categories, the training of postmen to classify streets according to postal delivery rounds, the drilling of 'town and river' type of factual information in geography and the classification of animals and plants in biology. It is a drill format that is suitable for the development of most concepts, especially groups of interrelated and easily confused concepts.

A simulated gambling game as a drill exercise

Yet another drill game, supplied by Wida Software, is based on the simulation, on the computer screen, of a 'one-armed bandit' fruit machine. Three cylinders can be rotated, by pressing appropriate keys. Imaginary 'coins' have to be fed to the computer to make it play. At certain, randomly occurring moments in the game, the player gains the right to hold or to 'nudge' one or more of the cylinders, just as in the real-life gambling machines. However, instead of attempting to line up three lemons, or other winning combinations of fruit, the player must line up three words that, together, 'make sense'. Once again, the content of the drill may be changed at will. The game was originally designed to drill correct sentence formation in foreign language teaching. However, with a bit of ingenuity, other applications may be invented. One history application involved the matching of WHO did WHAT and WHEN.

This last example, although a simulation of a sort, is not the type of CAI mode referred to as 'simulation'. The type of instructional process is very much a drill of a specific skill or body of knowledge, that has previously been learnt. An essential aspect of the 'simulation' mode, is that the student learns experientially how a particular system behaves under certain conditions, by 'playing' with the model of the system, which is the basis of the simulation. In the fruit machine example, the player 'plays' the machine, but does not play with the model on which the fruit machine operates and learns nothing of a general nature about the probabilities of winning and losing, how they are controlled and how they are stacked in favour of the machine (at least, there is no intention on the part of the instructional designer to promote any such learning).

The SAKI keyboard instructor

However, simulators, based on computers, have many applications in the drill-and-practice mode. A classic and very illuminating example is Gordon Pask's SAKI (Self Adaptive Keyboard Instructor), developed in the early 1960s, but only recently generally available, due to the fall of computer prices. The SAKI machine was initially devised to train the skills of punching a numerical keypad of ten digits. A later, more sophisticated version was developed to train and drill typing skills. The trainee is presented with a keyboard that has no markings on the keys and two screens. One screen presents information to be typed and the other shows a picture of the keyboard layout on which the keys light up to indicate the position of the key to be pressed at a given moment. The trainee approaches the trainer after some initial instruction on correct technique, typing rhythm and related essential knowledge.

Randomly selected letters appear on screen 1 and the position of each appropriate key is 'cued' by the lights that appear on the simulated keyboard on screen 2. As practice progresses, the general speed of presentation of data to be typed increases, and the intensity of the lights on the simulated keyboard decreases. However, all this happens in function of the error pattern and response rate of the trainee. He/she gets longer to respond to the letters which give difficulty, gets more intense cues and over a longer period of time for these letters and, furthermore, these letters begin to occur at a frequency proportional to their level of difficulty. All this occurs in response to a particular trainee's individual pattern of errors. He/she gets more practice, more time to respond and more help, adjusted on a continuous, 'on-line' basis by the control algorithm built into the

(a) ODD MAN OUT

```
Which is the ODD MAN OUT?
    1.  RUN
    2.  SHOUT
    3.  FINE
    4.  STRONG
```

(Student types in — 4)

```
That is correct.
    1.  RUN
    2.  SHOUT
    3.  FINE
STRONG is from the group ADJECTIVES
The other items are from the group VERBS
```

```
Which is the ODD MAN OUT?
    1.  HIGHLY
    2.  VERY
    3.  HIGH
    4.  ALMOST
```

(b) JACKPOT

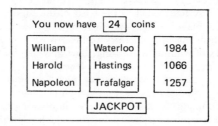

Figure 12.1 *Sample screens from the games*
'ODD MAN OUT' and 'JACKPOT'

trainer. This highly adaptive learning environment is extremely efficient, leading typical learners to achieve first-rate typing speeds in only a few hours of drill-and-practice.

Similar adaptive, computer-based, training devices now are fairly common for developing psychomotor and perceptual skills, such as radar tracking, missile firing and aircraft gunnery. Many arcade and home video games also develop, perhaps not all that intentionally, certain basic coordination or manipulation skills. It is more difficult to classify such devices as the full-scale flight simulators, or Moore's 'talking typewriter' used for the automated teaching of writing and reading skills. Basically, they too are simulated environments that provide the opportunity for intensive practice of some basic skills. However, a lot of initial learning (including of basic principles) may also take place during the experience.

12.3 The programmed tutorial mode

Many people equate the programmed tutorial mode (or just the 'tutorial' mode) with computer-based programmed instruction. To a large extent, they are correct. It is useful, however, to distinguish between what we have called the PROGRAMMED tutorial and other forms of teaching that are sometimes also referred to as the tutorial method. There is a world of difference between the personal tutor in a 'cramming college', charged with the task of getting his students through a particular examination in the most efficient and 'painless' way, and the university tutor who sees his task as one of a catalyst for the individual learning process of his students in pursuit of their own personally defined objectives. One may use mainly exposition and drill-and-practice techniques in his tutorial, whilst the other may favour the Socratic Dialogue.

The distinction between programmed and dialogue tutorials

For this reason, we stress the difference between the PROGRAMMED tutorial and the CONVERSATIONAL (or DIALOGUE) tutorial. Both may warrant being called a tutorial as they occur on a personal, one-to-one, basis, but the classic image of tutor and tutee as two people sitting facing each other at opposite ends of a log, has changed to student and computer on opposite sides of one of the many types of 'interfaces' – the term speaks for itself. However, the operative part of the term is the adjective – programmed or conversational.

The programmed tutorial mode of CAI is indeed akin to 'computer-based programmed instruction'. All the most commonly used techniques of the early days of programmed instruction are now being revived and applied in the design of CAI programmed tutorial sequences. Both linear and branching structures are used, though many people argue that the use of a computer to present a linear instructional sequence is rather like using a steam-hammer to crack a nut. This may, in general, be true, though there are some instances in which computer-generated effects or the computer control of ancillary media or equipment, may justify the application.

The possibilities for creative use of branching

In relation to branching, the computer opens up a range of possibilities that would have been difficult to arrange in the scrambled text, or primitive teaching machine, days of the 1960s. Theoretically, the computer can be programmed to branch to any number of alternatives, not to a limit of four or five at most, restricted by the number of plausible choices that the instructional designer can invent for one given multiple choice question, and also by the difficulty of organizing the pages in a scrambled text when there are very many different routes through the book.

In theory, a large library of computer-based instructional materials could be cross-referenced and indexed in a way that permitted students to back-track to any prerequisite learning that has become necessary in a new context but has unfortunately been forgotten through lack of use. Branching decisions, to faster or slower streams of instructional presentation, can be taken on the basis of a cumulative analysis of several responses over time and not just on the last, possibly atypical, response made. A variety of other non-computer-based materials may be incorporated into the system, their use being prescribed by the diagnoses of individual student progress and needs, that are constantly being prepared and updated by the computer. These alternative materials may be visual, as in videodisc and interactive videocassette systems already on the market. They may be print-based reading assignments, annotated bibliographies for free study, laboratory or workshop-based experiments and practical exercises, or specific group-learning activities. They may be all of these, knitted into a complex, multi-media, computer-controlled, truly individualized instructional system.

The reality is not always that creative

Unfortunately, in practice, few real CAI systems exploit these possibilities to the full. Too many are totally computer-based when they could be more effective and more economical if they were partly based on other media. Too many are a mixture of linear and branching sequences, not very different from programmes that actually existed in print and paper form some 20 years ago. Too many are not even good examples of the instructional programming art. We have seen examples of ostensibly CAI systems that did not really differ much from a standard textbook – presented electronically. They did not even follow the three basic principles of instructional programming:

* Active participation, by learners, in the learning process.
* Immediate knowledge of results and corrective feedback.
* Avoidance of excessive errors on the part of the learners.

Still less did they show creativity and perspicacity in the analysis of typical students' errors and difficulties, and the design of appropriate instructional solutions.

The creative application of the principles of programmed instruction

For want of a better starting point, the basic principles and techniques of programmed instruction, which have stood the test of time, should be systematically and creatively applied in the design of programmed tutorial CAI. Depending on the basic route of curriculum design being followed, one should perform an appropriately deep analysis of the structure of what is to be taught. This analysis may be behaviourally based (eg mathetical S-R analysis) when tangible performance objectives are being aimed for, or it may be subject-matter based (eg Rul-Eg/Eg-Rul) when the starting point of the design process is a given body of knowledge. Other approaches may be used. Landa's approach of mapping out the decision-making process to be learned, to its smallest component 'mental operations', may be attempted in relation to problem-solving skills, perhaps using algorithmic flow charts as a visual means of presenting the results of analysis. In the case of the design of a simulator for complex physical skills, Seymour's skills analysis methodology may be used to understand better the learning difficulties to be overcome.

The development of specific sequences of screens may well follow the principles laid down in Chapters 5 and 6, for the development of programmed exercises for specific objectives. After all, these techniques have proved to be effective for the design of many forms of individualized instruction. The detailed approach of mathetical exercise design has already been demonstrated, earlier in this chapter, to be eminently suitable for the design of the presentation screens of drill-and-practice exercises. It is just as appropriate for the development of expositive, 'demonstrate-practice-test' sequences in programmed CAI tutorials, when specific objectives are clearly definable for the sequence. When the aim is rather to explore a body of subject matter, the content analysis approach, resulting in a *rule-set*, may be used as a starting point. The information on separate screens may then be developed by a combination of the Rul-Eg (from rules to examples) and Eg-Rul (the opposite) tactics, depending on whether expositive or experiential strategies are favoured for the specific content to be taught. Normally, however, the amount of information presented in one screen, would not be bound by the archaic principles of small-step linear programme development. We might present two or more (related) rules in one frame. We would in fact follow the advice of mathetics, in presenting all related but contrasting rules together, at the first possible opportunity. This will emphasize the similarities and the differences and will assist in the welding of these separate rules into one powerful cognitive schema.

In order to present several rules in relation to each other, we would look for help from diagrams, tables or other charts, that might supply a visual image of the structure (schema) of the relationships that exist in the subject. This visual treatment of the subject matter, may, incidentally, be enhanced by the judicious use of colour, graphics and other special effects that our computer system may offer. Some systems can present animated graphics, capable of showing dramatically how an idea grew or changed in time, with much more effect than would be possible by means of a series of still diagrams.

As the complexity of our subject matter increases, the students may feel the need to look back at key concepts, or look forward to a more general, but simpler view of where they are heading in their studies. We may build into our control algorithm the facility for database interrogation, initiated by the student. We may need to prepare screens that define prerequisite concepts in more detail, or overview screens that show where we are heading. All these must be indexed in such a way that they are readily accessible, by means of key words or some other technique, when required by the individual student.

As we get more ambitious in our attempts to individualize the instruction process, our programming technique becomes ever more eclectic. We begin to incorporate many of the tactics of structured writing, in order to organize large bodies of information in readily available formats. We may use Eg-Rul tactics, to promote the discovery of some basic principles which underlie our subject. We get ever more conversational, in exploiting the flexibility that is possible in CAI. When exploring WHY a given student responded in a particular manner, we may start to employ some of the tactics of structural communication. Before we know it we may have crossed the hazy boundary between PROGRAMMED tutorial and some other mode of CAI.

Methods for the development of programmed tutorial CAI

There are several detailed methodologies that have been specially developed for the design of programmed tutorial materials. Most have a close affinity to the above-mentioned 'eclectic' style of programmed instruction materials development. Some are based on the direct coding of the detailed presentation plans into a computer program, by means of some higher-level computer programming language, such as BASIC or PASCAL. Others are based on the use of specially developed 'authoring languages'. We shall deal a little more closely with the differences in this final stage of materials preparation later on. The earlier stages, however, are all very similar, following the basic steps we defined before, of developing an overall plan, or 'structure chart', followed by the detailed design of the 'screens' that will be seen by the student and the 'control algorithm' that states which screens will be displayed to the student who gives a particular pattern of responses.

The main aspect that seems to vary enormously, is the technical jargon used to denote essentially the same stages and design tasks. Dean and Whitlock (1983) refer to the control algorithm by the (perhaps more straightforward) term of 'programme flowchart', while Godfrey and Sterling (1982) refer to essentially the same document as a 'local structure'.

As a general illustration, we present, in Exhibits 12.1 and 12.2, a very summarized view of the approaches outlined in the two books just mentioned. We shall refer to these exhibits again, in the later parts of this chapter. However, readers are recommended to study the originals, in order to gain a fuller grasp of the similarities and the differences between the two approaches (and also the above-mentioned problem of technical jargon differences).

Exhibit 12.1

The first stage, after the organization of a rule set and sequence, is the writing of the main body of material to be presented – the MAINLINE CHART. Note that in screen M40 the author has shown three alternative comments that might appear, depending on the student's response at this point.

The second stage is the mapping of the PROGRAMME FLOW CHART, shown here, indicating the incorrect and other responses that will be commented on and the routing paths between them.

The third stage will be a RESPONSE ANALYSIS, in order to plan the comments and hints to be given to the students who make errors or give unrecognized responses.

The fourth stage will be the planning of the layout of the screens, on squared paper.

The fifth stage will be the coding of the courseware, using an author language.

Exhibit 12.2

Note the differences in terminology. An early stage of the process defines the MAPPING or sequence of the main objectives to be achieved. Later, the treatment of each objective is developed in the form of a flow chart showing the LOCAL STRUCTURE. Note that this sequence is partly programmed tutorial, partly drill-and-practice and partly testing mode.

Exhibit 12.1 *Methodology for the development of programmed tutorial materials, as suggested by Dean and Whitlock (1983)*

From M20

M30

Right, it will be useful to record both what sort of documents will be typed and the method used to produce them.

Words like 'TYPIST' include a wide range of activities and methods. In this case the job advertisement would have been much more effective if it had mentioned the method of typing to be used.

Incidentally, how many kinds of typing can you think of? Type in as many as you can.

M40

Here's my list:

Copy typing   Audiotyping   Shorthand   Speedwriting   Wordprocessing

You will see that you got all/some of them and perhaps you thought of some others.

This simple example shows how a job title can include many different meanings.

M50

It also helps those looking for jobs to decide their suitability for a vacancy if the advertisement tells them something about the products of their work.

So a question asking about the kind of document typed would be useful. Careful questioning about these details could produce this information on the card:

'Copy typing of letters, proposals, reports and sales literature.'

Suppose an employer notifies a vacancy for a caretaker, what specific duties do you think might be mentioned in the advertisement?

M60

You will remember that the employer's complaint about the applicants for his typist job mentioned a number of drawbacks. Press the START button on your recorder to hear that part again.

Can you now list the capabilities or experience you think should have been recorded by Charlotte and added to the advertisement?

To M70

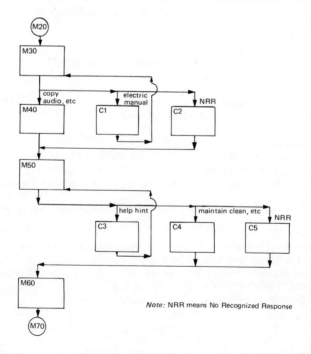

Note: NRR means No Recognized Response

Exhibit 12.2 *Part of the methodology for courseware development suggested by Godfrey and Sterling (1982)*

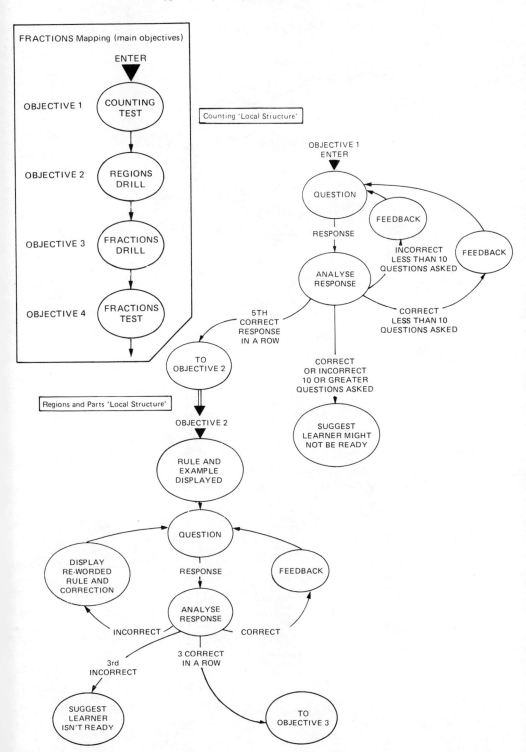

Exhibit 12.2 *(continued)*

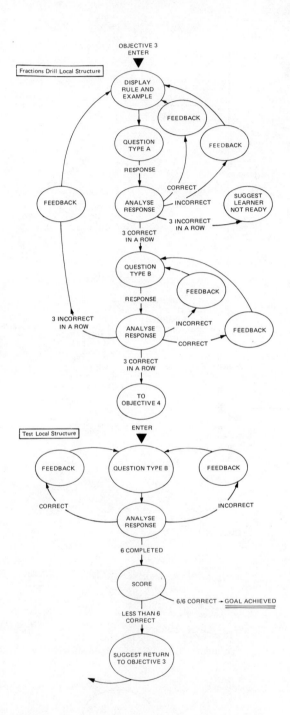

12.4 The conversational or dialogue mode

The chief differences between the *programmed* and the *conversational* (or dialogue) tutorial modes have already been mentioned. However, just as there are several approaches to the programming of instruction (and its application in CAI), so there are various techniques, or even basic theories, of development of automated conversations, or dialogues. Many authors have used these terms, in the context of CAI and in the more general context of instruction, to convey, perhaps the same general meaning, but with various specific differences of methodology and approach.

The common elements
The common element in most authors' approaches, is a striving to attain higher order cognitive objectives; abstract concepts and conceptual schemata, principles and the resultant strategies that may be formed by their combination. All tend to espouse experiential learning as essentially superior to expositive, programmed learning, for the achievement of these categories of objectives and all favour a student-directed, information-searching-and-analysis, learning situation. This, however, is strongly supported by previously designed materials that are tested to verify that the proposed objectives are really mastered by the students. To this extent, therefore, we are still in the general domain of 'instruction'.

Different approaches
As used by Bitzer (1976), in relation to the PLATO system of CAI, the DIALOGUE mode is seen as little more than a complex branching structure, in which the student's responses to a problem posed are not directly classified as 'right' or 'wrong', but are used as the springboards for further questions and problems that explore the implications of a particular decision taken. We can see the potential overlap here, with programmed learning, as, for example, the programmed text on 'The Waterloo Campaign' (Thornhill, 1965) adopts exactly this approach in its use of branching. This and other examples of the 'interactive case study' type do engage the student in a sort of dialogue and analysis/evaluation of the probable consequences of alternative courses of action. However, very often, after such analytic/evaluative comments and interchanges on the student's opinion, the materials end up by presenting the author's own opinion, or the actual historical decisions that were taken and the consequences that resulted.

The Socratic Dialogue
Another, still more open-ended approach is the attempt to simulate, in a branching sequence of frames (or 'screens') a 'true' Socratic Dialogue. This approach is not very often used, in fully automated instructional systems, as it entails never commenting directly on the student's own suggestions and decisions, but always following up with another question, that leads the dialogue yet further and deeper. As it is not possible to predict all the possible conclusions and observations that a student might make, it is hard to devise a suitable list of further questions to ask. The truly open Socratic Dialogue is not likely to be transformed into CAI by any of the currently used programming methodologies. It is indeed doubtful whether the 'true' version is an example of 'instruction', as the objectives that will be achieved by the learner are not generally specified in detail before the exercise. However, some 'close attempts' at simulating at least some of the aspects of Socratic Dialogue technique, exist in the field of CAI.

Structural communication as a conversational CAI technique
We would classify structural communication (discussed in Chapter 7) as one of these close attempts, at least in some of its applications. The student's responses are limited to the selection of any number of response components (key words or phrases that summarize the main ideas of the subject being studied). In selecting any number of components from a response-matrix of 20, 24 or even 30 items, the student has literally thousands of different responses open to him. If the response components are well designed, in the light of a deep understanding of the subject and much experience in discussing it with students, just about any plausible solution to the problem posed will be capable of being constructed (as well as many implausible ones). The method of response analysis (on the basis of what was included or excluded) and the resultant comments (if phrased in an open-ended and questioning manner, rather than in an autocratic or corrective manner)

serve as a very close simulation of the Socratic Dialogue. Once a given problem and its implications have been exhausted, the analogy breaks down somewhat, as the true Socratic Dialogue might then continue with new, previously unplanned problems or questions, arising from the discussion so far, whereas the structural communication materials continue with another, perhaps only loosely related, predetermined problem. We recommend the study, once more, of Exhibit 7.1, in Chapter 7, with special attention to the structure of the materials and their suitability for partial presentation by means of a CAI system.

There is no point, with present costs of computer memory and programming time, in presenting on the screen a 'Presentation' that runs to three or more pages of straight text. However, after reading the text (possibly in a library), the student interacts with a computer terminal that presents him with the first problem and the response matrix. The student makes his choice of response items at the computer and immediately receives a 'personalized' comment on his response, which is composed of a fusion of the special comments deemed valid for the items that he has included or excluded. He no longer has to wade through pages of comments to identify those that aim to deepen his own understanding of the problem presented. The presentation is slicker and easier to use. Unnecessary comments are not at the student's disposal. Furthermore, the system can analyse the response components actually chosen, and their frequency, thus identifying any need for the inclusion of further, originally unanticipated, comments.

We suggest that at this stage, the student might be given a choice of a second problem, from a varied selection of possible problems, thus giving the student the responsibility of deciding what appears to be the next most pertinent problem for discussion, in the light of the dialogue that has taken place so far.

The 'Caste' system

If we visualize the structural communication approach as the presentation of new information and the investigation of the resultant knowledge structures that are formed in the mind of the learner, the approach of Gordon Pask (named 'Caste') is almost diametrically opposite – the *presentation* of a model of the structure of a body of knowledge and the *search* for information that will make this structure meaningful to the learner.

Pask (1976) builds up a 'Conversation Theory' based on his own work and the work of Piaget, Vygotsky, Papert and others, in which learning is seen to result from *agreements* between people on the meanings to be ascribed to particular concepts. These agreements lead to *understanding*, which is seen as the ability to explain the concept 'from first principles', or to '*reconstruct*' it from previously 'understood' concepts. Thus the process of conceptual learning is an interchange of information that leads to the discovery of a new relationship between this and other, previously learned information. The interchange of information goes both ways – the search by the learner for necessary information to make sense of what he is experiencing, and the communication by the learner of the new discoveries being made by him – hence, 'conversation'.

Space precludes a fuller description of this theory and what it implies, in terms of the 'conversational transactions' that should take place between 'teacher' and 'learner'. The reader is directed to the original works (Pask, 1974b; 1976; 1984). The practical application of this approach, both in its CAI and human-tutor-based versions, commences by presenting the student with a 'map' or 'entailment structure' of the subject area under study, with all the topics and concepts interrelated, down to the simplest, already fully 'understood' concepts. The student then *explores* certain relationships marked on the map, asking the instructional system for 'descriptors', generally in the form of examples and non-examples of the relationships that interest him. This leads to the formulation of an *aim request* – a statement of the topic on the map that the student will, at this moment, attempt to make sense of, or rather to demonstrate understanding of, by deriving it from simpler, previously mastered topics.

The instructional system then questions the learner, by means of a series of multiple choice questions, to check whether he is really ready to derive the topic chosen. If the result is unsatisfactory, further 'explore' transactions are engaged in, to clarify better the structure of the interrelationships that exist between the various topics. If the learner shows 'readiness' he may ask the instructional system about the possible *learning paths* from currently known topics to the topic chosen as the current 'aim'. Given this information, the learner identifies intermediate topics, on one of the paths to learning, about which he needs to receive more information. The instructional system checks the learner's readiness to study this intermediate topic (termed a *goal*) by checking that all immediate prerequisites for at least one of the ways of deriving this goal, have been previously mastered. The learner then attempts to derive the unknown topic, from previously known ones, which involves him in model-building and both verbal and non-verbal explanations (the latter are demonstrations). These explanations and demonstrations (and the conceptual models formed) are evaluated by the instructional system to check that they have indeed been derived and explained by the learner through his own efforts to understand the topic and are not some previously seen, not fully understood explanation.

In this way the student gradually masters one topic after another, in any sequence that he sees fit, provided that the sequence is justifiable by his present state of understanding of lower-level related topics. The learner learns largely by seeking out the information that he needs to complete his understanding and the instructional system instructs by evaluating the learner's readiness to attempt a given learning task and indicating the sources of relevant information that the learner may use. These are generally stored on paper or other media and not in the computer memory. In the CAI implementation, the learner really 'converses with himself', asking himself the questions that the human tutor would ask, under the guidance of the entailment structure and other data systems stored in the computer.

This approach has been used experimentally in several applications (eg reaction kinetics, meiosis and mitosis, probability theory) but is not yet used very extensively.

12.5 The simulation mode

Whereas the conversational, or dialogue, modes of CAI are perhaps among the least practised on a large scale, SIMULATION is probably the most used and fastest growing mode. This is probably due to a series of factors:

* Computer-based simulations are sometimes the only way of developing certain types of learning experiences.
* They use the particular advantages of the computer as an ultra-rapid calculating and data processing machine, to the best advantage.
* The computer may often play a part in a more elaborate, not entirely computer-based simulation-game.
* Unlike the conversation mode, simulations (although possibly difficult, or at least, time consuming to programme for the computer) are often quite simple to plan from the instructional viewpoint. Thus, the instructional designer may delegate a major part of the production task to specialist computer staff.
* The use of simulations is usually an adjunct to normal teaching procedures – it does not disrupt traditional organizational practices, upset conventional teachers by threatening the jobs of anyone, or tie up the computer on a nearly permanent basis for instruction.
* Many computer-based simulations are in the science area, where staff are most receptive to the new technology.
* There are, however, ample valid applications for this mode in almost any educational or training context.

There are, indeed, several types of different applications for computer-based simulations. Dean and Whitlock (1983), present a useful classification into four basic types.

System facsimile. This is the training of staff who actually use computerized systems

1. System facsimile

to operate them correctly, under simulated 'safe' conditions. Examples include most business applications of computers that are used to assist interactions with customers or the general public. The staff who operate the system may affect the organization's productivity, or goodwill, immensely, so they must be well trained. However, training on the job is likely to introduce the sort of errors that lose custom and goodwill. Training off the job is not sufficiently realistic. Thus the use of the real system, but with simulated data and customers, is the most effective answer. This category is analogous to the use of real workshop machinery for the training of technicians, performing special exercises. The design of such training has little to do with the computer as such and much more with the analysis of the administrative procedures that are performed on it. The training designer need not, generally, get involved in the design of special CAI sequences, but simply uses the existing computer facility and software as a 'workshop' environment for the necessary training. The grading of learning difficulty can usually be handled by selecting suitable example situations. However, one can utilize the computer as a more active part of the instructional system by getting it to supply enhanced feedback of errors to the learner and manage the rate and difficulty of the examples presented for solution. In other words, the computer can take on a *drill-and-practice role*.

2. Apparatus operation

Apparatus operation. This is another category of simulation, based on computers, which we have already mentioned. The SAKI keyboard instructor described in the drill-and-practice section of this chapter is an example of a simulator of an item of equipment, which requires a computer to make it work. We also mentioned radar tracking simulators and the like. The computer is coupled to the equipment on which training is to be given, in order to enhance the feedback, or the guidance (or both), received by the learner. It may also be used to monitor learning progress and select appropriate levels of difficulty of the tasks set, appropriate length of practice sessions, etc.

3. Use of the computer as a database

Decision-making exercises. This is yet one more category that we have already discussed, at least in part. Large-scale simulation games, such as management games, war games and some social simulations, may require a fairly complex model of the real situation under study, that reacts accurately and quickly to the decisions and actions taken by the participants. Very often, an exercise coordinator and a store of paper documentation cannot keep up with the pace required to produce new data that simulates the consequences of particular decisions made by the players. The use of the computer as a data processor or as a calculating machine renders the game playable. In more sophisticated applications, the computer may accurately simulate a whole organization, or even a national economy.

4. Process modelling

Process modelling. This category has been left to the last, as it is probably the one that most people think of when referring to the SIMULATION mode of CAI. We have seen that the other three classes involve computers and some form of simulation, but other factors are involved (such as drill-and-practice) and the computer is generally a part of a more complex, multi-media, training system (involving, perhaps, paper documentation systems, equipment and apparatus linked to the computer, interactions with other managers in group decision making, etc). The process modelling, or 'pure' computer-based simulation of a phenomenon, involves only a computer and its software, in order to demonstrate the effects, or the characteristics, of a system under certain forces, constraints or influences. The learning intended is purely cognitive, in the knowledge domain; the forming of new concepts or cognitive schemata with respect to a phenomenon or process which, itself, cannot be readily studied in such depth. This type of computer-based simulation is the 'laboratory for learning' that enables the student to carry out difficult, perhaps dangerous or even 'fatal' experiments, without incurring the danger of the real-life situation. It enables many students to obtain, over and over again in a short time, the experience of a reality that occurs only rarely during a lifetime. This is the type of simulation that often gets educators excited about the real possibilities of the computer in the classroom in the short term.

The applications abound. One area in which dozens, perhaps hundreds of successful simulations have been developed, is medical education. Trainee doctors learn how patients with diabetes react to the intake of sugar in various quantities. Other simulations teach about the reactions of the body to different drugs or medicines. Yet others deal with environmental factors that promote epidemics, or the control of a common disease by the vaccination of different proportions of the population. In all such simulations, the student may vary some of the critical variables and observe how other variables are affected.

Another popular area for process simulations is in the sciences. Ellington, Addinall and Percival (1981) list dozens of examples, including a simulation of the process – of discharge of a capacitor, the flow of fluids through nozzles, gaseous diffusion, gravitation, interference and diffraction patterns of light waves, motion of satellites in orbits, and many others.

Yet another interesting area of application is economics and finance, where the effects of interacting market forces, tax laws or inflation rates may combine to give surprising effects under certain conditions.

Complex multi-faceted simulations

There is one other application to be mentioned here – the flight simulator. Simple navigational simulators (you can buy one down the road, for operation on most makes of home computer) are a form of *decision-making exercise*. They teach, more or less accurately, the factors that must be taken into account when flying and navigating a plane. In playing with these simulators, one learns when to drop the undercarriage, in relation to airspeed, when and how much to trim the flaps, how to allow for a crosswind when approaching an airfield, and so on. Such a software package, sold as a game for recreational purposes, does therefore give practice in the decision-making aspects of flying. As it presents the player with up to three visual displays – the instrument panel, the view in front of the cockpit and a map of the area in which the plane is situated – it is a simulation of the navigational and instrument-flying tasks in the pilot's job. When this same decision-making package is incorporated in a model of a real plane, that also supplies the correct physical feel, guidance and feedback for the manual skills of control, we have introduced the element of *apparatus operation*. Yet I have spent many happy hours deliberately 'crashing' the plane on my own copy of a flight simulation package, not so much to learn how to get the smoothest landing or the fastest route from A to B, but to experiment with the extremes of the variables – questions like, 'how do the flaps influence the stalling speed?' or, 'what is the relation between angle of climb and throttle opening that will give a constant airspeed?' In doing this, I feel that I have learned some of the principles of operation of a plane and the effects of changing certain variables – surely this can be classified as *process modelling*, the last of Dean and Whitlock's four classifications. It thus appears that complex simulations may fit into more than one category.

POLUT – an example of a typical simulation

In order to illustrate the structure and the design principles of a typical simulation, we shall choose a well known, quite 'elderly' (1971) example, called POLUT. We choose this example for it is one of the most characteristic of a bunch of simulations produced by the Huntingdon Projects in the late 1960s and early 1970s. These were the first systematic attempts to develop usable computer-based educational simulations of the process-modelling type. The POLUT simulation deals with aspects of water pollution – a topical subject in this day and age and one that may appeal to all readers, whether 'turned on' by science or not. The effects of pollution on the condition of a body of water and its ability to support life, depend on a great many interacting factors. The most obvious variables are:

* The kind of body of water (static/moving/big/small).
* The water temperature.
* The kind of waste being dumped.
* The rate of dumping.
* The type of treatment, if any, that the waste receives before being dumped into the water.

Exhibit 12.3 *The POLUT water pollution simulation*

Developed at the Polytechnic Institute of Brooklyn, New York, 1971, as part of the Huntingdon II project. Note that the computer system used did not have any special graphics facilities and was geared to producing a paper print-out of the tables and graphs. Hence, the rather simple, crude graph (by modern 'micro' standards).

INITIAL
INSTRUCTIONS

> IN THIS STUDY YOU CAN SPECIFY THE FOLLOWING CHARACTERISTICS:
>
> A. THE KIND OF BODY OF WATER
>   1. LARGE POND   3. SLOW RIVER
>   2. LARGE LAKE   4. FAST RIVER
>
> B. THE WATER TEMPERATURE IN DEGREES FAHRENHEIT
>
> C. THE KIND OF WASTE DUMPED INTO THE WATER
>   1. INDUSTRIAL
>   2. SEWAGE
>
> D. THE RATE OF DUMPING OF WASTE, IN PARTS PER MILLION (PPM) PER DAY
>
> E. THE TYPE OF TREATMENT OF THE WASTE:
>   0. NONE
>   1. PRIMARY (SEDIMENTATION OR PASSAGE THROUGH FINE SCREENS TO REMOVE GROSS SOLIDS)
>   2. SECONDARY (SAND FILTERS OR THE ACTIVATED SLUDGE METHOD TO REMOVE DISSOLVED AND COLLOIDAL ORGANIC MATTER)

DATA INPUT
BY STUDENT

> BODY OF WATER? *1*
> WATER TEMPERATURE? *50*
> KIND OF WASTE? *1*
> DUMPING RATE? *8*
> TYPE OF TREATMENT? *0*

OUTPUT OPTIONS
AND SELECTION
MADE BY
STUDENT

> DO YOU WANT A
> TABLE, A GRAPH,
> OR BOTH?
>
> (T, G or B)
>
> B

Exhibit 12.3 *(continued)*

THE OUTPUT

(a) THE TABLE

AFTER DAY 1 THE GAME FISH BEGIN TO DIE, BECAUSE THE OXYGEN CONTENT OF THE WATER DROPPED BELOW 5 PPM.

TIME DAYS	OXY CONTENT PPM	WASTE CONTENT PPM
0	6	2.67
1	5.24	9.82
2	3.41	15.38
3	1.16	19.69
4	0	23.04
5	0	25.64
6	0	27.66
7	0	29.23
8	0	30.44
9	0	31.39
10	0	32.12
11	0	32.69
12	0	33.13
13	0	33.48
14	0	33.74
15	0	33.95
16	0	34.11
17	0	34.23
18	0	34.33
19	0	34.41
20	0	34.46

(b) THE GRAPH

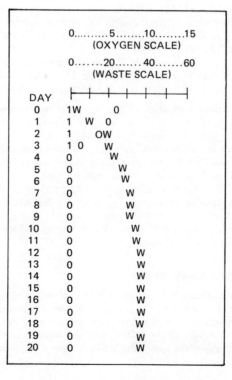

```
              0..........5.........10........15
                      (OXYGEN SCALE)

              0.......20....... 40.......60
                      (WASTE SCALE)

       DAY   ├──┼──┼──┼──┼──┼──┤
        0    1W        0
        1    1   W   0
        2    1       OW
        3    1 0     W
        4    0         W
        5    0           W
        6    0           W
        7    0             W
        8    0             W
        9    0             W
        10   0             W
        11   0             W
        12   0             W
        13   0             W
        14   0             W
        15   0             W
        16   0             W
        17   0             W
        18   0             W
        19   0             W
        20   0             W
```

However, there are many other, not so obvious, variables and constants that are involved in the process. Among these are:

* The rate of natural decomposition of sewage.
* The rate at which the water absorbs oxygen (this depends greatly on the type of body of water, being nearly eight times as fast for a fast river as for a small pond).
* The maximum oxygen content of the water (this depends in quite a complex way on the water temperature).
* The rate at which industrial waste decomposes naturally.

All these factors interact in complex ways with each other, resulting in a set of several difficult equations that must be solved to calculate the change in oxygen levels and pollutant levels in a given body of water, from day to day. To study the phenomenon thoroughly would involve many hours of calculation, in the normal way, to investigate just some of the infinite combinations of variables. Using the computer-based simulation, a group of students can, in one hour, visualize the effects of some 40 or 50 combinations of variables – enough to gain a very thorough understanding of how the principal factors interact – and without the need to perform any mathematics. A typical run, and the accompanying printout is presented in Exhibit 12.3.

The *instructions* indicate the five variables that may be altered by the students and the alternatives that are available for each of them.

The students select values, at will, for the five variables, in response to the *prompts* received from the computer. If, by chance, the students step 'way outside the bounds of reality', they are mildly chided and asked to choose another value, within more reasonable limits. For example:

* A value of water temperature above 90 degrees Fahrenheit gets the response from the computer – THE WATER TEMPERATURE IS HIGH ENOUGH TO DESTROY MOST LIFE. TRY A NEW TEMPERATURE.
* A value below 32 degrees gets the message – YOUR BODY OF WATER IS A BLOCK OF ICE AND CANNOT ACCEPT ANY WASTE. TRY A NEW TEMPERATURE.
* Best of all, an excessively high dumping rate elicits the comment – NEW YORK CITY ONLY POLLUTES ITS WATER AT A RATE OF 12 PPM/DAY. MAKE YOUR RATE BETWEEN 0 AND 14.

Once the variables are selected, the computer performs some necessary calculations (very complicated ones at that) and offers to present the results as a table or a graph, or as both. We show both in Exhibit 12.3.

Note that for the variables selected, we have managed to kill off just about all life in our pond by day four, when the oxygen content drops to zero. From then on, there is no hope for our pond, until we change the dumping and treatment variables. Note also that although we keep dumping at the rate of 8 ppm/day, the waste level stabilizes at about 40 ppm, after a few days. This immediately gives us an insight that, with other variables, we may be able to stabilize the waste and oxygen levels at acceptable values. Further trials will lead us to discover these levels, for each body of water.

12.6 The inquiry mode

The final classification in our schema of CAI only just qualifies to be included as a category of 'instruction'; it is really a mode of *information retrieval* from some form of organized bank of information, or *database* However, if the database is organized to supply all the information necessary to achieve certain groups of educational objectives and if the learner is given (or formulates for himself) a set of specific objectives to be achieved, then we have a form of instructional system – indeed, a very flexible self-instructional system.

There is, on the other hand, the unstructured, 'browsing' use of a computerized database, analogous to browsing in a public library, which is not aimed towards specific learning objectives and is therefore not 'instruction' in the sense that we are using the term. This is what Nicholas Rushby (1979) termed 'serendipity learning'. The planning of a bank of information for unstructured, serendipity learning, is not unlike the planning of the purchasing policy of a public library. We do not know exactly who will turn up to use the resources, or what they will be looking for. We do, however, have some information concerning the demand, by analysis of the requests we receive, through study of the bestseller lists, by careful analysis of the reports on new publications (from publishers or independent critics), and so on. How well we satisfy demand can be assessed through the study of user reactions, rather than the measurement of specific objectives achieved.

A very similar situation exists in education, in relation to general-interest, or orientation, courses, public lecture programmes, conferences and other such events. The organizers 'lay on' the events they think the public wants, or (through the analysis of some new trend or discovery) what they think the public ought to want. The success of such educational events is, quite legitimately, assessed by opinion polls or evaluation questionnaires designed to measure only the reactions of the participants to the content, approach, scope, methodology and organization of the event in question.

One potential growth area for the application of the computer in education is, no doubt, the creation of computer-based libraries, computer-based information banks on specific subjects, etc. One such application already in use in the UK is Prestel. Other videotext systems, available to the public in general and devoted to educational and cultural purposes, as opposed to advertising and newscasting, will no doubt appear in increasing numbers, throughout the world. Pundits already draw mental images of immense national or even international networks linking dozens of specialist libraries into one huge system which, to use the words of Roy Jenkins (1983), 'can allow the contents of all the world's great libraries to flow through every living room in the land'.

This vision, to become a functioning reality, will need a lot of work on the organization of the information of these 'great libraries' in such a manner that the not-too-skilled potential user can find his way about, locate what he wants, or what he might want if he only knew that it existed. The task of organizing such immense databases is not all that well understood. Many people are now interested in the topic and much research is under way. In relation to educational, scientific and other specialist use of such systems, the need is felt for a technology of subject matter analysis and organization that can produce a database on a given subject that is as helpful to the learner/researcher, as a human expert on the subject would be, if available for a personal interview. This has led to the concept of EXPERT SYSTEMS – computer-based information banks which emulate a human expert in a given subject area. Such an expert system should be able to respond to a variety of styles of interrogation. The normal approach to looking for information in a library – by searching through author and subject indexes – is but one possible style. What if the interrogator is not able to formulate his questions in these terms? Some computer databases (eg the ERIC service – Educational Resources Information Centre) will accept a key-word search that goes into much more detail than a conventional subject index classification. However, the user still has to specify the key words that interest him and, indeed, must do this with some skill if he is to avoid receiving a lot of irrelevant information as well as what really interests him.

A well organized database on a specialist subject is, however, little more than a well organized and cross-referenced library. It is an important and useful tool in the hands of an expert. However, it does not take over the role of the expert. The important aspect of a true expert system is that it takes on at least part of the problem-organization and problem-solving roles of the human expert. It is capable not only of furnishing the information solicited, but should be able to interrogate the user for relevant input information, prompt the user in the steps to be followed in solving a problem, or even take over the problem-solving process as a whole.

The special-purpose scientific software that solves a set of predetermined equations has been with us for a long time – any computer program is a 'problem solver' in this sense. But the solution of complex problems that require one to weigh various factors in relation to each other and make a decision on incomplete data – that is, *heuristic* problem solving – has only really been successfully computerized. The 'expert system' is one form of a computerized heuristic problem solver.

The two essential components of an expert system are:

1. *A structured 'knowledge-base'*: the information that is used by human experts is solving a particular category of problem, organized in a hierarchical or other appropriate manner.
2. *An 'inference engine'*: the rules, strategies, priorities, etc, that human experts apply when using this knowledge to solve problems or take decisions.

Both these components are assembled by interviewing acknowledged experts and transforming their insights into computer-based routines. Although this may sound simple, it is in fact a difficult and very time-consuming process. However, it may pay off, in that many computerized systems have sometimes shown themselves to be more effective and more reliable problem solvers than the human experts on whose accumulated knowledge they have been based. And, of course, they can be replicated and used in many locations by many people.

Such an expert system no doubt teaches the user something about the problem-solving process involved. Indeed, many such systems are constructed to explain to the user the logic of the steps that were followed in solving the problem. Thus, although their principal function is as a practical problem-solving tool, they do have an indirect educational function. However, they are not really instructional systems in our understanding of this term.

It is possible, though, to conceive of an expert *teaching* system, which is designed with the explicit aim of *solving the learner's learning problems* in a specific subject.

Approaches to the design of expert instructional systems

Several lines of research are being followed up in this area. The work of Gordon Pask, on conversational CAI, is of relevance here, as it is based on the identification of a student's readiness to study and understand a particular topic and the identification of prerequisite topics that should first be mastered. In the applications so far made of this approach, the diagnosis and guidance systems are (sometimes) computer-based, but the study materials then recommended are usually based in other media (especially print and slides). In the future, however, the study and reference materials could well be computer-based as well and, perhaps, available at a distance through videotext-type information distribution systems.

Another technique , of relevance here, is the structured writing approach as used in the information mapping (TM) service. We have already described this approach in Chapter 8. One reason for devoting a whole chapter to this technique is that, apart from its use as a means of writing clear communications and preparing manuals that may serve as learning or reference texts at the same time, it may be used for the analysis and organization of information for inclusion in a computerized database on a given topic. Robert Horn's 'maps' describing the technique, which we presented in Chapter 8, describe quite clearly how it is envisaged that such maps would be stored in a computer. We can see, from this description, that it is possible, with this technique, to build in a certain modicum of interaction between the user and the database. Instead of presenting a complete map on a given concept, which contains a definition, three examples, two non-examples, a visual schematic representation, an explanation of its use and a list of related concepts, the computer, when asked about the concept, could reply by asking, effectively, 'what do you want to know? I have a definition, example...'. The user of the system would then choose just what he thinks he needs in order to fulfil his current learning or reference requirements. He may ask for 'another example', if the one given was not sufficiently clear. If the definition includes some other unknown term or idea, he need only inform the system (for example by typing in the word, followed by a question mark) in order to be immediately presented with a definition and choice of other information. As soon as this doubt is cleared up, he would return to the previous point in the database and continue his study.

So far, the system only reacts to user-instigated queries, by offering 'menus' of information options and responding to the user's decisions. The building-in of system-instigated queries, in the form of questions that check the understanding of the topic being studied, may allow the system to pinpoint the information that the student/researcher needs, thus turning it into an instructional system, as opposed to merely a bank of information, organized for easy reference.

Using the 'expert system' approach, it should be possible to interview a range of 'expert' teachers on a subject which has previously been organized in the form of 'maps'. These teachers would relate the decision-making process which they apply in order to diagnose specific students' learning problems, in order to select appropriate knowledge content, in order to organize the message, the number of examples, the test questions and the remedial actions that would be taken in order to 'solve the student's learning problem'. This body of expert teachers' heuristics on a given subject (say, digital electronics) would act as the *inference engine*, which would select presentations from the *knowledge-based*, structured in the form of carefully written maps.

The interesting aspect of this approach is that one can start by structuring a knowledge-base which, by itself, is already usable, either in text form (a learning/reference manual) or in computer-based form (at a distance, for example). Then, as time and resources allow, one may develop the inference engine: the computer-based expert instructor that embodies all the heuristics and insights that are used by the best teachers of the subject. We have started a project along these lines, on digital electronics. The knowledge-base, or at least a large part of it, is complete and is being used as learning and reference material in conventional instructor-led courses (Arce and Romiszowski, 1985). By observing how the instructors use the knowledge-base, how they 'get inside the heads' of the students and how they take their instructional decisions, we hope, in time, to build the inference engine that is necessary to complete our 'expert electronics instructor'.

The design and development of such systems is still in its infancy. It is, however, an area which is showing much promise for the future.

13. Coding the Courseware for the Computer: Use of Standard Languages and Authoring Systems

The last stage in the preparation of a courseware package is the transformation of the detailed plan (the screens and the control algorithm) into a set of instructions for the computer. We do not see it to be the function of this book to go into great detail as regards this task. We have, indeed, stressed at several points, that if the initial design is well and fully documented, an instructional designer who has little expertise in the computer language concerned can, nevertheless, plan all the details of the courseware, leaving the coding to be performed, as a sub-contracted task, by a specialist programmer. The instructional designer must, however, have a thorough knowledge of what is possible/impossible/easy/difficult to accomplish with the given language, in order to plan a package that the programmer will be capable of transforming into a working system.

Two approaches to coding

Basically, there are two possible approaches to the coding of the package:

1. Use of a general purpose computer language.
2. Use of a special purpose courseware-authoring tool.

Each approach has its advantages and disadvantages. Also, there are many alternative languages to choose from in each of these two principal categories.

13.1 The use of a computer language

General purpose computer languages are often classified in terms of 'level', the lower levels referring to machine-like languages and the higher levels to human-like languages. In the early days of computing, programmers worked almost exclusively in machine code; languages that specified, step-by-step, the changes in state that specific positions (addresses) in the computer's memory should undergo.

High and low level languages

Later, a series of 'higher-level' languages were developed to simplify the task of programming, rendering it less mathematical and more like the processes of logical communication in natural human languages. These higher-level languages depend on the presence in the computer of an intermediate 'translator' language that transforms the human-like commands into the appropriate machine code routines. This transformation process is not always that efficient, in terms of speed and use of available memory, so, although most programming is carried out these days by means of one or other of the higher-level languages, there are some applications which still require the skills of programming directly in machine code. This is true in many games and other applications that present complex and possibly moving images on the screen. Some educational applications may well fall into this category.

The choice of a computer language

A skilled programmer should be consulted for advice on the language to use for a given application. Of course, in practice, one may be restricted in the choice of languages by the specific system to be used. A Sinclair microcomputer, for example, has its own version of BASIC permanently resident in memory (this type of memory is called ROM – read only memory) and there is no readily available facility for changing to some other language. An Apple II computer, on the other hand, with the installation of appropriate extension boards and the purchase of appropriate software, may have a variety of languages loaded into memory as and when required. Both computers can, of course, be programmed directly in machine code, if so desired. Of the available higher-level languages, not all are equally suitable for the preparation of courseware. COBOL, for example, was developed for business data processing applications and is not all that convenient for anything else. FORTRAN add ALGOL were developed for scientific and caluculation applications so are rather clumsy for dealing with large bodies of text, although they are sometimes used for courseware development. The most popular language, especially on microcomputers, is BASIC. As the name implies (Beginners All-purpose Symbolic Instruction Code), the language is easy to use and suitable for a variety of applications. There are, however, almost as many 'dialects' of BASIC as there are manufacturers of computers that use the language. Some versions are more useful than others for courseware development, principally in respect of the graphics facilities that they offer.

There are, however, several other general-purpose programming languages that are contenders for first place in popularity among courseware developers. Most of them offer substantial advantages over BASIC in terms of flexibility and versatility when applied to the design of courseware. The strongest contender at the moment is probably PASCAL. Other interesting languages include LISP and SMALLTALK. The first of these is the basis for the LOGO language, developed by Seymour Papert at MIT, now used widely for the development of logical analysis and programming skill in young children. The second was developed by the Xerox Corporation as an experimental language that would enable non-experts to write sophisticated programs. From the courseware developer's viewpoint, the system offers many interesting features, for example the capacity to display several overlapping 'pages' of information on the screen at once, like sheets of paper stacked on a desk. By pointing at a particular sheet with a special pointing device, one brings it to the 'top of the pile', revealing all its content, rather like pulling a given sheet out of the stack on the desk. This, and other special facilities, open up courseware design possibilities which would be difficult or impossible to implement with, say, BASIC. However, as usual, easy availability wins the day. Thus BASIC, followed by PASCAL, are the most common choices.

Have a go yourself   For those interested in learning to code their courseware packages, we recommend the 'suck it and see' approach. Get hold of a suitable computer system, take a beginner's course in programming if possible, or else just study the manual and have a go. If possible get hold of some programs already prepared, key them in, run them, then analyse them carefully, in order to see how they work. Try to apply some similar strategies to courseware of your own design. Progressively get more ambitious. It is not all that difficult, but it IS time consuming. As in all skills, practice makes perfect. However, do not forget that programming is not a totally reproductive skill, but a very highly productive, creative, skill. Many possible programming solutions exist to most relatively complex problems. Some are more economical, rapid, or just plain elegant, than others. The selection of good solutions depends on the development of strategies and cognitive schemata, which are strengthened as much through exchange of ideas with others as through sheer practice. So try to arrange ample opportunities to share your experiences with other programmers, preferably more experienced than yourself. If at all possible, arrange to work with others, as a team. The task of programming is tedious. You will need all the help and encouragement that you can get from sharing your ideas and results with the other members of the team. Also, the division of a large package between several programmers means that there is something substantial, ready to use with the students, much sooner than if only one author is involved. Do everything possible to avoid the 'loneliness of the long-distance programmer'.

13.2 The use of an authoring system

Courseware authoring tools currently available fall into two main categories – those that are structured in a way that is similar to a higher-level computer language and those that are 'menu-driven', in the sense that they present alternatives, in plain English, for the courseware author to choose from. We refer to the first group as courseware authoring LANGUAGES and the second as courseware authoring SYSTEMS.

A typical menu-driven rountine might run something like this:

A menu-driven system in action

* LOGON (some code word that, when typed in by the author, sets the procedure in motion, 'logging-on' the author, or in other words, registering the session and presenting the first 'menu':
* Do you wish to:

  A create a new lesson?
  B continue writing an incomplete lesson?
  C modify an existing lesson?
  D study a lesson?

  (There may be yet other alternatives. The author responds by pressing the appropriate key. In our case, let us press A.)

* OK. You wish to create a new lesson. What is its title? Please give it a code name that does not exceed 20 characters. Type in the code name and press ENTER.
  (The author types in an appropriate code name, say MATH14.)
* OK. The code name for this lesson is MATH14. Every time you wish to retrieve this lesson, use it or alter it, use the code name to call it up. Now you can give the lesson its full name and present any other information that should appear on the title page. Press CONTROL and C together. This will clear the screen and set it up so that you can type in your information, just as if you were typing on a normal typewriter. On the top line of the screen you will see all the information you need to plan your work – the line and column numbers that you are typing in. Remember that the screen has 24 lines, each made up of 40 columns. Press CONTROL and C to continue. Then press ENTER when you are finished.
  (When the author follows the first instruction, the screen goes blank and a flashing 'cursor' appears in line 1, column 1. Above this first usable line appears a message which says 'line 1/column 1'. As the author types, this message changes, to show the current position in which the next character will be typed. SPACE and NEWLINE keys allow the author to move the cursor to any position in order to type in the messages planned earlier on the screen planning sheet. When the screen is complete, the key ENTER registers it in the computer's memory and presents the next 'menu'.)
* Do you wish to create another screen of introductory text, such as a preface, index, contents list, etc? Press Y (yes) or N (no).
  (Let us assume that the author types Y.)
* As before, press CONTROL and C to start and ENTER when you finish typing the screen. Press CONTROL and C.
  (The author now types in the preface or whatever it is that is to be presented, then types ENTER.)
* Do you wish to create another screen of introductory text, such as a preface, index, contents list, etc? Press Y (yes) or N (no).
  (This time the author types N.)
* You may present a textual message to the student, followed by a question. The question may be open-ended short-answer type (up to 50 characters) or else multiple-choice, with up to eight alternative responses. When ready to type in this screen, press CONTROL and C.
  (The author does this, receives a blank screen, as before, with a flashing cursor, and types in his first exercise.)
* What is the correct response that you expect from the student? Type it now, then press ENTER.
  (The author types in the answer expected.)
* What message should the student receive after giving the correct answer?

Type it in and press ENTER.
(The author types in his confirmatory/congratulatory message.)
* How many incorrect, or anticipated alternative responses, do you wish to comment (maximum 10)? Type in the number and press ENTER.
(The author may, for example, type 4.)
* Type in the alternative/incorrect response number 1. Press ENTER.
(The author types in the response.)
* Type in the comment associated with this response. Press ENTER.
(The author types it in.)
* Type in the alternative/incorrect response number 2. Press ENTER.
(This cycle of response and comment repeats four times as the author has requested four incorrect answer comments.)
* You may present a textual message to the student, followed by a question. The question may be ....
(The cycle repeats, as shown above.)

The example just shown is a highly simplified sequence from a hypothetical menu-driven authoring system. It illustrates the main characteristic, however: the author needs very little skill in using the computer, in order to input his lesson plans. It also illustrates the characteristic of limited choice open to the author: in the case illustrated, to a 'herringbone' type of branching sequence. Real author languages are generally not quite that restricting. Most allow for sub-sequences following on from incorrect responses. Some allow the student to return and review sections at will. Usually, some form of answer-matching facility is included, that allows the acceptance of slightly misspelt words, abbreviated names, 'key words' as indicators, or partly complete responses. Most systems also allow for the unanticipated response-comments and directions for those who give a totally unrecognizable answer. Some offer built-in options 'on the menu' for the preparation of graphic material, use of colour or other special effects. Others give the option of including sub-routines written in other computer languages, to supplement the limited options available to the author directly from the 'menu'.

13.3 The use of a courseware authoring language

The structured, or coded, authoring languages, are rather like a special computer language. They contain commands that summon the available authoring facilities just when the author wishes to use them. The basic facilities are much like those just described for menu-driven systems, but there is no predetermined format to the control algorithm. The author decides at each point just what branches or loops are to be incorporated and informs the computer accordingly. The author is much less restricted, but the authoring task becomes more dependent on skill in use of the particular language. Some authoring languages are quite complicated, necessitating the use of sizeable reference manuals and many hours of practice.

The mastery of an author language of this type is not, however, a difficult task. A good description of what is involved in coding a lesson plan with a typical author language, is given by Dean and Whitlock (1983). This reference also describes the characteristics of most of the authoring courseware currently available. In practice, however, unless you are starting with a 'clean slate' to design a CAI system for your own needs, the choice of author language will be limited by what is available to run on the type of hardware that you own. It is as well to check carefully, before committing yourself to buying new equipment, just what packages exist that will help you to code your CAI lesson plans. As this is a rapidly developing field, it is essential to obtain up-to-date advice.

The range of authoring languages on the market
Dean and Whitlock (1983) provide a listing of about a dozen authoring languages that are available on the market, ranging from the ultra-sophisticated PLATO to the very much simpler PILOT. Most of these languages are tied to one particular make (and often, model) of computer, though some, like PILOT, exist in several versions for different computers. Systems like PLATO are designed

to be used on large mainframe computers (in the case of PLATO, on CONTROL DATA equipment) which makes their implementation rather expensive unless the particular computer system is already installed. Even then, the authoring language software is often leased, rather than sold, and the costs are high. This limits the application of such systems to intensive, large-scale, projects that will make sufficient use of the resulting courseware to justify the financial outlay.

The simpler packages, capable of being implemented on relatively cheap and widely available microcomputers, like for example the Apple II, are probably of greater interest to the majority of readers who might wish to get started in a small way. These packages are more limited, especially in terms of the CML facilities they offer, as they run on systems with a much more restricted memory capacity and thus cannot keep such sophisticated records of individual student progress.

The main facilities offered

However, the minimum facilities that *most* of them offer are:

* The presentation of text and questions on the screen, in an organized fashion, using the space to best advantage.
* The acceptance of student responses through the normal input devices available (usually the keyboard).
* The analysis of the responses received, to check whether they are correct (this may be simple or sophisticated, involving the analysis of synonyms, spelling mistakes, etc).
* The recording and filing of the responses made, in order to supply data for materials revision or student guidance.
* The ability to branch to other parts of the material in the courseware package, depending on the response patterns of a given student.
* The provision of immediate feedback, on the screen, in the form of comments on the responses made by student.

Other facilities that may be available, depending on the system being used, are:

* The creation of high resolution graphics in order to illustrate the presentations on the screen.
* The provision of colour (often up to 16 or more different colours) in order to highlight the information presented.
* The creation of sound effects, ranging from simple 'beeps' included to focus the student's attention, to sophisticated music and even voice synthesizers.
* The possibility of building in new commands, in order to extend the versatility of the package.
* The possibility of including sequences of courseware prepared in computer languages, such as BASIC or PASCAL.
* The possibility of creating item banks of test questions, for the automatic generation of tests.
* The possibility of using special input devices, like light pens, touch-sensitive screens, graphic pads, paddles, etc.
* Much greater sophistication in the management of the learning process, in record keeping and in student guidance.

Early planning

The field of CAI/CAL is fast developing, especially in relation to the courseware authoring tools available on the market. As these become more versatile and sophisticated, new opportunities open up for the creative instructional designer. However, every authoring tool has its limitations. It is necessary, therefore, to make a careful decision, early on in the ID process, on the specific hardware and software tools to be used, so as to avoid lost time and effort in replanning unrealistic screen designs and control algorithms.

One should also bear in mind that the computer screen is not always the best presentation medium for the message. Often a paper-print or an audiovisual presentation offers distinct advantages. Do not be afraid to mix media, using the computer for what it does best and other materials as support. How this can be done effectively will be discussed in the next section of the book.

PART 4.
Techniques for Audiovisual Materials Development

Overview

The section of this book which deals with audiovisual materials design may appear to have been better placed earlier on, if we are to adhere to the 'historical' sequence of discussion; after all, the use of film in education and training commenced before the Second World War. So did educational radio. Thus, scriptwriting for instructional purposes would seem to have a much older history than computers. However, a closer examination of the bulk of early educational/training films or radio (and indeed early audiovisuals employing slides and tapes, filmstrips, etc) reveals that these were, in the main, informational rather than instructional. They were designed as unidirectional messages on a given topic. It was up to the teacher to build them into the lesson, developing introductory and follow-up activities, evaluation tests and transfer tasks. Very seldom were these materials planned as part of a true instructional system and even less frequently would they 'stand up on their own' as a complete self-instructional system. The techniques of scriptwriting and of production were 'borrowed' from the commercial documentary and entertainment tradition, with very little adaptation to the special needs of instruction. Only after the development of instructional programming in the print medium did it become common to approach the design of audiovisual materials in an objectives-based and 'systems' manner.

Many of the techniques of materials development, first invented for the preparation of print-based programmed instruction, were adopted or adapted for the development of audiovisual 'instructional packages'. This transfer of techniques was quite successful in some cases and not so successful in others. The audiovisual medium is, by its nature, linear and sequential (or at any rate, that was the case until quite recently). Thus, where a linear, step-by-step sequence of instruction is appropriate (as in many procedural tasks) an audiovisual (or just audio, or just visual) sequence of presentations, followed by appropriate practice, proved to be highly effective (often much more effective than a print-based instructional text, for example). Where, however, the content to be learned is highly structured, complex, or interrelated to other topics which may or may not have been previously understood, the linear approach to instruction breaks down (in the audiovisual medium as much as in print-based self-instruction).

Successful instructional designs, when dealing with such content (examples include mathematics, sciences and most higher level conceptual learning), have used audiovisual materials as the 'explanation' or 'demonstration' component in a more complex, multi-media package which allows branching, repetition, alternative presentations, even group discussions. These were often coordinated by a teacher acting as 'resources manager', but sometimes by a printed workbook or study guide. Thus, the scriptwriting techniques for the audiovisual components continued to be based essentially on models for the transmission of pre-designed messages and continued to resemble the techniques used in commercial, documentary, advertising or entertainment scriptwriting.

Suddenly things have changed, for the interlinking of video and the computer has produced a new audiovisual medium, which is capable of as high a degree of interaction and individualization as any medium so far available to the instructional designer. It is for this reason that we have chosen to place the audiovisual materials section after the section on the impact of the computer. The early chapters in this part of the book examine the different uses to which audiovisual materials are 'traditionally' put to in instruction. We will note that, in relation to the individualization of the instruction process, these media have contributed very little. Apart from the individualization of learning rate, by putting the audiovisual in a carrel under learner control, the only other significant aspect of individualization was the occasional offer of 'alternative media' for the learner to choose from. The same message was made available in print form as a text, in audio form as a tape, or in audiovisual form as a videotape, for example. This was supposed to cater to 'individual learning styles' (learners may be 'visual learners' or 'aural learners', 'serialists' or 'holists', etc). However, the theory on this aspect is not very developed and in reality the cost of providing such redundancy in the course limited the practical application of this approach. However, the appearance of the interactive videodisc as an instructional programming medium shows signs of changing all this. Many new possibilities for the interactive and individualized structuring of audiovisual instruction are now available. In time, the costs are likely to become very accessible indeed. It will not be long before an hour's instruction on interactive video will be cheaper than an hour's instruction totally based in a microcomputer – the videodisc is potentially a very cheap information storage medium. And many say that today, an hour's CAI is much cheaper than an hour's conventional instruction at any reasonably large scale of usage (numerous target poulations).

However, as has been the case with so many 'new media' in the past (educational radio and television, programmed instruction and 'conventional' audiovisual instruction are all examples), the designers of instruction have not been prepared to exploit the benefits and opportunities offered. Despite the many excellent pilot studies and the many success stories, the large-scale use of new media has not generally offered the benefits expected. After early enthusiasm and some success, the popularization of the media has inevitably led to disappointment and gradual reaction and rejection. I believe we are about to witness a similar situation in relation to computer-based instruction as we have already experienced with programmed instruction, educational television and the like. A large part of the blame for this is due to those who misused the media in question, for applications where they were not suitable or, even when in principle they should have been suitable, using inappropriate instructional design methodologies. As, at the time of writing, the interactive video movement is just beginning to 'ride on the crest' of the first wave of enthusiasm, it is opportune to sound a warning. At least one of the purposes of this section of the book is to outline some of the dangers and suggest possible ways in which they may be avoided.

Chapter 14 presents a case study – a very simple and straightforward instructional objective, to be achieved by means of audiovisual materials (in this case, normal videocassette). This (true) case study serves to illustrate some of the points we have made in this introduction. It also serves to define quite clearly the three principal roles of audiovisual in an instructional (or indeed any teaching-learning) system. These are: motivation, information and instruction in its true sense.

Chapter 15 continues on to develop a classification schema of the forms of audiovisual materials and systems commonly used in instruction. Examples are given of each category in the schema. We see that in some of the categories, the audiovisual material is but a component of a more complex, multi-media package, whilst in other cases, it is possible to construct audiovisual self-instructional systems.

In Chapter 16, we examine a variety of approaches to the design and scripting of audiovisuals and develop a model for the design of instructional audiovisuals which is coherent with the principles of our generic ID model. We also analyse some excerpts of audiovisual scripts prepared for a variety of specific purposes – motivation, pure information and true instruction. We note, however, that the 'true' instruction that is possible with the 'conventional' audiovisual media such as slide-tape or videocassette, is limited to basically expositive instruction, following a more or less linear sequence of pre-planned study. Some repetition of the last learning exercise is possible when the package is studied on an individual basis under learner control. Some skip-linear sequences are possible when a learner demonstrates prior knowledge of a topic. But flexibility is limited and is often awkward to implement in practice, especially with less sophisticated learners who may 'lose themselves' when trying to wind a tape back or forward.

In Chapter 17, we finally come to a discussion of the higher degree of interactivity and individualization that is possible through computer-control of a videotape or videodisc. However, for the reasons outlined above, we proceed warily, suggesting that this too may turn into a 'passing fad' unless we are more careful to utilize it more appropriately than has generally been the case of other 'new media' in the past. Indeed, modern-day interactive video has some predecessors that slipped rapidly into oblivion in the 1960s and 1970s. We analyse why this happened and what there is about the 1980s' variety that may make interactive video a 'survivor'. We also review, albeit briefly, some other 'new technologies' which are threatening to revolutionize the way we live and learn (or so we are told by the futurologists). We conclude that interactive video and some of the other upcoming communication technologies do indeed offer the potential to design mass instruction in ways that hitherto may have only been possible in the very privileged interaction, on a one-to-one or small-group basis, of highly motivated, goal-seeking learners, with 'the best experts in the land'. However, in order to achieve this, radical changes will be needed in the common approaches to the design of instruction. The basic principles of learning and the generic ID model are capable of absorbing and accommodating these new techniques. But are the instructional designers capable, or indeed willing, to develop and use them? Are we prepared to design materials that are highly interactive and operate effectively under learner-control, rather than rigid system-control? Are we equipped to make intelligent and efficient use of the novel and very-high-frequency capabilities for the changing of the communication mode (print/graphic/video/ audio/animation/etc) which interactive videodiscs offer us? Will our instructional design skills keep up with the rapid progress of electronic technology, or will these new media turn into 'passing fads' because we fail to develop ID methodologies that utilize them in a truly cost-effective manner?

14. Audiovisual Media: Informational, Motivational or Instructional?

A case study

A few years ago, in 1979 to be exact, I was responsible for the organization of a programme of practical training in the 'production of videocassettes for training'. The client organization was a large telephone company that had just invested in video equipment for its new training facility. The trainees were a mixed bunch. Some were technical personnel with extensive experience of work in commercial television studios, but no background in training. Others were trainers, with qualifications in psychology or pedagogy and one or two years' experience as planners of training for the personnel of the organization, but no experience with the use or the production of video.

The training programme was organized in two 'modules', each of a week's duration, separated by a month or so, to allow the studio facilities to be completed. During the first week, we had at our disposal a rudimentary studio equipped with portable cameras and recorders and full editing facilities. Much of this first week was devoted to the planning of video for specific instructional applications. The trainees were divided into four teams and each team had the task of planning, scripting and then shooting and editing two videocassettes. Topics were chosen from our immediate environment – how to operate a Sony portable video camera, the use of lighting in a small video studio, identifying long, medium and close-up shots, how to use the editing desk, and so on. All the project tasks were instructional in nature, with clear overall objectives. Given the rudimentary nature of the equipment at our disposal, the resultant tapes were of surprisingly good technical quality, probably due to the experience of the technical personnel with studio experience. The same could not be said of the 'pedagogical' quality. All four groups had got carried away by the medium and its new possibilities. The use of zoom and quick cuts from one scene to another was so overdone as sometimes to lose entirely the sequence and overall context of the message being 'put across'. Musical background was used with such artistry as to distract all attention from the commentator and what he or she was saying. Special effects were used 'because they were available' and thus presented a challenge to the production team to see what could be done with them. Dialogues, discussion and dramatization were employed for no clearly discernible purpose. In the tape demonstrating different camera shots, a class of kindergarten children was used as the subject being filmed and the team got so involved in making good artistic use of the children's spontaneous reactions and play activities, that few of us could remember at the end of the showing what the overall objective of the tape was, let alone describe the special characteristics of each type of camera shot demonstrated. All were quite unanimous, however, that the tape was an enjoyable experience to watch.

When, at the end of these practical exercises, a group evaluation session was run, few of the participants showed real understanding of where, in their scripts, they had gone wrong, why their unbridled use of the special effects at their disposal had diminished rather than enhanced instructional effectiveness, or how they might proceed in order to remake their productions. One problem was that they had no real objective evidence to prove whether their productions would or would not teach effectively. They all had previously mastered the objectives set for the exercises and were not therefore able to act as a real sample of the target audience. As 'peer evaluators' they had not mastered the concepts of instructional design in this medium sufficiently to give useful feedback to their colleagues. Many indeed felt that the instructors were exaggerating the defects. We therefore decided to change our strategy for the second module of the training programme. Rather than giving guidelines for video planning to be followed by practice exercises and feedback (the expositive strategy of the first week), we adopted an experiential approach – each team would be given a specific training task to be accomplished by means of video and would find out, by testing on a real target audience, what works and what does not work. Then the group members would formulate their own guidelines and principles, to be employed in further exercises. We hoped, by adopting this strategy, to overcome the resistance and the rejection (based on disbelief) of principles presented by the instructors. In addition, this approach would simulate more exactly the working environment in which the trainees would find themselves, as video producers in the training department of an organization.

The first task, given to all four groups at the start of the second module of training, was to teach someone who had no prior relevant experience how to dismantle and assemble a typical telephone, of the type installed in most households and offices. This was seen as a typical training situation, which could form part of the training of factory assembly personnel and/or field installation and maintenance technicians. As, in our case, the exercise was also an experiment in the value of video as a medium for this type of training, we were somewhat stricter as to the 'rules of the game'. Only the video medium was to be used. No subsidiary textual or graphic material was allowed. To ensure that these rules were followed, and also that no prior learning would facilitate the mastery of the new task, the target audience was made up of men and women of a countryside background, who had never worked with complex mechanical implements, never been employed in a factory of any kind, never owned a telephone of their own and were indeed largely illiterate. Each of the four groups had the same task to perform, with an ample supply of 'target trainees' on which to test their productions. Each group had first to analyse the task, learning as they went along, how a telephone is constructed. The telephone set used was a modern design, which is made up of seven principal sub-units, that slot together and join up rather like a jig-saw puzzle, requiring only the use of a screwdriver and quite a considerable amount of dexterity.

Once the task was analysed, each group scripted and produced a videocassette to teach the necessary skills and procedures. Not one of the groups succeeded in teaching anyone of the target audience to perform the whole task. This led, therefore, to an analysis of learning difficulties and revision to the structure of the programmes, rescripting and reshooting, followed by try-out on fresh members of the target audience. It took the groups at least two remakes to achieve the objectives. One group managed to teach the task effectively with their third version, whilst the other three groups had to make a fourth version before they achieved success. The second stage of the week's training was therefore based on the analysis of a total of 15 different versions of a video to teach the same objectives, each version supported by objective data on its real effectiveness.

It was interesting to note that the four final versions, that actually did teach all the task effectively, seemed to have been produced from the same script. The differences were to do with style of language and treatment of shots, but the overall structure of the tapes, in terms of sequence, content and position of each shot, content of the commentary accompanying each shot and instructions to the viewer on what to do, were almost identical. In contrast, the earlier versions, that did not succeed in teaching, varied widely with respect to all these factors. It is possible to list the changes that occurred, as follows:

What the case study illustrated

1. Early versions were based on a complex shooting script, using two or more cameras, various shooting angles, frequent cuts to new camera angles, much zoom-in and zoom-out, etc. The last versions all used only one camera, positioned over the shoulder of the person who was demonstrating the task

(that is, in the position of the eyes of the performer of the task). No trick angles were used, no overall scenes showing the operator from a distance (the eyes of his supervisor perhaps), no sudden cuts or changes of continuity. Only some restricted use of the zoom-in and zoom-out facility, to identify small detail or overall context of the workspace, remained in the scripts, in relation to camera shots.

2. Early versions had long and complex commentaries, explaining the overall objectives of the task, when it is necessary to perform the task, the technical terminology used to name the various components and so on. Furthermore, the style of the commentaries gave the impression that they were written before the shots were planned – the visual content was there to enhance or illustrate the audio message. The final versions of all four groups were quite different. The commentary was cut to the bare minimum. Often there were long pauses as the hands of the demonstrator performed a particularly intricate step. Sometimes, difficult steps were repeated without any commentary at all. Technical jargon was cut out and replaced by more common terms whenever possible. In particular, the style of the commentary gave the impression that it had been recorded, live, during the filming of the demonstration, that it was the *demonstrator* who was talking. Much of the need for linking phraseology was eliminated – the script would say 'now pick up THIS part and hold it like THIS' while a finger would point out the relevant part, then a hand would pick it up and turn it to show, without a shadow of a doubt, what 'THIS' meant.

3. Early versions were invariably in the form of a 'straight run' through the task, demonstrating all seven steps of dismantling and all seven steps of assembly. Only one of the four groups built in a break, between assembly and disassembly, with a suggestion to the viewer that he or she might try to 'have a go' before continuing. Other than this, very little direct instruction to the viewer, or planning to involve the viewer as an active participant, was included. By the second version, all four groups had commenced to involve the viewer as an active participant in the learning experience. Stops were built in after one or more steps and the trainee was required to practise what was demonstrated, step by step. Failure to perform correctly was handled by allowing the trainee to rewind the tape and watch the relevant demonstration again. The improvement in learning effectiveness was immediately apparent, but much use was made of the rewind option and the overall learning time was therefore quite long. Not all the trainees managed to learn, although the group that built practice stops most frequently into the presentation (after each one of the steps) got closer to achieving its objectives. Then the breakthrough – one of the groups realized that the assembly and disassembly tasks could help each other, so maybe it was not a good idea to teach them one after the other in almost separate sessions. Why not teach them together? Once a step of dismantling had been done (for example the removal of the cover) the final assembly step could be demonstrated and practised. Then one could return to disassembly, taking the process a stage further (remove the cover and detach the dialling disc) and now the assembly of these two components could be demonstrated and practised. This was the sequence adopted in all final versions of the tape.

What we can learn from the case study

It is interesting to analyse the experiences of the four groups on this course in the light of some of the principles that we have already outlined in earlier chapters of this book. The objectives of this particular video presentation were instructional. Early attempts of the four groups could be seen to have treated the exercise as more informational or motivational than really instructional. Effort was put into explaining why things were done and justifying why they were done that way. This led to the over-use of heavy and detailed commentary and of trick visual effects and camera angles, that did nothing to improve the communication of the task that was to be learned.

Furthermore, the instructional objectives are clearly to do with the mastery of algorithmic procedures and the necessary physical skills to perform them. There is little theoretical justification for asking trainees to derive the procedure, logically, from 'first principles' (as the early scripts of some groups tried to do), as opposed to just applying the predetermined procedure. In order to learn to apply the procedure, however, the trainee requires a clear demonstration and ample opportunity to practise, with appropriate corrective feedback as and when necessary. We note the appearance of our familiar demonstration-practice-feedback exercise design model, already encountered in our discussions of lesson planning in general and of the mathetics model of programming in particular. Due to the nature of the objectives, the demonstration should be visual as much as is possible and the practice must be 'hands-on' execution of the task. Because of the limitations imposed on media and method choice in this exercise, the feedback and correction must be supplied by the trainee himself, comparing the actual results of practice, with those shown on the tape (no supervisor is peering over his shoulder to stop and put him right as and when required). Therefore the task must be practised in manageable steps, so that action and the evaluation of its results are always contiguous in time.

Finally, the breakdown of the learning task may be performed in may ways and some are perhaps better than others. Among the tactics that were attempted by the groups were the 'PARTS' method and the 'PROGRESSIVE PARTS' method, the latter showing itself to be superior in this case. Why this should have been so is difficult to analyse with certainty, from a theoretical viewpoint. However, given the very special characteristics of the target audience and the rather large number of steps in the procedure to be learned (14 if we include assembly and disassembly as two separate tasks) the 'progressive parts' method (which provides more practice in all and also builds the steps together into a whole activity) could be expected to offer advantages. The same may not, however, be so true for the case of more sophisticated trainees learning a new task that is in many ways similar to other tasks they have learned before. Such a target audience may well be successful in following a step-by-step presentation rather akin to the structure of a workshop manual (albeit in video), using prior experience to judge success and knit the separate steps into a whole meaningful task. To illustrate better the contrast between the two sequencing tactics adopted, we present a part of the sequences developed (see Figure 14.1).

The 'PARTS' approach	The 'PROGRESSIVE PARTS' approach
1a. Demo step 1 of disassembly. 1b. Practise step 1 to standard. 2a. Demo step 2 of disassembly. 2b. Practise step 2 to standard. 3a. Demo step 3 of disassembly. 3b. Practise step 3 to standard. 4a. Demo step 4 of disassembly. 4b. Practise step 4 to standard. 5a. Demo step 5 of disassembly. *(continue through 14 steps)*	1a. Demo step 1 of disassembly and the 　　last step of the assembly procedure. 1b. Practise these steps, disassembling 1 step 　　and then assembling again. 2a. Demo step 2 of disassembly and the 　　penultimate step of assembly. 2b. Practise disassembly of steps 1 and 2, 　　and assembly of last 2 steps. 3a. Demo step 3 of disassembly and . . . *(continue through 7 such steps)*

Figure 14.1

We have spent so much time analysing this one case, as it reveals most of what must be kept in mind when planning audiovisual (and indeed purely audio or purely visual) instructional packages. The case illustrates that, although there may be many new and special production techniques that a new medium offers (allowing us to do things that we previously could not do), the overall basic principles of instructional design still hold good. To exploit the new facilities at our disposal, we must identify a real need for them, and not just use them 'because they are there'.

The four-level ID design approach in practice – Level 1

We have seen, in this case, yet one more illustration of a multi-level decision-making approach to instructional design. Firstly, at *Level 1*, we define the problem and the constraints on our choice of solutions, satisfying ourselves that instruction is indeed the appropriate solution, and that the overall methodology to be employed is capable of really solving the problem. In our case study, this stage of decision making was imposed 'from above' by the designers of the exercise for their ulterior motives (to create an experiential exercise for the course participants). In a real-life situation, one would have considered the other possible solutions, such as not trying to train totally inexperienced personnel to do the telephone assembly task (a recruitment and selection-based solution alternative) or selection of a 'live' instructor/demonstrator from the shop floor and training him to give small-group training sessions on the shop floor (a different alternative of delivery media). There are many possible variations of overall approach to the problem and many of them could indeed be successful. The choice of the video medium did prove, however, to be a viable alternative, even if not necessarily the most cost-effective or rapid in all manner of practical situations.

–Level 2

Secondly, at *Level 2*, the overall objectives were identified as being principally INSTRUCTIONAL. The project exercise said nothing about motivating the target audience to take up employment in a telephone company on such assembly tasks, or about explaining to them the importance of performing the task correctly. If these had been the principal objectives, the resultant videocassette would have been very much different. This is not to say that in a real-life situation similar to the one simulated in this case, the motivational and informational objectives may not be important. However, they should not be confused with each other in the same video presentation, or the probable upshot will be that none of the different objectives will be achieved. In this particular case, it is probable that motivation is intrinsic in the execution of the task. In reality, people would arrive with a long-term motivation to earn a living by carrying out the task and, in the short term, would be continually motivated by their success in mastering successive steps of the task. This indeed was quite apparent in our simulated case. Although the sample used for try-out was not composed of potential recruits, they were quite obviously deeply motivated to continue working on the task when the videocassette led them to sense success. Motivation was very high on the final versions and not at all high on the early versions. Thus, it is a fair theoretical position to concentrate only on the achievement of the specific instructional objectives in the scripting of such a videocassette. If necessary, other informational and motivational objectives should be achieved in other exercises.

Once identified as to their nature, the objectives should be specified more clearly, in performance terms, on the basis of an analysis of the task. This analysis reveals the nature of the task as reproductive, based on the application of a standard algorithm, using simple psychomotor skills. This leads to the adoption of a theoretical position in favour of an expositive instructional strategy, of the 'demonstrate-practise-evaluate and correct' variety. Furthermore, as the algorithm to be applied is composed of many steps and the target population has little experience of learning similar tasks, some form of step-by-step presentation will be required. It only remains to be decided how to break the overall task into steps – what size of step to adopt and what sequence of steps to follow

–Level 3

At *Level 3*, decisions should be taken as regards the detailed sequence and breakdown of the learning into manageable steps. We are at the stage of lesson planning – and so far we have not really got to consider the special characteristics of the video medium that had earlier been selected. Now we begin to take these into account, but always in the context of our general instructional design and development principles that apply equally to any medium. We consider the need to go through the cycle of demonstrate-practise-feedback for every step. We note that observational control, by an instructor or other 'expert' in the task, is precluded. We therefore consider how the trainee may best be provided with feedback information so that he may take his own corrective actions. We come up with the observation that, apart from the obvious need for the trainee to repeat any demonstrations that did not communicate completely, feedback can be enhanced by chaining the separate steps more tightly to one another (an observation based on the application of some of Skinner's findings) and this can be achieved by adopting the 'progressive parts' sequence, in which each new step is practised in conjunction with all previously learned steps. Analysis of the nature of the task shows us that the progressive-parts methodology can only be easily put into practice when the trainee learns assembly and disassembly at one and the same time. Otherwise, someone has to prepare a series of part-assembled or part-disassembled telephones for each trainee to practise on.

The adoption of this tactic pleases from several other points of view. From the psychologist's viewpoint it is more conducive to the formation of an idea of the task as one whole assembly-disassembly procedure (this pleases the Gestalt and cognitive field fans). The hard-and-fast Gilbertian instructional programmer observes that although the disassembly steps are learned in a 'first to last' sequence, the assembly steps (which are arguably the more important and more motivating) are practised in the reverse 'last to first' sequence (so that an element of 'backward chaining' has been built into the exercise). The subject matter expert is pleased because the trainee is learning in a way that more closely simulates the real job situation, where telephones (at least in the job of field maintenance) do not always have to be completely dismantled and assembled. As the real job often involves partial disassembly followed by assembly from that starting point, the selected training procedure is a closer simulation of reality and offers more practice runs over the steps that are more frequently practised in reality. The reader may study the 'progressive parts' sequence shown in Figure 14.1 to check this. Finally, the video expert is also pleased, as the progressive parts approach seems less repetitive and less fragmented as regards the style of the visual message, requiring only seven breaks for practice, rather than the 14 breaks necessary in the linear 'parts' approach.

—Level 4

It is at *Level 4*, however, that the media designer really gets into the decision-making process as a principal partner. Now that all the major structural decisions have been taken, it is the video expert's job to decide how best to implement the lesson plan. It is here that knowledge of the language of visual communication plays its part. It is now that the role of any audio commentary is put into its proper place. It is now that decisions should be made on whether appropriate camera shots and special video effects will contribute to the overall effectiveness of the product as an instructional package (or a component in such a package).

Other uses of the AV medium

Of course, we have been dealing here with a very specific type of instructional package, aiming at a specific category of objectives. It should not be surmised that this type of videocassette package is the only one we recommend, or even that it is the most common application of the video (or more generally the audiovisual) medium to instruction. It is probable that the role of television in education is, quantitatively, much more in the informational and motivational camps than in the 'hard' instructional applications exemplified by our case study. The traditional role of educational television in many countries has been limited to an enrichment of the other instructional activities performed by the teacher. The British Open University has found that even in their situation of constraint to three principal media of communication (radio, TV, and printed materials), the media of radio and television play a supportive and enrichment role to print, which bears the principal burden of instruction. However, as we are dealing principally in this book with the development of instructional materials, we shall not devote much space to these other very important roles of audiovisual media. The interested reader is directed to the extensive and rich literature on media design and development, written with general educational applications in mind. This literature varies from practical how-to-do-it books (for example, Garrison, 1970; Wilkinson, 1971; Romiszowski, 1974; MaCrae et al, 1981; Wittich and Schuller, 1973; Brown *et al*, 1973; Heinich *et al*, 1982) to more theory-based and research-based treatments of the audiovisual field or of specific media in particular (see Schramm, 1972; 1973; Bates and Robinson, 1977; Dwyer, 1972; Baggaley, 1976; Baggaley and Duck, 1980).

15. Audiovisual Instructional Packages: A Review

We shall in this chapter concentrate our attention on the analysis of audiovisual packages that can, at least to a certain degree, 'stand on their own' as self-instructional systems or as components of such systems. We shall therefore link our discussion and our choice of examples to the general philosophy expounded in this and the previous volume of our instructional development series. We shall also limit our discussion to the domain of audiovisual media. This will not exclude the mention of purely audio-based packages or of purely visual packages, but will not include some training simulator packages that rely on the tactile, kinaesthetic or other senses. (Some such devices have indeed been mentioned in earlier chapters on simulations and games.) Using the matrix in Figure 15.1 as a reference, we shall therefore be concentrating our discussion on examples that would classify in the lower half of the first three columns – audio, visual and audiovisual media that are largely 'self-presenting' rather than relying on a teacher to present them as part of the lesson. This implies that the package, whatever its other characteristics, is designed and developed on the basis of a detailed lesson plan.

15.1 The functions of AV media in an instructional system

We already have a model for the development of such lesson plans. This was originally developed in the previous volume, but has been used at various points in other chapters of this book. It is characterized by having a three-column format, which corresponds to the three forms of intercommunication that occur between teacher and learner in a 'true' instructional process. Figure 15.2 uses this lesson-planning model as a framework on which to organize some of the applications that exist for audiovisual media in instructional systems.

AV media as 'information transmitters'

Perhaps the most obvious, most common and most 'conventional' use of the audiovisual media is in the information transmission role.

Indeed, perhaps the most frequently quoted justification for the use of audiovisual materials in teaching is to bring into the classroom stimulus material which otherwise would require the learner to go out to seek the experience (or, perhaps, could not otherwise be experienced at all). This is the justification for geography films, dramatizations of historical events, descriptions of industrial and other processes, interviews or debates with well known or controversial figures, etc. Such descriptive stimulus material may indeed be built into an instructional design, although the scripting and production of such material may not itself follow strict instructional design principles. For this reason, it is possible to make use of a vast range of existing audiovisual material as components in a purpose-designed instructional plan. Not all materials that one comes across are, however, equally effective at communicating their intended message. One should be quite clear firstly about the overall purpose in the mind of the author of the material. Was it to inform? If so, to inform whom about what? Have the characteristics of the target audience and the informational content been adequately taken into account in designing the communication? Was it to influence or to motivate or interest the target audience? If so, to interest or motivate whom, and in what? Once more, have important audience and content characteristics been taken into account? Or was the author's real intention to teach something? Are there some discernible specific objectives that learners should achieve? If so, what are these

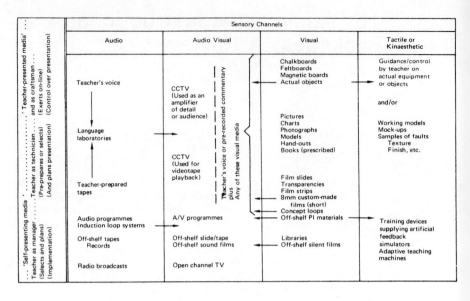

Figure 15.1 *A classification of instructional media
(Romiszowski, 1974)*

Teaching activity (Stimulus)	Learning activity (Response)	Control activity (Feedback)
Pre-instructional or opening activities — AV overview, preview, 'advance organizer' — 'Motivational' AV presentation		
Instruction — Transmission of new information — demo, explanation, etc — Presentation of 'models' of desired performance — Presentation of case material for analysis, etc	— AV as a practice exercise — AV 'job aids' — Use as a guidance or timing device during learning	— AV recording of learner's performance for evaluation — Comparative presentations of correct and incorrect responses
Post-instructional or closing activities — AV summary — AV enrichment		

Figure 15.2 *Possible functions of AV in a lesson*

objectives and what aspects of a total instructional system are really present in the material being analysed?

Very often, the answers to such questions are unclear, ambiguous, or else show that the original authors of the materials did not really consider them at all. No discernible purpose or target audience can be identified. Even so, this does not preclude the use of such material, if the instructional designer can impose valid objectives.

AV media as presenters of performance models

Of course, not all stimulus-materials are descriptive presentations of events, phenomena, or people. In cases like the one analysed in the previous chapter, the stimulus material is a careful demonstration, to be studied and imitated by the learner. Other examples of such use of audiovisual media as 'models of desired performance' include the 'role-models' presented by means of film or videotape as a first step in an expositive strategy geared at the development of interactive skills. One specific example, often applied, is in the training of office personnel and sales personnel in the procedures of handling customers and clients. Another similar application is the presentation of videotaped 'models' of good examples of specific classroom-teaching skills, as a first step in a 'microteaching' programme of teacher training.

AV media as presenters of cases/ examples

Yet another form of stimulus material is the presentation of good and bad, or appropriate and inappropriate examples, as the first step in a sequence designed to develop specific discriminations or concepts. In such a case, a series of short snippets of audio, visual, or audiovisual material may be presented as examples for criticism, comment or classification. One application in the training of submarine crews involved the presentation of the sounds of different types of ships, as heard in a submarine below water, to develop the skills of auditory recognition of friendly and enemy ships in situations where the use of sonar equipment is likely to give away the submarine's position. Another application, in medical education, involves the presentation of short excerpts from patients' interviews, as a stimulus to the practice and development of diagnostic skills associated with the interpretation of *what* patients say and also *how* they say it.

Also, of course, we have the stimulus materials that may be used as pre-instructional or post-instructional presentations, aiming to motivate, to supply 'advance organizers' or to indicate wider implications.

AV media as practice – guides – aids – opportunities

Coming to the central, or learner activity (or 'response') column, we note certain highly specific uses for audiovisual media. One may use audiovisual devices as *practice-media* as in the case of the tapes actually recorded by the learner in a language laboratory exercise. One may also supply some forms of 'job-aids' to facilitate practice, by means of audiovisual devices. One example that springs to mind is the use of audiotape as a timing device. This is done in the 'Sight-and-Sound' type of typing training situations, in which the trainee who is looking at an exercise to be typed (presented in the normal way on a sheet of paper) is also listening to an audiotape that is reading the exercise in the appropriate rhythm for typing. Each character to be typed (including spaces and punctuation marks is read out to the accompaniment of a metronome, set to give a specific typing speed. In time, the typist graduates to progressively faster tapes, thereby being paced to work at more productive levels. A similar timing use was employed in the Swedish 'RITT' system for operative training, which we shall describe later on. Another form of exercise control function is performed by the audiotapes used in the rolemap form of exercise, described in Chapter 8. In these cases, an audiotape performs a timing function by just running on in silence for a prescribed time and then giving instructions to finish one activity and commence another. Visual or indeed audiovisual materials may also be used in such an exercise-control function, as in the case of a series of slides that change at a predetermined rate, showing the stages of assembly of a transistor radio on a production line. Such applications, used in the 1960s at the Philips factory in Eindhoven, perform the three functions of partial demonstration (stimulus material), control of the rate of practice (response-control material) and presentation of expected results, by which the trainee can evaluate progress (feedback material).

We have thus come to consider the third, or 'feedback' column of our lesson planning model. The slide sequence just described is characteristic of one type of feedback material, which, in a step-by-step procedure, presents to the learner a visual, audio or audiovisual unit with which to compare his or her own efforts. Such feedback material is especially useful when the learner must identify correct or incorrect procedures or decisions by their results and these results may be identified in reality through the auditory or visual channels. Sometimes, as in the case above, the expected outcome of a task-step is illustrated and the learner is left to compare the actual results achieved with the expected results. In other cases, a recording of the learner's behaviour is presented for self-evaluation, as when videotape is used to show a learner his or her own performance in a microteaching exercise, during an interviewing skills role-play session, during a simulated sales meeting and so on. In yet other cases, a somewhat more sophisticated feedback approach may be employed, which presents the learner's responses and contrasts them with correct responses. To some extent, the audio-active-comparative methodology of using a language laboratory does just that. In contrast to the simpler audio-active approach, which involves listening to and repeating a series of drills and exercises, the audio-active-comparative approach records the learner's own attempts in the spaces left on the tape presenting the exercise. When the tape is replayed, the learner hears his or her own attempts alongside the models of pronunciation presented in the exercise, thus facilitating the identification of any pronunciation errors. As the learner's responses are recorded on one track of the tape and the models on another, it is possible to repeat the exercise as often as is necessary, re-recording ever more perfect versions on the learner track, erasing the previous attempts at the same time.

A rather more complex approach, though basically similar in principle, was employed in the submarine training example mentioned earlier. This approach, developed in the British Royal Navy in the 1960s, used the audio medium in all three stages of the lesson design. The topic of ship recognition was first introduced by an instructor, explaining the principles of auditory recognition and classification of ships by the noise they make and developing a classification of 20 basic categories of sounds. To do this, the instructor would use an audiotape of characteristic sounds as stimulus material. Then the trainee seamen would pass to drill and practice on an individualized training exercise which in effect simulated the task of ship-noise recognition. Each seaman would listen to a range of noises that would be presented to him over headphones and would classify them by responding on a special keyboard. If he classified the sound correctly, he would know, because a new sound would be presented for classification. If, on the other hand, he were to classify the sound incorrectly, the two sounds – that originally presented AND the one that he thought it was – would be presented superimposed on each other, so that the difference was accentuated. The seaman would then respond again, trying to correct his original error, until the discrepancy between the sounds was eliminated, thus giving him knowledge of having correctly responded.

Similar tactics are employed on some automated devices for the training of musical perception. The technique is becoming quite common in music education microcomputer software. The micro generates sounds and supplies the corrective feedback. In the original Royal Navy example, which was implemented well before the era of the microchip, a bank of five four-track reel-to-reel tape recorders were employed to furnish continuous access to all 20 sounds to be classified. This was controlled by a special dedicated logic device, built of valves, resistors and bits of wire, that would select a sound at random from the 20 available and, when the learner responded, would either select another (in the case of a correct match) or else would superimpose the incorrect choice on the original sound, playing both through the earphones (in the case of an incorrect match). Today, the implementation of such training exercises requires rudimentary programming skills, as compared to the design of special 'teaching machines'. However, the principles of the design of such exercises have been around for a long time and are largely independent of the specific medium of communication being used.

We have analysed these few (some quite unusual), examples of the use of audio, visual and audiovisual media in the design of instructional exercises, to illustrate that our general instructional design approach holds good even in this field. We have shown that in some (perhaps only a few) cases, audiovisual media may supply all three of the essential components of communication in an instructional system. More often, however, they are restricted to only some parts of the whole system, most usually to the supply of stimulus material. We thought it useful, however, to illustrate all three possible functions, in order to alert the instructional designer to the possibilities that exist for the fuller exploitation of certain media in instruction. We meant also to illustrate, yet once more, the value of thinking carefully about the structure of an instructional system, so that we may not fall into the error of expecting too much from a media presentation. It is too often that one comes across a videotape, a film or a slide-tape presentation, used on its own largely, and described as an 'instructional system' or 'instructional package', when even a perfunctory analysis reveals that the package is only transmitting new information, or merely awakening the interest of the audience in a subject. The presentation may do this very well indeed, but that does not justify our passing it off as an 'instructional' package.

15.2 Types of AV packages and components

Now that we have an overall view of the role and the potential of audiovisual media in self-instructional packages or systems, we should have a look at some of the great variety of such packages that can be produced. In the analysis of this variety, which is summarized in Figure 15.3, we have included some categories of both visual and audio components in such packages. In order to give coherence to our classification, we have included REALIA (real objects and situations) as a source of principally visual (but also audio, in some cases) information. We have defined a number of horizontal rows for different types of (mainly VERBAL) audio – notably the 'commentary' type of audio component, often in the form of voice-over by a presenter who is never seen, or who only appears infrequently in linking sessions, and the 'conversational' type of audio component, which we take to include two or more people debating or discussing a topic, the dramatized presentation of plays and playlets, attempts to converse with the learners or to involve them in the presentation in ways other than just watching, listening and possibly completing some form of workbook at key moments. This distinction, though not too strict in all cases, is meant to discriminate between media presentations that are purely expositive in nature, and those that might be thought to lead at least in part to experiential learning. Finally, we have also included 'blank' rows and columns, to be able to classify pure 'audio' or 'visual' packages and also, on the visual continuum, a column for 'print' – this last in order to be able to include in our schema such hybrid types of package as the audio-workbook which is dependent on the interaction of the audiotape and print media. We have not included, however, a listing of examples of purely print-based self-instructional packages, as the variety of these is so great as to overload our schema. In any case, we do not propose to deal with print-based packages in this chapter, as these have been dealt with at some length in preceding chapters.

The examples listed in the cells of Figure 15.3 are but a sample of some well known and some lesser known varieties of audiovisual package, selected to give a coverage representative of the possible variety and aimed at stimulating the reader's imagination to invent yet other forms. As we have stressed so often, the instructional design and development process is heuristic and highly creative in nature.

We shall now give a short description of some of the examples mentioned in the schema presented in Figure 15.3. These descriptions are quite short, as they

VISUAL COMPONENT

	1. No visual component	2. Print	3. Realia	4. Still visuals	5. Moving visuals
A. No audio component	Outside the scope of this chapter		Exercises programmed around the use of real objects – 'fault museum' – on-the-job practical training exercises	Instructional packages using sets of visuals only – radio assembly training at Philips – the game of 'RUST'	Silent moving pictures, as the main instructional component – 'concept-loop' 8mm films
B. Expositive commentary/ instructions	Audio-instructional packages/programs – the 'audio-active comparative' language laboratory tapes – audio-discrimination training – submarine crews, musicians, etc	Audio-workbooks – generally the stimulus material is on tape and practice/feedback in the workbook	Audio-directed practice with real objects/situations – the 'RITT' system – 'Walkman' directed practice/drill	Audiovisual packages – slide/tape – disc/filmstrip – tape/cartoon book which include practice/question and feedback	Video instruction – telephone assembly – on-job, video-based training Instructional television (some) Interactive video (some)
C. Dramatized/ conversational (possibly experiential) script	Radio programmes and audio-tapes with stimulus material designed to create experiences – case material – plays (post-listening de-brief required)	Audio-workbooks aiming at attitude change, etc – 'Glaxo' medical programmes – rolemaps – group programs	Audio-directed exploratory discovery – 'Walkman' directed nature walk – audio laboratory experiment guide	Audiovisual systems that attempt to generate group activity – radiovision – narrow band video	Audiovisual/video systems that promote group activity/interaction – TV-executivo – instructional television (some) with group audience participation

AUDIO COMPONENT

'Hybrid' or 'Eclectic' systems
– the 'audio-tutorial' approach developed by Postlethwait
– the 'resource-based' multi-media approach

Figure 15.3 *A schema of audiovisual package types*

are intended to illustrate the conceptual schema which we are presenting in this chapter. Readers interested in learning more about a specific type of package are directed to more complete sources. We shall work horizontally across each row of the schema, commencing with the first one called 'No audio component' (that is, with purely visual materials).

Cell A-3: the 'fault museum'

The first cell (the intersection of 'No audio' and 'No visual') is necessarily empty. So is the next ('No audio-Print') cell, for the reasons we gave above, of limiting our schema to the main area of interest of this section of the book. In the third cell, we have included one specific example of a whole range of possible packages which depend on the presentation of a real situation or object, to be 'played with' or 'used as a tool', the result being the achievement of specific instructional objectives. Of course, just about all practical and many simulation exercises could be classified in this category. Also many on-the-job practice exercises can be set up so that just by doing the job, the learner gets feedback of sufficient quality and with sufficient frequency and immediacy to guarantee the achievement of the desired learning objectives. The example mentioned specifically is one often used in the training of quality control skills. Very often, a person must develop specific perceptual skills in order to identify, visually, and deviations from expected standards of quality – surface finish, colour, texture, etc. A 'faults museum' is assembled, of real examples of the product ranging from 'perfect' to quite obviously 'defective' for a particular characteristic (say shade of colour in a stocking). The range of examples in the 'museum' must include many 'in-between' as well as 'obvious' good and defective examples, so that the learner is exposed to levels of 'fine-grain' discrimination, that initially are beyond his or her capabilities to recognize. If several different characteristics are important, the museum should include graded examples for each of the characteristics on its own and in combination with others. Training depends on using the 'museum' as a job-aid. First, a range of standard examples is presented, the trainee attempting reached, training continues 'on the job'.

Cell A-4: still visual packages

The next column of the first row deals with visual instruction packages, using *still representations* of some form or other. Any form of still visual media could be used: pictures, photos, slides, etc. The two examples are of 'true' instructional systems, in that both use still visual media as the main means of demonstration, practice and feedback. Both have already been mentioned earlier on. The slide-controlled, assembly-task training, as used at the Philips radio factory, breaks the task into such distinct and small steps, that a series of slides are sufficient to show the trainee what to do next, with what components, and how it should look when finished, thus providing both 'up front' information and guidance and then 'end of step' feedback information. In between, the rate of presentation of the slides, which is variable, may be used to control the development of appropriate production speeds. The presentation speed in the case described was under the control of the learner, who would decide when she was ready to progress to a faster rate of production.

The other example mentioned in this cell uses a different form of still media presentation: colour photographs of the leaves of coffee plants in various stages of infection by a disease colloquially called 'rust'. These photographs were used in the form of a card game (based on the game of 'snap'). The game provides stimulus information to the trainees (plantation workers), provides practice exercises and feedback of results. (Thus it is a complete instructional system.)

Cell A-5: the 'concept loop' film

In the last 'no audio' cell, we mention the 'concept loop film', as an example of highly successful instructional use of silent film, very highly acclaimed in the late 1960s and early 1970s. Its use seems to have gone into decline in recent years, probably as the result of the general displacement of super-eight film by the more versatile, more convenient and, in reality, cheaper videocassette. In its heyday, special loop-film projectors were developed, using cassettes of film that could not be longer than about three minutes and, in most models on the market, without a soundtrack. These apparent limitations, however, proved to be the strengths of the medium, as they forced the designer to plan short video clips that were self-contained and which would use the visual-communication potential of the content to its best advantage. The medium forced the designer to think in terms of specific objectives to be achieved by the visual communication and to make the best use of the short communication time available. Hundreds of effective loop films were developed in subject areas ranging from physics and biology in the formal 'education' area, to nursing, medical diagnosis, engineering and sports in the 'training' area.

Since the advent of more convenient and cheaper methodologies, the use of single concept super-eight films has waned and production of new films has almost stopped. This has, however, led to many designers forgetting how much could be achieved with the more limited medium, or rather, how many instructional situations may actually benefit from very short self-contained presentations and from a reduction or indeed total elimination of the audio commentary. There is nothing to stop the instructional designer of today from implementing the same design principles that were so effectively applied to the development of the concept loop films. However, as the case study presented in the previous chapter demonstrates, one tends to get carried away by the more sophisticated capabilities of currently available media, often to the detriment of the package's effectiveness.

Cell B-1: audio-instructional packages

The second row of our schema presents examples of packages that use an audio commentary (as defined above). In the audio-only cell, we list two examples previously mentioned in this chapter. These are language laboratory tapes, of the audio-active-comparative type (a 'true' instructional system) and the audio-discrimination exercises as used for the training of submarine crews and musicians. We have not included the host of expositive audiotapes that present descriptive information, as these do not, by themselves, constitute instructional systems. They may, of course, be readily incorporated as part of an instructional design – usually the information-transmission, 'stimulus material' part of the package.

Cell B-2: the 'audio-workbook'

This brings us to the consideration of the 'audio-workbook', a combination of an audio recording as the principal stimulus material, with a printed workbook that supplies guidance on use, practical exercises and usually the feedback component. Sometimes, however, the feedback may also be included on the audiotape. Such audio-workbooks, when well designed, are true instructional packages, providing all the system-learner interaction that is necessary to ensure the achievement of the predefined objectives.

Many authors and instructional designers have advocated and implemented the use of audio-workbook systems. They are essentially quick and easy to design and produce and, when appropriately used, are quite effective means of individualizing instruction. Very often, however, due to the limitations of audio as a medium of instruction in certain subjects, the audio-workbook is used in conjunction with other types of exercises. Langdon (1978) mentions three such uses – as a review or revision of course content previously studied in other ways, as an adjunct used in conjunction with other learning materials (for example a book which is referred to as reading assignments at specific points in the audiotape) and as 'update programmes' designed to keep previously trained people up to date on new developments in their job or the subject area related to it. One very widespread application of this type is the provision of audiotapes and audio-workbooks to the medical profession. Doctors have a constant need to update their knowledge. This has led to the offering of various services of an update nature, to which doctors may subscribe. One such service, based on audiocassettes, was launched in the 1960s in the United Kingdom by the Grays, a husband-and-wife team of medical practitioners. The Grays offered a constantly growing library of audiotapes on new developments in the medical field, on a subscription basis. Many thousands of doctors subscribed to the service. One big advantage of audiocassettes, for the doctor, was that, at least in the United Kingdom, the typical medical practitioner spends several hours a day in his or her car, travelling to visit patients in their homes. This, otherwise lost, time can be spent listening to 'update' tapes, rather than to music, on the car's audiocassette system.

This use of pure audiotapes does not, of course, qualify as a complete instructional system. Some services similar to the Grays', therefore supply a simple question and answer workbook, enabling the doctor, at some time that is convenient, to evaluate comprehension and retention of the information presented on the tape. This may lead to further study and review of certain parts of the tape.

Langdon describes a method of indexing the tape in such audio-workbook applications. One side of the cassette tape is used for the audio message, whilst the other side has a sequence of numbers recorded at five-second intervals, rather like the page numbers in a book. When working on questions in the workbook, the user has index references given in the text, so that the appropriate section of audiotape may be rapidly located. The cassette is slipped into the player so as to play the 'B' side with the numbers recorded on it and rapid wind forward or back soon locates the appropriate number. Then one just flips the cassette to start listening at just the right point of the commentary. This simple trick overcomes much of the inconvenience associated with the use of audiocassettes for 'random access' review purposes. According to Langdon, this indexing method is independent of the equipment used and can locate any desired section of the tape in no more than 30 seconds. Further details on methodologies for the development and use of audio-workbooks may be found in Langdon (1978).

Cell B-3: audio-directed drill-and-practice exercises

The next call presents examples of audio packages involving the use of realia – real objects and situations – to supply the necessary visual stimulus information. An early and 'classic' example is the audio-training RITT system, first developed at the SKEFCO company in Sweden, in the early 1960s. This method uses an audiotape to describe and to pace the performance of a cyclic industrial task. It was first applied to the training of operatives working with semi-automatic turret lathes and other such-like machines in a ball-bearing factory. This was batch-production work and labour turnover in the factory was high, so that usually, when the time came around to produce another batch of a particular type of product, the previously trained person was no longer available and someone new had to learn the job. The typical learning curve on these jobs was such that full-speed production usually took four weeks or more to be achieved. By then the batch was produced and a slightly different job had to be learned. The RITT system was in essence an audiotape recording that explained the sequence of operations to be carried out. The new trainee would listen to the recording, while a previously trained person, also listening to the tape, would carry out the task as instructed, thus providing a demonstration. Then the two would change places, the new trainee attempting to execute the job while listening to the instructions on the tape. The experienced worker would watch and, if necessary, correct the novice's actions (physically, by new demonstration, rather than by explanation). Once the novice was managing to complete the tasks to a reasonable (but very slow) standard, the experienced worker would leave. The novice would now work under the control of a series of tapes, each one recorded at progressively faster rates of work. This, of course, meant that more and more of the explanations had to be cut out. Eventually, the trainee would be working to a tape recorded at 100 per cent of EWS (Experienced Worker Standard), which would give only the bare minimum of guidance as to the sequence of actions to be performed. The trainee would move from one tape to the next at his own discretion, when he felt ready to try a faster speed of work. However, the typical four working weeks to reach EWS, was cut to between two and four days.

The system, without the pacing element, also proved itself to be most effective in a variety of situations. The present author used it, in comparison with printed self-instructional materials, for the training of engineering apprentices and found it about 30 per cent more effective (in terms of retention) and much more efficient (in terms of learning time) (see Romiszowski, 1974). The British Royal Navy used a similar system, some years later, for teaching new sailors how to find their way about an unfamiliar ship. The trainee would simply don a 'Walkman'-type of portable cassette player and headphones (or better, a single earphone so as not to cut out all environmental noise) and would follow the recorded itinerary, instructions and explanations, to learn the locality and method of operation of all the ship's important systems. As compared to the original Swedish RITT equipment, the current 'Walkman' is simplicity itself. The Swedes developed their methodology before the audiocassette age and therefore used heavy, reel-to-reel tape recorders, located in a separate room out of harm's way and beaming the taped instructions to the trainees by radio-waves.

Compared to today's equipment, the RITT system was clumsy and expensive from a 'hardware' viewpoint, but the instructional design principles which it employed were first class. Effectively, it was an automated and much more efficient version of the TWI (Training Within Industry) approach to skills and procedures training. It has many applications and some interesting side benefits (some related to drastic cuts in labour turnover on the jobs in which it was used). One very special side benefit reaped by several Swedish firms was related to the use of seasonal labour from Southern Europe, to overcome labour shortages in certain sectors of Swedish industry.

These seasonal workers would stay only a few months in the year and thus created special training problems, especially as most of them did not speak any Swedish on arrival. It was found possible to record tapes with two tracks in parallel presenting a synchronized set of instructions in two languages. This enabled a Swedish worker (who was not a qualified instructor) to supervise the training of an Italian or Spanish trainee (who did not speak a word of Swedish). No words would be exchanged. The audiotapes saw to all communication needs, apart from the visual demonstrations, needed to show the trainee how to work and to show the supervisor that learning had taken place. Further details and references on the RITT system and related audio-instructional approaches may be found in Romiszowski (1974).

Cell B-4: the 'traditional' audiovisual package

The discussion of 'audiovisuals that don't use visuals' in the more accepted sense, brings us to the category that, to many, is the only meaning of the term 'audiovisual package' – a package composed of an audiotape (or disc) and a set of slides (or a filmstrip). Many people use the term 'audiovisual' in this restricted sense – one can sometimes come across statements like, 'Should I use an audiovisual or a videocassette for this lesson?'. We prefer the more descriptive terms of 'slide-tape', 'disc-filmstrip', or whatever, reserving the term 'audiovisual' for the whole field under discussion. We shall not dwell very much here on this category, as we shall return to study examples of slide-tape and similar packages in the next chapter. We should stress here, however, that such packages may, or may not, be 'true' instructional systems. Many (indeed the majority) are not. The teacher must build them into his or her own instructional design. The danger, however, is to use them in isolation, as if they were complete instructional systems and expect to achieve objectives which actually require the use of further practice exercises, discussions and so on. Many commercially available packages include teacher guidelines on how to build them into an overall design. Some supply supplementary print materials that in effect make the total package into a 'true' instructional system. But most are merely *information packages*.

Cell B-5: moving audiovisuals

The last comments are equally valid for the vast majority of audiovisual packages that are based on *moving visual presentations* – film and videotape in particular. The majority of these are also basically informational or motivational materials, which may be used as components in an overall instructional design, but do not by themselves classify as 'true' instructional packages. The problem is that these, as indeed the previously discussed categories, often are labelled 'instructional' and are taken to be such by their users. In the following chapter, we shall be careful to distinguish between audiovisual scriptwriting for informational or motivational material and the design of truly instructional sequences.

Among the examples in our schema we list productions like the one described in the last chapter's case study. This is by no means unique. The South African company, Iconotrain, has developed such video-based operator training systems for many large organizations. Much of the training in the South African gold mines is carried out by means of on-the-job use of short video snippets, similar in style to the telephone assembly example described earlier (and also similar to the 'concept loop' methodology in some respects). The implementation is however similar to the RITT system, in that skilled colleagues are used to provide corrective feedback when the trainee applies to the job what was demonstrated in the tape. The skilled employee has a smaller role to play, for all the demonstration is performed by the videotapes but, in the final feedback-giving stages, his role is important. As with the Swedish experience, it has been found quite easy to produce multi-lingual versions of the videotapes and thus overcome previous difficulties of communication with migrant workers from neighbouring countries and territories. Iconotrain claims that whole training centres have been replaced by on-the-job training based on short video segments.

Of course, this operative-training application is not the only type of video that would classify as instructional. In the conceptual learning field, short video sequences followed by questions and discussions can be acceptable instructional designs. Sometimes, the questioning and discussion element is instigated by the video script. Viewers are asked to respond to questions or to solve problems, then to return to the videotape, where a response is given or various approaches to the problem are discussed. Such video-with-questions would not be classed as fully 'interactive video'. It is, however, a step towards the conversion of purely informational material into an instructional package.

Some 'hybrid' or 'eclectic' systems

Finally, we come to the last row on our schema, devoted to packages or systems that employ a more involving, partly experiential, use of the audio medium, allowing some form of involvement or interaction by the viewer/listener. On the 'border' between this and the previous row, we have listed a few difficult-to-classify examples. These are difficult to pin down as they are basically integrated multi-media and multi-method approaches, which could be classified in any combination of the cells on our matrix, depending on the exact way in which they are implemented in a particular case. The 'audiotutorial' system (Postlethwait, 1972) was conceived as a system of learning activities that could take any form and utilize any media combinations, the whole system being controlled by instructions and explanations contained in an audiocassette. The audiocassette may be considered to be a formal commentary, but the learning experiences that it 'knits together' may be both expositive and experiential in nature. The 'audiotutorial' system developed and spread in the USA from the mid-1960s to the late 1970s and continues to flourish. In Great Britain, during the same period, an approach termed 'RESOURCE-BASED LEARNING' developed. In essence, there is very little difference between this and the 'audiotutorial' approach, except that 'resource-based learning' does not necessarily use an audiotape as a study guide. This task may be performed by an audiotape, but is more usually contained in a printed study guide. Once again, however, the resources used may contain both expositive and experiential materials, utilizing a variety of presentation media. Group activities and research projects involving the community may also be included. The 'GROUP-PROGRAMS' and 'ROLEMAPS' are other approaches that use a variety of, often experiential, group learning activities, interspersed with some expositive textual or other materials, the whole package being controlled by an audiotape. This tape often acts not only as a study guide and source of orientation, but may include essential study material, task and role descriptions, rules of play and so on, as well as often acting as a timing control for the various activities. Further details may be found in books devoted to group-programs (Thiagarajan, 1977) and rolemaps (Dormant, 1980).

Cell C-1: audio programs as experience generators

In the category of experiential audiovisual packages, it is somewhat more difficult to draw the line between what we have called 'true' instructional packages and others (not so 'true'). This is because experiential learning aims to achieve general, attitudinal or cognitive skill objectives, which are not always directly measured by the results of a given exercise, but rather by observing: (a) the process by which the results were achieved and/or (b) the long-term changes that occur after the learning experience, and/or (c) by engaging in a relatively free 'debriefing' discussion that aims to generalize the principles discovered or explore the grounds for the attitudes formed. All this implies that experiential learning exercises are rarely, if ever, composed of just a media-based package. Rather, the media presentations used must be incorporated in an overall instructional design that includes other learning activities.

For this reason, we have been rather more free in our inclusion of examples, on the assumption that they are, or will be, incorporated into a carefully designed instructional lesson plan. We therefore include, in the 'audio only' cell, radio programmes and pre-recorded tapes of dramatized situations, plays, debates and discussions, that are to be used as the stimulus material for specific learning activities on the part of the listeners.

Cell C-2: audio-workbooks aiming at attitude change

Similarly, in the 'audio and print' cell, we mention the form of audio-workbook which is sometimes used in semi-experiential packages dealing with the analysis of dramatized cases of human behaviour. These are sometimes used as parts of management or salesman training programmes, as well as in school study areas such as literature and theatre. One practical example mentioned in the schema concerns some packages developed by the present author for the updating of doctors, as part of a service offered by the Glaxo pharmaceutical company. The content of these packages dealt with controversial issues, on which British doctors held diverging views; one of these was to do with the issue of measles vaccination. Some doctors felt that the official Ministry of Health position in favour of mass vaccination of all children would cause more work and trouble than the 'normal' treatment of measles cases as and when they occurred. The opposite argument hinged on the possible complications (in a small minority of cases) as a result of contracting measles. The package was therefore developed as a dramatized discussion between two doctors, one in favour and one against the mass vaccination programme. This audio-dramatization was accompanied by a booklet, mainly of graphs and other statistical displays reproduced from medical journals, which the doctor in favour would use as visual support for his arguments: 'Now just look at Chart 3; these statistics were quoted in the *Lancet* of May 1968; don't you agree that they prove my case?...' In addition to the charts, the booklet contained instructions on exactly how to give the measles vaccine (indications and contra-indications, dosage and procedure) in the form of an algorithmic flow chart that could be detached and used as a job-aid, if and when the doctor decided to vaccinate. We feel justified in considering this audio-workbook to be partly 'experiential' in nature, as the arguments for and against the scheme were not all logical and one-sided. There were good points to be made for each position on logical grounds, as well as some weaknesses in the arguments. The prime deciding factor was, in the final event, emotive: the main objective was that, through the experience of participating vicariously in the somewhat emotive arguments presented, the listener would 'come round' to agreeing with the proposed scheme.

Cell C-4: audiovisual systems for group activity

In the 'experiential audio and still visual cell' we have included two very different examples. The first – RADIOVISION – is quite well known, at least to British readers. It is a method quite extensively used by the BBC in its schools broadcasting. The method consists of supplying schools with a package of visual materials, usually slides, in advance of the broadcast of a particular radio programme. The teacher, or some other member of the group, then presents the visuals at appropriate moments, indicated by the programme's presenter. In some applications, the total package differs little from an expositive slide-tape presentation, except that the audio message is being received over broadcast radio – a mere technicality. In other, more creative applications, however, the radio programmes are conversational in style, partially dramatized and the visuals do not only illustrate the audio message, but also guide subsequent group-learning activities in the classroom. This form of radiovision package may indeed have some experiential learning component built into it. It may also be quite highly structured, including plans of the pre-broadcast and post-broadcast activities to be carried out, thus qualifying as a 'true' instructional system.

The other example, 'NARROW BAND VIDEO', is something completely different. It is a relatively new technique for the transmission of still video pictures over narrow wavebands (as used for telephone transmission). A link-up of this nature, using television cameras at each end and a rented telephone line to connect them together, may, with the suitable interface equipment now available, enable teleconferencing with 'live' video transmission for a price that is comparable to the older 'audio-only' telephone link-up systems. This example is not, of course, a type of package, but is rather a system that may enable rather novel means of communication and instruction over long distances. Such systems may be of particular use in their capacity to arrange an exchange of ideas between different people and if properly utilized should be capable of promoting both expositive and experiential instructional strategies. Little has so far been done with these new systems. More details about them may be found in an article by Bretz (1984).

Cell C5: television and instruction

The examples in the final cell of our matrix reflect to some extent those just described. They do, of course, include *moving* image transmission. Also they have been selected to illustrate that not all practical large-scale developments are the preserve of the 'developed world'. The TV-EXECUTIVO system is an on-line teleconferencing system in use since the mid-1970s throughout Brazil. It is similar in conception to many other teleconferencing systems available, perhaps even earlier, in other parts of the world. However, the Brazilian system is actually in regular daily use. Long-distance communications in Brazil are operated by a national company, EMBRATEL, that operates both international (satellite) communications and national (microwave) communications. The microwave system consists of a network of booster stations linking all large towns, which transmits both telephone, radio and television signals from one region to another. As EMBRATEL already had the capacity to transmit television signals between all large towns, it was relatively simple to install television studios in several of the chief centres of population and reception/study centres in all the other large towns, all these facilities being installed on EMBRATEL property. The system operates by renting time to any user organization that is interested in communicating with groups in various cities. It offers live video transmission from the studio centres to all other centres and live audio communication between all centres. It is therefore possible to hold a seminar employing real-time live presenters and/or sophisticated audiovisual presentations, to a geographically scattered audience, who may participate with questions, or comments, and may engage in cross-talk conversations and small group discussions as part of the experience. Apart from its most popular usage – to transmit the messages of head office to outlying branches (this usage alone paying for the investment in the system in little over one year), the TV-EXECUTIVO is used by many non-commercial groups, such as universities, research groups, special interest groups and also for the training of all categories of personnel. Many of the applications are indeed experiential in nature, encouraging discussion, arguments, the use of dramatizations, case studies and so on.

The other, and final, example is instructional use of television. Once more, we should distinguish between the 'truly instructional' uses of television and the 'enrichment' uses. In the USA, all school television tends to be labelled ITV (instructional television), while the same mix of various programme types is referred to in England as ETV (educational television). It is quite clear, however, that some uses of television in education do really qualify as examples of 'instruction', whilst others do not. We go back, as always, to examine the methodology by which the programmes and the system of use destined for them were designed. Were they based on relatively clearly stated objectives? Was a system designed that could be expected to achieve the predetermined objectives? Was this system in some way evaluated, and improved if necessary, until the objectives were being achieved with a reasonable level of certainty? If the answers to these three questions are positive, we have an instructional system. Television, all by itself, is unlikely to constitute a complete instructional system, except in extreme cases, like the case study in the last chapter. However, television as the principal medium of information dissemination, supported by other means of controlling practice and supplying feedback, can be the basis of highly effective instructional systems. Many such systems exist throughout the world. We choose to single out yet again, however, a system that has been operating in the State of Maranhao in the north-east of Brazil. This state had no effective state secondary education system before 1968. Then, one year later, over 20 per cent of the 15 to 18 age group were in full-time secondary education participating in a television-based system of instruction. This operates without schools as we know them, without teachers as we know them, and uses few support media other than some printed materials. The system has stood the test of time, though, and is operating, 15 years later, in much the same way as originally, but on a yet larger scale. This system is designed to lead students to achieve the normal range of school-leaving objectives. By all accounts, it succeeds as well as most of the 'conventional' systems. We have put the Maranhao system in the experiential/conversational cell of our schema, although in reality quite a proportion of the broadcast lessons are mainly expositive in style, aiming as they do at the achievement of typical secondary school curriculum objectives. There is, however, another component in the system. The original conception of the system included a substantial element of self-help activities to be developed at local community level. The group leaders in the villages were often social workers and their role included the organization of supplementary activities related to the implementation of social change by the direct application of the TV course content to the local problems. Thus, in reality, a large proportion of the significant learning which took place was through group discussion and action, motivated and fuelled by the media lessons, but organized and executed by the group leaders and members in general. It is true to say that the learning which took place in one locality, as a result of a given course, was quite a bit different from another locality with different problems and inhabitants.

Similar, television-stimulated group discussion and learning may be encountered in many recent adult education projects in developing countries. Agricultural education in India uses a model which is based on the broadcasting of short, often controversial, programmes on specific farming techniques, which aim to stimulate locally organized discussion and decision making.

Some North American distance-education systems adopt a similar philosophy. Both in Canada and the USA, there is a swing towards the adoption of distance-education systems that beam programmes from a central source to scattered groups of students, working with trained tutors/animators, whose job is to get the groups working experientially in discussion or application of the 'seed' information that was in the programme. This model of distance education, although somewhat more costly than the individual-learner-at-home models, seems to offer both pedagogical and economical advantages over the 'conventional' university model.

Conclusion   We have dealt in this chapter with many aspects of the nature and the variety of instructional media packages. We have shown that the design of instructional packages is based on the application of general instructional design and development principles (heuristics), that have already been amply demonstrated and discussed in earlier chapters of this book and in earlier volumes in this series. We have seen that the variety of possible designs for media packages is about equivalent to the variety of instructional design problems that may present themselves. We have also seen, in the last section of this chapter, that the variety of media combinations is also very rich. We hope to have illustrated, by means of presenting this variety, that the range of designs is not exhausted and that any instructional developer may be able to add to this variety by the invention of new forms of media or new ways of exploiting old media. However, such invention should be guided by the application of well tested instructional design principles.

In the next chapter, we shall have to restrict our sights to some extent. We shall be dealing with the Level 4 considerations of scriptwriting and development of audiovisual packages. However, in the light of the rich variety of options, we must restrict ourselves to only certain types of scriptwriting and certain forms of package. In general, we shall confine ourselves to the planning of *self-instructional audiovisual packages*.

16. Developing Audiovisual Instructional Materials

16.1 Some common approaches to the development of audiovisuals

Audiovisual instruction owes much to the arts of theatre, cinema and commercial television – areas where most of the techniques, special effects, traditions and non-verbal language of the audiovisual medium were developed. Much of this inheritance is valuable, but some is also a source of trouble.

Contrast between the theatre and instructional message design

Instruction shares, with these other areas, the need to communicate a message. However, the precision of the communication that is necessary for instructional purposes (and also for documentary information, as in news programmes) is much greater than in the theatre, for instance. The playwright usually has some general, perhaps moral, message to communicate, which he does by means of a story. The detailed information transmitted, sentence by sentence, as the story unfolds, may however be of secondary importance. It is possible for a viewer to miss a sentence, or even a whole scene, yet gain the overall message that the play seeks to transmit. Playwrights deal in impressions and emotions, rather than in specific facts and concepts. The authors of instructional or 'hard' informational messages are more concerned with the detailed content of the message. This is not to argue that education and training never get involved in the shaping of impressions and emotions – the affective domain depends heavily an this, as does the stage of 'warm-up' or motivation to learn, that often opens a new lesson. However, the bulk of instruction in the cognitive and psycho-motor domains relies on the clear and unambiguous communication of specific information – not always verbal information, hence one reason for the importance of audiovisual media.

Contrast between the design of instructional and purely informational messages

Instructional messages are ones which seek to promote learning, and purely informational messages are ones that seek to promote communication without necessarily being concerned with long-term retention or concept formation. They have many aspects in common. They also differ in some critical ways. Both types of message design are concerned with the clear and unambiguous communication of specific information. Both are concerned with the choice of language and words that will be understood by the intended audience. Both are concerned with economy of words or illustrations, so as to communicate the intended message efficiently. However, the purely informational message is seldom designed with due regard to all the characteristics of the audience that may influence comprehension and, especially, effective learning. The writer of informational news or documentary material is concerned with style, with idiom, with analogies or metaphors that *might* make the message more meaningful to the audience. The writer of an instructional message is concerned with all of this and in addition such matters as the previous learning experience of the intended audience and the learning skills they possess; the structure of the information to be communicated and how it relates to information already learned by the intended audience; the number and choice of examples that should be included in the message in order to ensure that all (or at least most) of the intended audience will understand and learn; the size of 'step' of new information that the intended audience is capable of taking before it becomes necessary to engage in practice and reinforcement activities; the amount and type of practice that should be provided in order to facilitate the desired learning; the criteria that should be applied to judge whether the learning tasks have been adequately performed; the remedial messages that must be prepared in advance to cater for those who do not demonstrate complete learning success straight away, etc.

It is largely because of these differences that exist between a truly *instructional* audiovisual and other forms of audiovisual communication that it is necessary to develop a specific approach to the scriptwriting. Many approaches do exist and some of these owe more to the theatrical/cinema traditions or to the journalism/documentary traditions than to a systematic application of the principles of instructional design. The aim of this chapter is to develop such a systematic approach. We shall commence, however, by analysing some approaches quite commonly encountered in the production of audiovisual packages. From these analyses, we shall attempt to synthesize an approach which fits in with our general 'systems' approach to instructional design and development.

The audio-based approach

The 'audio-based' approach is often encountered in the production of audiovisuals for institutional clients. The client brings a typed 'script' which is generally not so much a script intended to be spoken, as a typed message on some topic, written by someone accustomed to producing technical reports. The initial meetings with the audiovisual producers are taken up in trying to find out what the client really wants to communicate and in transforming the typewritten document into an audio script, capable of being spoken in a natural manner. Some ideas may spring up at this stage as regards the visual content of the final product, but often the client has no real idea of what is worth illustrating, what needs illustrating or whether indeed any visual content is essential. Very often it becomes obvious that the client wants an audiovisual simply because it is fashionable to use audiovisuals. On other occasions the client is aware that the final product should impress or motivate and believes in the emotive power of the AV medium. He is quite happy to leave the planning of the visual content to the specialists but has strong reasons for controlling the verbal content (for example to protect himself against criticism from superiors, subordinates, competitors, etc).

We may summarize this approach as follows:

We have experienced this approach in many projects. When the client's intention is purely motivational, promotional or informational, the approach can be made to work. An example that springs to mind is the request received by an AV firm from a jewellery firm that had just won a contract to set up shops on a fleet of cruise liners carrying tourists to various parts of the world. As only one salesperson could be put aboard each ship, it seemed a good idea to make use of the already existing closed circuit video systems to attract the interest of the passengers. Rather than producing straight advertising, the client wished to screen informational programmes. For example, on a ship going to Brazil, they would screen a documentary about Brazilian gemstones, with some low-key promotion at intervals. The client arrived armed with a 20-page text about 20 different types of gemstone and a request to produce an audiovisual within a week, as the ship was due to sail. The AV producers could do nothing but cut and modify the typescript until it resembled an audio script capable of being delivered in the 15-minute limit specified. This took one day. Then they went to the client's premises and shot a few hundred slides, of all the gemstones to be described under all manner of 'artistic' lighting conditions and mounted as all manner of jewellery. This took another day. They then put the show together during the remaining three days. The result was quite successful, in terms of its objectives (motivational and informational). However, we dread to think of the result if any *instructional* objectives had been envisaged.

We have often experienced requests to produce 'instructional' audiovisuals, which have commenced along the lines of the case described above. It is often quite difficult to get the client to agree to 'back-pedal' and review the work that went into producing the typescript, packed solid with technical information at 'seven concepts to the paragraph', which he is blithely expecting the learners to understand, assimilate, remember and apply, after one exposure to an audio–visual.

The video-based approach

Another quite commonly encountered approach is almost the exact opposite of the one described above. We may call it the 'video-based' approach, as it starts with an existing set of visual materials – video clips, film shots, or just a collection of slides. These may have originated as a coherent set of shots, taken with a specific purpose in mind (although not according to a carefully prepared shooting script), or they may 'just happen to exist' in some firm's photo archives, in a commercially run slide or video clip library, or even as a collection of 'holiday snaps'. For some reason, someone decides to make an audiovisual presentation from the collection of images available. This is often unavoidable in the field of journalism, for example as in the case of the preparation of an obituary for screening on television. It may also occasionally be the only way to proceed in instructional message design, for example when we are dealing with the communication of historical events. It is however all too often the approach used in order to avoid the stages of careful, objectives-based, instructional design which should precede the scripting and production of an instructional audiovisual.

The producer is not so much an instructional designer as a photographer-artist, who first goes out in search of a collection of interesting and pleasing shots of the subject in question, and only later, on a light table, starts to consider which shots to use, which to discard, the sequence of presentation of the shots selected and so on. These decisions are guided mainly by the quality of the shots, their relative brightness, the visual attractiveness of their content and so on – not by the requirements of the message to be communicated by the content, as no detailed plan of audio/video interaction exists at this stage. At most, there is a rough 'storyboard' to guide the organization or selection of the images, but the detailed script is not yet written. Only when the images are selected and ordered and an estimate of the time exposure of each sequence of images is made, does the producer pass to the stage of writing a detailed audio commentary to accompany the visual sequence already planned.

We may summarize this approach as follows:

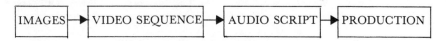

Most low-budget documentary films follow this sequence of design and development. They are forced to do so by the restrictions imposed by low budget, portable equipment, one-man crew and so on. Also, the film producer (say in the case of a film that involves interviews) may not have a clear idea of what the final structure of the film will be until he has amassed the separate shots and evaluated their relative interest value or newsworthiness. The final film usually stands or falls on the creativity with which the producer collects useful material when in the field and the skill with which he knits this material together on the cutting table. The audio content, if at all recorded at the time of shooting, is often redone after the final cutting and generally bears no resemblance at all to the original; what gets included and what gets left out is governed by a mixture of expedience, luck and artistic judgement – not necessarily the best situation for the design of instruction.

Many so-called 'training' films and much 'educational broadcasting' is produced in a similar way. I was involved in the production of a series of BBC programmes for managers, which presented New Developments in Training. Each programme, of 30 minutes' duration, was devoted to a specific training methodology or technique. The treatment given to the technique was a mixture of interviews with practitioners and location-shot film of the technique in action. Usually the half-hour programme would present four or five variations or case studies of the technique in question. Initial planning was limited to research, selection and invitation of the participant contributors, visits and interviews 'on location' together with the shooting of location shots and the preparation of an overall 'timetable' (a sort of storyboard that established the sequence and timing of the component parts of the programme).

On the day scheduled for production, the BBC would allocate the use of studio, camera crews and equipment for one hour only. All the morning could be used for rehearsals and 'mock interviews'. These were not closely scripted, however. The interviewer worked from a set of key questions which would be asked. However, in order to maintain naturalness and spontaneity, the questions would be asked in a slightly different manner every time and, naturally, the answers given would vary considerably from rehearsal to rehearsal.

At the appointed hour, we all went into the studio and started shooting. In less than one hour, all shooting was completed. Nothing was repeated unless things went so wrong that 'cutting and pasting' could not cover it up. More often than not, anything that went really wrong during shooting would simply be eliminated from the programme. In general, the final programmes bore a 70 to 80 per cent resemblance to the original storyboard outline, as regards the content and sequence of shots. As regards the audio message accompanying the shots, the resemblance between the final version and the original intentions of the interviewed participants, was much less still. The programmes were, however, well received by managers (the target audience) and were held up by the BBC as 'good educational television'. All that just goes to prove that 'you don't know what you don't know'. A viewing audience seldom misses what was not presented in a programme, unless they are already informed about the topic and thus, by definition, are not the intended audience for an *instructional* programme.

The integrated audiovisual approach

A much better approach to the design and development of audiovisual instruction is the 'integrated audiovisual approach': the planning of the audio and the visual content of the message at one and the same time, before either one is produced. This approach usually follows a sequence of message design which we may summarize as follows:

| STORYBOARD | → | AUDIOVISUAL SCRIPT | → | SHOOTING SCRIPT | → | PRODUCTION |

This approach has many variations and we shall examine some of these briefly, in the next section. Before we go on, however, we shall define the differences between the three types of 'script' mentioned above.

Storyboard. A storyboard is a general outline of the programme or presentation that we intend to produce. It consists of a list of key points we wish to make, key items of information we wish to communicate, or key images we wish to present. The storyboard may thus be mainly visual (sketches of images and captions to indicate how they will be used), or mainly verbal (items of information to be communicated and notes on the types of images that may be used as illustrations). The storyboard is very much a working document. As such, it is often drawn up on separate cards, which may be reordered or replaced with ease during the planning stages. The non-specialist 'client' would normally be involved in the planning and/or the approval of the storyboard.

Audiovisual script. The storyboard, once approved, is transformed into a more detailed, step-by-step script. This is usually prepared in two columns, labelled respectively 'audio' and 'visual', where both parts of the message are carefully planned and written out (or sketched). The audio part of the script is the prototype for what will be said and what sound effects will be recorded. The 'visual' part of the script shows (or describes in words) every shot that will be used in the final production. The audio and video components are designed alongside each other, to ensure that the two media complement each other in the most effective manner possible. The changing of the visual images is tied to a given quantity of text or other audio content, so that correct timing is ensured. The language and all other aspects are carefully evaluated, not so much by the client this time, as by the specialists in audiovisual media production.

Shooting script. The audiovisual script is a clear presentation of the instructional developer's intentions, but is not necessarily all that easy to follow in the studio or on location. We are then dealing with camera crews, lighting technicians and so forth, who have their own jargon and for whom timing and camera angles, etc, must be defined with much greater precision than is necessary for the designers, developers and validators. A shooting script may have many columns, one devoted to each camera used or to each member of the studio crew. Thus, during shooting, each person involved knows just where to look on the script and may take his/her cues from the immediately preceding actions of the other members, as listed in the other columns of the script.

Simplified procedures for special cases

The sequence and the three stages of scriptwriting described above are typical for productions which involve the filming of video sequences which employ more than one camera. In the case of other types of productions, the sequence may be slightly different. If using only one camera, in the style of 'snapshots' (a technique often used in documentary film production and sometimes in low-cost video produced with portable equipment) a much simpler form of shooting script will suffice. It may even be possible to do without a shooting script at all, working directly from the audiovisual script, orienting the cameraman verbally, shot by shot. There is no need to script the continuity during the shooting in such cases, as the production is really put together later, on the cutting table. Instead of a shooting script, we will need to produce a *cutting script*, once the shots have been taken and examined, in order to direct the technician who does the cutting.

In the case of slide-tape audiovisuals, it is generally unnecessary to produce a shooting script or a cutting script. The producer can work quite adequately from the audiovisual script. It may be useful, however, to produce a separate version of the audio component, in a form that is more convenient to read and use during sound-recording sessions. A text cut up artificially into chunks that accompany specific images does not necessarily give a clear idea of pace, emphasis, or even the meaning, to a speaker recording the soundtrack.

There are, of course, special types of audiovisual productions that require yet other variations to the production sequence we have outlined. A case in point is the production of multi-screen, or multi-media presentations, which require the careful orchestration of the audio message and several parallel visual tracks.

16.2 Special case: the development of instructional audiovisuals

Poor aspects of much existing audiovisual instruction

In the previous section, we mentioned some of the differences between the planning of instructional audiovisuals and other forms of audiovisual programmes/presentations. We have seen that the 'audio-based' approach has serious limitations for most types of instructional presentations, as well as for many informational and motivational packages. The 'video-based' approach, despite its obvious value in making use of historical or other already existing images, also has severe limitations as a basis for the design of instructional messages. The integrated audiovisual approach overcomes most of these weaknesses, ensuring that the audio and visual script components are designed to complement each other, both contributing to the achievement of the intended instructional objectives. But, of course, the mere integrated planning of the script's components does not assure that instruction will occur. It is fair to say that most slide-sound 'instructional' packages are developed by methods similar to the 'integrated audiovisual approach' described above, but a study of the products available commercially on the market will quickly convince us that the majority of existing slide-sound packages are unlikely to be effective or efficient instructional materials. This is because they are not clear about the objectives which they hope to achieve, or the informational content/presentation/pace are not well suited to the achieving of the stated objectives, or the objectives cannot be totally achieved by means of an audiovisual package, or the presentation is too long without any planned breaks or practice exercises. In short, the audiovisual presentations are seldom planned as part of a systematically designed lesson. It remains for the teacher – the user of the package, to build a lesson plan around the audiovisual – not always an easy task when the package is poorly designed in the first place.

Audiovisual design as part of the overall instructional development process

The way to avoid the production of *purposeless, pseudo-instructional packages*, is to ensure that the audiovisual message design process is part and parcel of a systematic instructional design and development process. We wish to be certain not only that our audiovisual has some specific objectives but also that these objectives are really worth achieving. We should therefore go through the steps of Level 1 design, to ensure that the problem we are trying to solve can really be solved by means of instruction. We should then go through the stages of Level 2 design, which defines the specific objectives of instruction and, among other things, leads to the selection of specific media for specific objectives, in a rational manner. Then, at Level 3, we should plan the details of the lesson which will utilize the audiovisual as one of its components. Thus the later need to plan how to make use of a given package is eliminated, for it comes with a ready-made utilization plan. In the case of self-instructional packages, which in effect constitute the whole lesson, it is still a good idea to first write out the 'dynamics' of the proposed package as a skeleton lesson, so that we may check that all the essential ingredients of an instructional system may indeed be 'packaged' effectively in the medium or media selected. Finally, using the lesson plan as a guide, we proceed to Level 4 – the detailed planning and scripting of the audiovisual presentation. In summary, the total process is:

Level 1: Decision to develop instructional system.

Level 2: Overall instructional plan, including the selection of specific instructional media.

Level 3: Detailed lesson plans for the partially or totally mediated lessons.

Level 4: Audiovisual message design and development.

Suggestions for the systematic development of instructional audiovisuals

Several suggestions for the systematic development of audiovisual instruction can be found in the literature. Brown, Lewis and Harcleroad (1977) go no further than to suggest that one should establish the objectives and content of an audiovisual carefully, before commencing the message design stages. They do not specify exactly how this should be done, suggesting that the teacher should follow the methods he or she is accustomed to using for the planning of other forms of instruction.

Jerrold E Kemp (1980) is more specific, suggesting that all the planning may be done by means of cards, somewhat like the Dacum Process described elsewhere (Romiszowski, 1981;1984) for the general planning of curricula. He suggests a three-column table to be mounted on a board or wall (see Figure 16.1) on which the specific instructional objectives may be organized in sequence. The informational content to be communicated may be layed out alongside each objective and appropriate media alternatives may be selected and posted in the third column. Once this overall plan is completed to satisfaction, the selected media for a given objective are planned in more detail on a separate set of 'storyboard cards' (see Figure 16.2).

Kemp then suggests that the next stage should be a 'first attempt' at the audio content of the script. This should be written with reference to the storyboard cards and checked for accuracy, completeness and so on, before the final, audiovisual script is developed. At this stage, timing considerations lead to the modification of

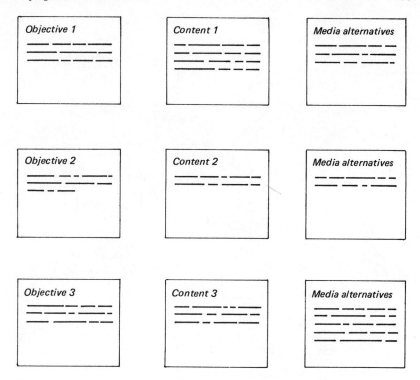

Figure 16.1 *Planning the overall structure of a media package, using cards on a wall*

the audio script and the more detailed specification of each visual image or sequence, resulting in a fuller audiovisual script (see Figure 16.3).

We see some similarity between Kemp's suggestions and our own. His first stage is a somewhat simplified, special purpose version of our Level 2 overall instructional plan. However, Kemp then goes straight to the storyboard planning stage, which for us is the first stage of specialist (Level 4) audiovisual message design. Whereas, in some projects, it may not be relevant to devote much time to the consideration of how an audiovisual package is to be used in a given lesson, we feel it is valuable to impose the self-discipline of deciding consciously whether it is or is not relevant to plan a user-guide (another term for a lesson plan) to accompany the package. We therefore recommend that the Level 3 stages of detailed lesson planning be always at least considered. For similar reasons we would encourage the execution of the Level 1 analyses which may indicate: that no instruction is required; that the cost of developing audiovisual instruction is not justified by the cost of the problem; or that suitable sources of ready-made materials exist and should first be investigated.

Non-instructional audiovisuals as components of instructional systems

There are, of course, many cases when an overall instructional plan specifies an audiovisual presentation as part of the means of attaining a specific objective, but the audiovisual itself is not an instructional package. Some examples are:

* The use of a video clip or a filmed dramatization as case material for a group discussion.
* The use of a video clip of a skilled performer as a 'model' of required performance standards (as in microteaching).
* The use of audiovisual materials as 'openers' in order to interest the audience, motivate them to participate, etc.
* The use of 'informational' film material as a means of reviewing a subject to

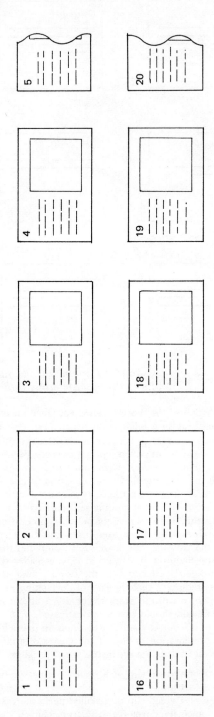

Figure 16.2 *Planning the detailed sequence and tactics of a media package using a 'storyboard' layout*

COMMENTS	VISUAL	AUDIO
1. The title	OPERATING THE SWITCHBOARD	Operating the switchboard An XYZ Company production (music)
2. Narrator introduction		Hello. This is the first lesson on the operation of the type of switchboard installed in our offices (pause)
3. General view of machine and operator		All secretarial staff take their turn on the switchboard and therefore . . .

Figure 16.3

Stages and steps	Completion			
PLANNING	Unit 1	Unit 2	Unit 3	Unit 4
1. Express your idea				
2. Develop the objectives				
3. Consider the audience				
4. Get some help				
5. Prepare the content outline				
6. Select the medium				
7. Write the treatment				
8. Make a storyboard				
9. Develop the script				
10. Prepare the specifications				
11. Schedule the picture taking				
PRODUCTION	////	////	////	////
12. Take the pictures				
13. Process the film				
14. Make work copies				
15. Edit the pictures				
16. Edit narration and captions				
17. Prepare artwork, titles and captions				
18. Record narration				
19. Mix sounds				
20. Prepare final copies				
FOLLOW-UP	////	////	////	////
21. Write the instruction guide				
22. Prepare for use of the materials				
23. Use the materials				
24. Evaluate·for future use				
25. Revise as necessary				
26. Copyright materials				

Figure 16.4 *A checklist for controlling the audiovisual media development process (adapted from Kemp, 1980)*

be treated later in more detail, or summarizing a subject that has been dealt with before.

In any of these, or similar, cases, it may not be necessary for the audiovisual message designer to follow through the whole four-level instructional design process. A short briefing, as to the specific purposes of the presentation, may be sufficient to get the designer started. In such cases, a simplified procedure, like that suggested by Kemp (or even simpler when there are no specific objectives to be achieved) may be adopted. It may be possible to start immediately at the storyboard stage of the message design process, proceeding from there to the development of a step-by-step audiovisual script.

The specific techniques of scriptwriting for non-instructional audiovisual presentations (motivational, promotional, etc) are too many and varied to be covered in depth here. We shall restrict ourselves in the next section, to the analysis of one principally motivational example and one principally informational example, as well as several instructional examples. These examples of scripts will serve as case material for the illustration of the principles outlined so far and the more specific tactics that may be employed in special circumstances.

Tricks of the trade   In relation to specific tactics, or 'tricks of the trade', suggestions abound in the literature. Some of these are 'blanket' general rules (for example the limits on the maximum exposure time of a slide) which should be taken with a grain of salt. Many of these rules have been developed in the realm of motivational or informational message design and tend to be indiscriminately applied in the realm of instructional package design. Some authors do, however, present useful checklists developed specifically for instructional audiovisual design. One such list of suggestions (Kemp, 1980) is presented in Figure 16.4. Another list (Langdon, 1978) is presented in Figure 16.5.

Figure 16.5 *A checklist for controlling the development of audio-workbook packages (from Langdon, 1978)*

	UNIT 1	2	3	4	5	6	7	8
1. Project team assigned								
2. Content conference								
3. Questions and ans written — SME								
Reviewed and edit — Coor								
4. Objectives written — Prog								
Objectives approved — SME								
5. Obj and questions reviewed — Coor								
6. Record content — SME								
Transcribe recording — Sec								
7. Script edited — SME								
8. Questions, cues, etc added — Prog								
9. Criterion test written — Prog								
Criterion test approved — SME								
10. 1st draft program review — Coor								
11. Script typed — Sec								
Script proofed — Sec								

Figure 16.5 *(continued)*

Task										
12. Workbook typed — Sec										
Workbook proofed — Sec										
13. Script recorded — Coor										
Recording proofed — Prog										
Recording approved — SME										
14. Workbook xeroxed — Sec										
Criterion test xeroxed — Sec										
15. Test draft compiled — Coor										
16. Developmental testing (date)										
17. Test scored — SME, Prog										
Test results on matrix — Prog										
Test analysed — Prog										
18. Test conference — Coor										
19. Revisions complete — Prog										
Revisions edit — SME										
20. Workbook revisions retyped — Sec										
Script revisions retyped — Sec										
21. Script recorded — Coor										
Recording checked — Prog										
Workbook xeroxed — Sec										
22. Tape and workbook check — Coor										
23. Prepare										
— Tape _____										
— Workbooks _____										
— Tests _____										
24. Field test (date)										
25. Test scored — SME, Prog										
Test results on matrix — Prog										
Test analysed — Prog										
26. Test conference — Coor										
27. Revisions complete — Prog										
Revisions edit — SME										
28. Final Okay — Coor										

KEY:

Coor	—	Project coordinator
SME	—	Subject matter expert
Prog	—	Programmer (instructional designer)
Sec	—	Secretary

16.3 Some examples of scripting for audiovisuals

In this section, we present a series of excerpts from audiovisual scripts, in various stages of preparation. Some of them are being designed with instructional objectives in mind, whilst others are principally motivational or informational in intent. To a large extent, the excerpts, presented in Exhibits 16.1 to 16.4, speak for themselves. We shall, however, make a few comments on each of the exhibits, in order to focus on their principal characteristics.

16.3.1 A motivational audiovisual

Exhibit 16.1 The first exhibit presents part of a script (much modified in order to shorten it) of an audiovisual originally prepared for an organization in Brazil, as part of an instructional system for the in-service training of typists and secretaries. The audiovisual was intended to 'kick off' the course unit for typist training that dealt with the quality of work-aesthetic layout of the page, proper use of language, correct grammar, application of the company standards and rules in relation to official documents, and so on. This was one of the chief groups of instructional objectives identified in earlier analyses. The need for a motivational element in this course unit was indentified as follows:

Level 1. The performance of typists in relation to the quality aspect was extremely deficient. This was shown to have serious repercussions on the organization's image and, therefore, this performance deficiency was seen as a priority problem. Job-aids already existed and were deemed sufficiently well structured to resolve most typists' doubts, if they were to be systematically used whenever the typists were in doubt. However, in general, typists made no use whatever of the job-aids provided. Further analysis showed that this was partly a 'knowledge deficiency' (most typists did not know how to use the more specialist job-aids, such as organization standards and norms) and partly 'execution deficiency' (most typists knew how to use the common job-aids, such as dictionaries for checking spelling, but in practice never used them). There seemed, therefore, to be an element of the problem that would require instruction and also an element that required other forms of action.

Various alternatives were considered. The most obvious would seem to be the strengthening of positive consequences for those who use the job-aids and maintain better quality. Job reorganization, promotion or disciplinary measures were also considered. The nature of the organization, however, made it impossible to implement any such measures in practice. The organization was part of the civil service (the government of one of the states) and civil service employment policies made it impossible to use salary or other bonus incentives, to promote or demote for reasons other than years of service, and so on.

The only alternative that seemed open to the instructional designers, was to attempt to develop higher levels of motivation in relation to the quality of the work produced, as well as demonstrating and teaching the use of the existing job-aids. Thus, the Level 1 proposals included some *motivational* objectives as part of the project specification.

Level 2. The overall instructional plan was next developed. This included the motivational objectives as well as the more specific instructional objectives identified at Level 1. A more detailed task analysis, now undertaken, suggested that, in relation to the motivational objective, typists may gain a better working climate and personal relations with their superiors, if they maintain 'first time off' quality in their work. The more specific formulation of the objective took this aspect into consideration.

The content of the motivational aspect was thus defined as 'the personal benefits of maintaining high quality in your work'. The methods were to be 'lighthearted demonstration of the benefits of improved work quality and the unpleasant consequences of poor quality work' and the media selected were 'slide-tape audiovisual package'. This selection was motivated by the need to repeat the course frequently, to small groups in various localities (which imposed a Level 1

Exhibit 16.1 *Part of an audiovisual script prepared for a slide-tape presentation to be used as initial motivation on an in-service course for typists*

VISUAL	AUDIO
(An initial series of slides, illustrating the work of the typists in the organization.)	(An introductory section which explains the importance of typewritten documents to the proper functioning of the organization.)
	Now we come to the question of QUALITY! . . . we don't want to give our customers a heart attack every time they receive a letter from us, do we?
	. . . nor is it a good idea to give a bad impression of our organization and its products. It's quite natural to make generalizations, from poor typing to poor dressmaking, for example.
	The quality of the typing, whatever the document, is ultimately dependent on *you*. It is no use relying on the boss to correct your every letter . . .
	. . . he's much too busy to do it all the time . . . and he won't thank you for the extra work either!

Exhibit 16.1 *(continued)*

	There are plenty of tools in the office to help you maintain and improve the quality of your typing — dictionaries — collections of model letters — special typists' reference books. Learn how to use them . . .
	. . . and use them ALWAYS! . . . when appropriate, of course.
	You'll be surprised how much retyping time and effort you will save, just by checking up on what you are not sure about BEFORE you start to type the draft.
	If you can't find the answer in the reference books, then ask someone more experienced . . . they'll be happy to help you out . . .
	. . . much happier BEFORE you make a mistake than afterwards.

Exhibit 16.1 *(continued)*

Remember . . . you win too! By maintaining quality in your work, you can change this sort of working environment . . .

. . . into this!

(musical exit)

THE END

decision to make the course as individualized and self-instructional as possible) and by the limited range of media alternatives available to the organization. Further considerations were: cost and the ability to use impersonal and exaggerated situations as illustrations, using artwork, rather than real-life photography or filming. These last considerations really formed part of the Level 3 lesson planning stage.

Level 3. As we are not in this case planning an instructional sequence, it would be inappropriate to use a three-column lesson planning format (information-behaviour-feedback). We are in the business of 'telling a story' at this point. Furthermore, our choice of media limits us to the 'transfer of information'. Our lesson plan therefore takes the form of a step-by-step outline of the messages to be communicated and the means to be used to communicate them. We reproduce part of this plan below.

> *Step 1*: Review the position and the work of the typist in the organization. Stress the importance of the typing function for the efficiency of the organization. Use real examples and real-life illustrations to establish the real context, already well known by some of the trainees and soon to be experienced by the others.
>
> *Step 2*: Illustrate the harmful effects (both on the organization and the individual typist) and the potential benefits (for the typist) of work quality. Explain that it is easy to improve work quality and emphasize that it is the typists' ultimate responsibility to ensure adequate work quality. Use fictitious, exaggerated examples of good and bad typists, pleasant and unpleasant bosses or working environments (in order to avoid personal indentification with specific characters).
>
> *Step 3*: Illustrate the usefulness of each of the job-aids that are available. Use typical examples of errors that are commonly committed in the organization. Use a mixture of artwork and photography, as the topic requires, for clarity and impact.
>
> *Step 4*: Etc, etc.

The excerpt of the script presented in Exhibit 16.1 is the part that deals with Step 2 only. Before being fully scripted, as shown in the exhibit, the plan presented above was analysed by photographers, artists and audiovisual experts, in order to produce a storyboard more detailed than the above-mentioned outline but less fully worked than the script. This storyboard (not shown) was the document approved by the client.

16.3.2 A storyboard for an informational audiovisual

Exhibit 16.2 The sketches and notes presented in Exhibit 16.2 were not prepared as part of a self-instructional package design exercise. They were intended to form part of an introductory audiovisual on 'media selection' to be used to start off, or summarize, a teacher-led session on the subject (part of an in-service programme of teacher training developed in Hungary, under the auspices of UNESCO). The idea for this section of the audiovisual (there were many other topics dealt with, which we shall not list here) was adapted from a similar sequence of visuals, originally prepared in the UK (under the auspices of the Council for Educational Technology) as a multi-overlay transparency that accompanied a lecture package on media in education. A study of the visuals will reveal how they could be superimposed on each other, to create a series of images (a story) on the overhead projector.

In the context of the audiovisual project, however, the idea was adopted as the starting point for an INFORMATIONAL slide-tape presentation. We stress the informational aspect of the project, as this gives a clue as to the final structure of the script (not presented here as it was in Hungarian).

The ideas presented in each card of the storyboard were developed into a full script, taking care to maintain a fairly rapid flow of slides – thus each card of the storyboard was transformed into several slides, in order to accompany the quite extensive audio script that was developed. This led to several extensions of the initial idea. For example, our European child was related to several 'familiar' environments (countryside, cities, etc) and several 'unfamiliar' environments. We managed to squeeze in the reverse example (African child in relation to European environments) which pleased UNESCO, the sponsors of the project.

Exhibit 16.2 *A storyboard prepared to develop the theme of audiovisual media as means of enriching or improving the learning environment*

Exhibit 16.2 *(continued)*

② Schoolboy in his natural environment
(European / North American)
Establish the connection between the student
and the local subject matter.

③ Our subject may learn about his local
subject matter in two ways
a) by direct experience
b) by mediated experience
What are the advantages / disadvantages
of each alternative?

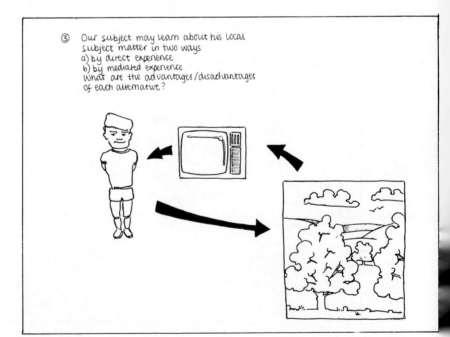

Exhibit 16.2 *(continued)*

④ Our student and a distant subject matter (tropical climate)
Identify the limitation of the natural learning environment.

⑤ Establish the use of mediated messages as overcoming limitations/enriching the natural learning environment.

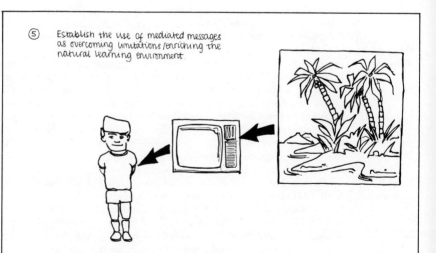

Exhibit 16.2 *(continued)*

⑥ Review the variety of media available
to the teacher, which may be used to
improve or enrich the learning environment.

16.3.3 A participative, 'semi-instructional' audiovisual

Exhibit 16.3 The script presented in Exhibit 16.3 is part of an audiovisual produced for use on training courses for industrial training instructors, offered by a UK technical college. The audiovisual was part of a multi-media pack designed to teach the principles of communication and the selection and use of instructional media.

No Level 1 analysis This project was not born out of a concrete performance problem observed in a specific job situation (as was the case of the typist example), but arose to meet the needs of a curriculum and course structure already developed, approved and in operation in the college (indeed, the basic curriculum of the instructor training course was fixed in accordance with the requirements of a national certificate awarded by the City and Guilds Institute to instructors and technical teachers). This being the case, there was no real opportunity to perform a true Level 1 analysis in order to establish that trainee instructors really needed to master the objectives set up for the specific course module in question. We shall take the need for the module for granted and proceed to Level 2 analysis.

Level 2 analysis *Level 2*: The overall instructional plan, prepared next, was based on a SUBJECT analysis (as opposed to a task analysis) which sought to interrelate the topics of the course as a NETWORK of learning activities organized in sequence and interdependence. The approach was based on the modified 'network analysis' approach pioneered by Tom Wyant (1973; 1974) and the final course was intended to run on a totally individualized, or small-group, basis in a learning resources centre specially set up for the purpose. Each learning activity corresponded to a file drawer containing a variety of materials, both instructional and informational, in a variety of media. Each 'node' or junction on the network was represented by a self-test, which could be used to check mastery of antecedent learning tasks and readiness for subsequent learning tasks (post-test and/or prerequisite test). Each learner could proceed at his own pace, along a track of his

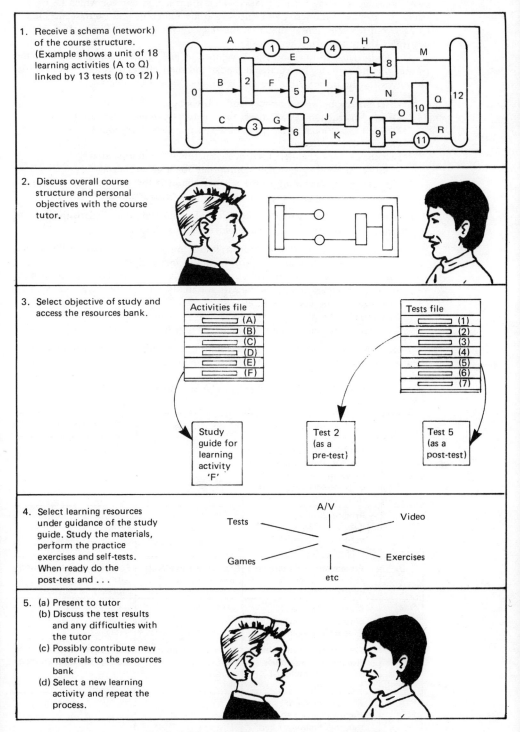

1. Receive a schema (network) of the course structure. (Example shows a unit of 18 learning activities (A to Q) linked by 13 tests (0 to 12))

2. Discuss overall course structure and personal objectives with the course tutor.

3. Select objective of study and access the resources bank.

   Activities file

   Tests file

   Study guide for learning activity 'F'

   Test 2 (as a pre-test)

   Test 5 (as a post-test)

4. Select learning resources under guidance of the study guide. Study the materials, perform the practice exercises and self-tests. When ready do the post-test and . . .

   Tests   A/V   Video
        Games   Exercises
                etc

5. (a) Present to tutor
   (b) Discuss the test results and any difficulties with the tutor
   (c) Possibly contribute new materials to the resources bank
   (d) Select a new learning activity and repeat the process.

Figure 16.6 *Procedure of study in an individualized, resource-based workshop for trainee instructors*

own choosing, selecting learning tasks which interest him and for which he is ready (by means of the tests) and selecting study materials and exercises to his liking (from the selection available in the relevant file drawer). Figure 16.6 illustrates the procedure schematically.

The audiovisual in question was developed as one of the media available to the learner for the activity 'Instructional media and their characteristics'. The specific objectives for the presentation were: The trainee should be able to...

* Discriminate between one-way and two-way communication.
* Define 'instruction' as a process that requires two-way communication.
* Define instructional *media* (as opposed to methods, techniques).
* Define 'sensory channel'.
* List the sensory channels used in instructional communication.
* Identify the sensory channels appropriate for a given message.
* Classify media according to the sensory channels they use.
* Identify media applications which qualify as 'instruction'.

The excerpt of script presented in Exhibit 16.3 is the part that deals with the fourth and fifth objectives only. An audiovisual was chosen:

(a) as a convenient medium to illustrate the highly visual content of the lesson – a demonstration of the use of media in instruction;
(b) as in a later exercise, trainees are asked to analyse the slide-tape from a variety of viewpoints, including its suitability for the objectives, its 'status' as regards true instruction, etc;
(c) because of the philosophy of the course structure that sought to offer a variety of media alternatives whenever possible.

Level 3 design

Level 3. Before proceeding to the scripting of the audiovisual presentation, a detailed 'lesson plan' was drawn up. This is reproduced below:

Figure 16.7 *Lesson plan for the 'instructional media' AV*

	Information	Learner activity	Feedback
1	Explain the objectives of the AV. Present a 'case' to illustrate the importance of these objectives for the instructor.		
2.	Present illustrations of 1-way and 2-way communication. Define each mode.		
3.	Present further examples. Ask the learners to clarify them.	Classify the examples given.	Feedback by comparison with correct answers supplied on tape.
4.	Present a formal definition of 'instruction'. Give a clear example.		
5.	Present cases of teaching situations as examples and non-examples of 'instruction'.	Classify all examples and non-examples correctly.	Comparison with correct answers supplied on tape.
6.	Give clear examples of the difference between method, technique and media. Define each.		

] **Figure 16.7** *(continued)*

7.	Give further examples and non-examples. Ask learners to classify.	Classify all examples and non-examples.	Comparison with correct answers supplied on tape.
8.	Illustrate the concept of 'sensory channel' by means of one or more examples.		
9.	Ask the learners to identify other examples of 'sensory channel' (include the kinaesthetic sense).	Identify the sensory channel used in each example.	Comparison with correct answers supplied on tape.
10.	Etc . . . etc . . .		

The lesson plan developed at Level 3 is already quite detailed and illustrates clearly every step of the proposed instructional sequence. The overall strategy adopted is quite clearly EXPOSITIVE. The sequence follows the 'Demonstrate/explain-Practise-Feedback' model of expositive instruction quite closely. There is nothing very unusual about the lesson plan. It illustrates the intention to use an audiovisual slide-tape package as the medium of delivery, but otherwise, it may serve as a plan for a teacher-delivered lesson, or a self-instructional text, just as well. The decision to use a slide-tape has cut down the variety of options, especially as regards the modes of learner activity and feedback to the learner. A live lesson could be much more varied in this respect.

One might also criticize the plan, as regards its status as 'true' instruction. There is, after all, no guarantee that the learner watching and listening to the presentation will in fact respond to the questions posed on the tape, before the tape itself supplies the correct answers. This is analogous to the classroom situation when the teacher asks a general question, for *all* to think about, and then invites *one* student to supply the answer – there is no guarantee that the others gave any thought to the question at all. It is for this reason that we have named this exhibit 'semi-instructional' – it has the appearance of two-way communication, but in reality we cannot be sure that it has occurred.

Level 4: the script

Level 4. The lesson plan went through the usual stages of storyboard to final script. At storyboard stage, the author came up with the idea of the 'two lovers' as a visual means of illustrating all sensory channels. Note that the two steps 8 and 9 are the only ones dealt with in the piece of script presented. At the storyboard stage this example was represented as two cards. In the final script, it has been expanded to a sequence of seven slides.

16.3.4 A fully self-instructional slide-tape workbook system

Exhibit 16.4

The last exhibit we present here is interesting on several counts. It dates back to the mid-1960s, having been produced as a research project in an investigation into 'audiovisual programmed instruction'. In the early 1960s, I was engaged on a project at the Pressed Steel Company, Oxford, which was responsible, together with a similar project at the Baker Perkins Company, for the first large-scale application of linear programmed instruction to engineering apprentice training in the UK (see Romiszowski, 1965; Dodd, 1967a). The project caught the imagination of many trainers at that time and the programmes were used in many other companies. The majority of the programmes written in these projects were later published, partly by the Heinemann Publishing Company in booklet form and partly by International Tutor Machines in linear teaching machine form. In 1966, the author, together with the staff of International Tutor Machines, performed a research project into the techniques of programming in the audiovisual medium. One part of this research involved the transformation of an already published and validated linear text on Industrial Safety, into an audiovisual version. The content, sequence, types and number of examples, etc,

Exhibit 16.3 *Part of an audiovisual script, prepared for a slide-tape presentation to be used on a training course for industrial training instructors*

VISUAL	AUDIO
(An earlier part of the slide set, showing the role of communication to instruction.)	(An earlier section of script, about as long as the part reproduced here.)
	Now, how about the SENSORY CHANNELS used for communicating our messages? Our message may use any one, or a combination of the senses, to reach its destination. For example, the girl here is receiving the boy's message by the AUDIO channel.
	But what about the girl's message to the boy? (pase — 4 seconds) The hankie she has dropped is a VISUAL 'come on' message to the boy . . . and a pretty old fashioned one at that!
	The boy however is a bit more up to date. He knows not only WHAT message content will be effective but also that sending it by both AUDIO and VISUAL channels, simultaneously, will be more efficient! (pause — 4 seconds)
	As our tale unfolds, the other sensory channels come into play . . . taste, smell and the tactile, or touch senses . . . . . . How many sensory channels do we use in communication anyway? (pause — 4 seconds)

Exhibit 16.3 *(continued)*

If you answered FIVE, you're WRONG!
Sight, sound, touch, taste and smell make up five,
but there's one more!
The SIXTH sense we're after is the kinaesthetic sense
of internal bodily coordination — the sensory channel
that tells us when our muscles are doing what they
should be doing — *and doing it right!*

(pause — 2 seconds)

Very important in most practical tasks!

OK! Just to sum up, here's another motorist driving
along . . .
What sensory channels does *he* use to receive the
information necessary to control the car?

(pause — 6 seconds)

Artwork:

VISUAL	— THE ROAD
AUDIO	— HORNS, ENGINE
TACTILE	— BUMPS ETC THROUGH WHEEL
KINAESTHETIC	— LOCATING CONTROLS WITHOUT LOOKING

In normal circumstances, four sensory channels may
be used — vision, to see the road, audio when strange
noises occur, touch to feel vibrations and bumps
through the steering wheel and, of course, the
KINAESTHETIC sense to grab the gear lever without
looking down, to apply the right pressure to the brake
pedal, and even to light a cigarette without taking the
eyes off the road.

were all maintained equivalent in the two media – only the media changed.

Version 1: the audiovisual linear programme

Originally, we produced a version that was entirely in slide-tape, following a style of scripting rather like that illustrated in Exhibit 16.3 (channels of communication). In effect, the original printed linear programme was read verbatim, with minimal changes on to tape and synchronized with a set of slides more or less equivalent to the illustrations in the original programme.

This initial version of the audiovisual programme was tested on groups, matched to groups that had already studied the printed version. In terms of immediate learning, there was no significant difference between the two programmes. In terms of acceptability (learner reactions) there was no significant difference. But the audiovisual programme took, in general, about 30 per cent less time to complete.

Version 2: The workbook-audiovisual programme

Later on, the second version, incorporating a workbook, was prepared. The reason for this was that studies of the earlier versions had suggested that, although immediate learning was near 100 per cent of expected criteria, later on-the-job performance was not as safe as had been hoped for. One possible reason for the lack of transfer may have been forgetting, and linear programmes (still less audiovisuals) were not convenient media for rapid revision. The second version incorporated a *workbook/job-aid* as a secondary medium. This had several functions:

* It provided spaces for learners to make overt responses to direct problem-centred questions (each slide that was analysed was repeated in the workbook as a picture) thus leaving the learner with his own permanent record of study.

Exhibit 16.4 *Part of an audiovisual script prepared for a self-instructional slide-tape-workbook package for teaching industrial safety to new apprentices*

VISUAL	AUDIO
 Fred	In this module, we will study the basic principles of industrial safety. Remember: SAFETY BEGINS AT HOME, when you PREPARE YOURSELF to come to work. Fred, here, is about to strangle himself with his own scarf. He should never have entered a workshop, full of fast moving machinery, dressed like that!
 Bill	What about Bill? He's just arriving to work. Look at his appearance! What's wrong with the way he's turned out? Switch off the tape for a moment and turn to page 1 of the workbook. Read the rules listed there and answer the questions about Bill. (pause with music, as signal for switching off) You should have written down three criticisms of Bill's turnout: — his untied shoelaces. — his loose, oversize jacket. — his long hair — Yes! Many people get scalped every year by getting hair caught in machinery.
	Now here's Fred again — without his scarf. But he's behaving in a potentially dangerous manner again! Oily rags seem harmless enough, but if the oil gets on the shop floor, someone may have a nasty spill. Remember: THINK AHEAD — PREVENT ACCIDENTS BEFORE THEY OCCUR!
	Here's another potentially dangerous situation. Who do you think is in danger? What is the thoughtless behaviour that might cause an accident? Stop the tape, read page 2 of the workbook and answer the questions. (pause with music) Well, of course, both Fred and Bill are in danger and both of them are acting in a thoughtless manner. Bill is running, and Fred cannot see where he is going. Both are breaking the rules of safe behaviour in the workshop.

Exhibit 16.4 *(continued)*

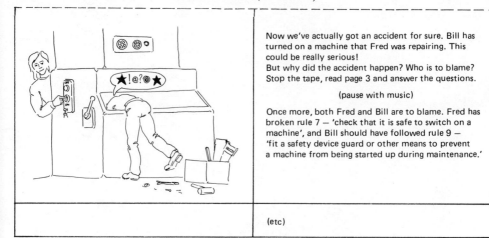

Now we've actually got an accident for sure. Bill has turned on a machine that Fred was repairing. This could be really serious!

But why did the accident happen? Who is to blame? Stop the tape, read page 3 and answer the questions.

(pause with music)

Once more, both Fred and Bill are to blame. Fred has broken rule 7 — 'check that it is safe to switch on a machine', and Bill should have followed rule 9 — 'fit a safety device guard or other means to prevent a machine from being started up during maintenance.'

(etc)

* It provided summaries of key safety rules and procedures discussed in the programme, in an ordered and indexed manner very suited to rapid reference.
* Rules listed in the workbook were not all spelled out and repeated in the programme (as was the case in the earlier versions) thus significantly reducing the length of the audiovisual.
* The rules and related problem-slides were reproduced on the same page in the workbook, so that workers using the booklet later as a reference job-aid, were apt to try the problem again, or at least relive the earlier study experience.

When tested comparatively, this new version of the programme proved itself superior to both the earlier versions on all counts: immediate learning, acceptability, rate of learning and transfer to the real workshop situation.

In practice, the workbook on its own is a form of self-study material as it contains the rules and examples of their application (or the results of breaking the rules) – it has a sort of 'Rul-Eg' structure. It is however not fully programmed in the sense of applying a systematic 'demonstration-practise-feedback' sequence to each instructional objective, until the learner reaches full mastery. This aspect is left to the audiovisual lesson.

This aspect of sharing the tasks of instruction, practice and reference among different media is one interesting aspect of this example and one which is often overlooked in the design of mediated instruction. One tends to get 'hooked' on the *principal medium* selected for the lesson and neglects to consider the use of simple *secondary media*, which could both simplify and enhance the overall instructional package.

We shall not go into detail as regards the earlier stages of design that preceded the script presented in Exhibit 16.4. The earlier exhibits have already covered these points adequately. A rigorous Level 1 analysis did not take place in this case, as the very concept of front-end analysis was unknown at the time. Overall objectives for the whole project were derived through very traditional approaches to the planning of training. Level 2 design was similarly skimpy, as we were preparing media for an already existing curriculum and course structure. Level 3 design adopted the 'Rul-Eg' approach that was popular at the time (see Chapter 5). Level 4 design is illustrated by the exhibit. The reader may, if so inclined, attempt to recreate the earlier stages of design by working back from the script.

Before going on to the next chapter, the reader may also like to have another look at the four exhibits, as we shall be referring to them again.

17. Interactive Video and the New Media: An Evaluation

17.1 What are the 'new media'?

In our discussion, we shall apply the term 'new media' to a whole range of new possibilities of information transmission, which have become, or are becoming, a reality due to the progress made in microtechnology. Many of the 'new media' are not really brand new concepts, but their capabilities have been extended by the addition of microprocessor control. The laser videodisc is an example of a totally new medium, now slowly demonstrating its potential. The multi-screen slide show, however, is not new, but, given microcomputer control over the projectors, allows designers to change slides with such speed and precision, as to allow the creation of totally new visual effects, impossible to achieve in other media. Thus the addition of the microcomputer to even the humble slide projector or audiotape recorder may produce, in effect, a new medium of communication with its own special characteristics. And the addition of the microcomputer to the new and revolutionary laser videodisc has created yet another new medium – the interactive videodisc, which will figure as the principal subject of this chapter. But before we proceed with that topic, let us review some of the other promising new media developments which have appeared as 'spin-offs' from the micro revolution.

Multi-screen audiovisuals We have just mentioned the relatively humble slide projector, which, when coupled up in a bank of two, four, six or more, under computer control, is capable of presenting a host of new visual effects. The computer controls slide changes, fading, dissolving, forward and return random access and whatever else is required, enabling one to program rapid cascades of slide changes that may simulate a moving image, may vibrate from one predominant colour to another at any desired frequency, or indeed to the rhythm and tones of a particular tune, may present a slide out of sequence, or offer random access library facilities, and so on. The programming of such a show is a specialist and time-consuming task, but the results are often well worth it. Once a particular slide-mix effect has been programmed, it can, of course, be transferred to videotape through a telecine chain, so that multiple copies or repeat presentations may be easily managed without the complication of always setting up a bank of computer-controlled projectors. The multi-projection systems now available may well, in the long run, be destined to work as special effects generators for video presentations, rather than regular projection systems. However, they have opened up a new dimension of creativity in the use of still images for dynamic displays.

Split-screen and special effects video Anyone who watches television is aware of the awesome variety of special effects now possible, even in routine low-budget productions. Moving pictures twist and fly across the screen, as if in three dimensions and stack, one behind the other, like a pack of cards. A split second later, they fly off again, distort, do a loop-the-loop, fill the screen and, for a moment behave like the normal TV image we are all accustomed to. Then, once more, the special effects take over. Most of these effects are now pre-programmed and are available to the programme director at the flick of a switch or the twist of a knob. Whereas the majority of such special effects may not have much direct application in terms of the improvement of learning, they do enhance the level of interest and motivation of the audience, when used with a modicum of judgement. Some, such as split-screen effects, may offer instructional advantages for specific categories of objectives – we mentioned the use of this effect in relation to the simultaneous presentation of two objects or situations in order to facilitate their discrimination.

Computer-generated graphics and text

Another development that now appears regularly on our TV screens, is computer-generated graphics of various types. Most economics graphs and forecasts are now computer-generated as are the weather forecast charts. Every other motor car advertisement seems to present computer-generated schematics, with three-dimensional effect sometimes, which somehow miraculously spin and turn on the screen and eventually turn into the real car. The sub-titles on the foreign film are computer-generated and superimposed on the video at the time of transmission so there is no need to print a special dubbed copy of the film. The pages of text and graphics stored in the Prestel, Ceefax and other videotext systems are all computer-generated. This means they can be corrected or updated, word by word, character by character, at any moment, without the need for expensive retyping and setting of whole pages and chapters. Computer-generated graphics and text pages are of very real use in the educational and training field. Quite suddenly, two processes which were very costly and time-consuming – the updating and revision of printed messages and the preparation of special artwork for visuals – have become much faster and relatively inexpensive. No longer do we have the same excuses for out-of-date books, with boring, uniform text and few illustrations. Computer-generated graphics and text, in a variety of fonts and sizes, may now be produced on relatively inexpensive equipment, using special applications software available from your corner computer store.

Speech compression

Another contribution of the computer is the technique of 'speech compression'. A normal audio recording is digitized and then re-formed in compressed form, allowing the same message to be transmitted in a fraction of the time previously required, without loss of clarity or distortion of tone. The principal use of this technique is for the more rapid transmission of messages across phone lines or the more economic storage of audio, but some research has shown that speech-compressed audiocassettes used in education are exceptionally effective instructional media.

Narrow-band video

Narrow-band video is another development which is capable of cutting message transmission costs and, as a spin-off, is opening up new educational possibilities. The transmission of a video image is expensive, as a relatively wide waveband is required to carry the signal. This also limits the number of channels that may be transmitted simultaneously. Narrow-band video overcomes these limitations, by the expedient of transmitting the image at very slow-scan speed, so that only one screen image is formed every half-second, second, five, ten or 30 seconds. Special micro-controlled equipment is required to link up a normal TV camera at the transmitting station to normal TV receivers at the receiving stations. The transmission occurs 'on-line' however and anything that the camera focuses on may be transmitted, almost instantly, as a still image. For many purposes, a series of still images are a satisfactory medium of communication. The big advantage of the technique is that the transmission costs go way down as the waveband required for transmission is very narrow – the slower the transmission speed, the narrower the waveband required. Most systems can transmit stills at about one every few seconds over a normal phone line. The possibility of small group, or individual, distance education, with 'on-line' audio and visual communication, both ways, is at hand.

Laser holography and 3-D video

A number of systems for the transmission of three-dimensional images are now in development. Laser holography offers the possibility of creating three-dimensional images in space, 'as if they were really there'. Other, perhaps simpler, systems may soon bring the *appearance* of three dimensionality to the images on our home television. No large-scale, across the board benefits are likely to be gained in the educational and training fields, in the immediate future, but no doubt, once such systems become cheaply available, they will further enhance the impact of video messages and may contribute in special ways to the achievement of specific categories of instructional objectives.

**Communica-
tion
satellites**

The communication satellites, now so common that they have already revolutionized world telecommunications, have yet to play a significant role in education or training. Early attempts, such as the projects sponsored by UNESCO in India, national projects in Canada and the USA, and so on, have just scratched the surface of possibilities. Most of these projects have concentrated on the transmission of normal television broadcasts (ETV) to a wider audience – the name of the game has been 'massification' of educational opportunity. However, the large number of communication channels offered for telecommunications, coupled with data transmission developments, low-cost media for data storage, home computers, narrow-band video and the like, may soon be all integrated to offer economical, individualized, distance education systems, operating semi-automatically but with instant tutor-tutee two-way communication and catering to minority interest groups in flexible, learner-selected time-slots.

17.2 What is 'interactive video'?

**The general
meaning of
the term**

The term 'interactive video' may be understood in a general or a highly specific way. In the general sense, interactive video simply means visual (or rather, audiovisual) communication which has the characteristic of interacting with the viewer/listener in some way. In this sense of the term, we could classify as 'interactive video' the system of education that was set up in New Mexico in the 1960s to cater for the Zuni Indian reservations. Apparently the Zuni Indians do not take will to rigid timekeeping and this created a special problem for their education. The solution tried some 20 years ago was to set up a mobile viewing station on the reservation, linked by a closed-circuit TV system and a telephone to an educational film library at a local university. The Indians, whenever they felt like it, would individually phone in and request the screening of films selected from an available menu. If they got tired, they could phone in to stop the film, which would be set up to continue from where they left off, on some other future occasion.

In the specific sense, however, the term 'interactive video' has recently come to be used to describe the latest and (to date) most sophisticated audiovisual communication system for individual use; the laser videodisc linked to a microcomputer with graphics and text generation capabilities, through a special interface, which allows the computer to be used to control the videodisc, transforming it into a random-access store of moving images, stills and text. This is the ultimate of current technology. There are of course simpler systems, such as user-controlled random-access videocassette players, which are also referred to by many as 'interactive video'. And why not indeed? It is video and it does interact with the audience. But it is the interactively controlled videodisc that seems to have grabbed the imagination of many media experts in the educational and training fields.

The reasons for this are not hard to identify. The videodisc, as a storage and playback medium, offers a number of distinct advantages over its rivals. Some of these are:

* a very high storage capacity for its size (a single laser videodisc about the size of a gramophone record, may store over 100,000 still images or pages of text, or about 500 hours of audio recording, or about two hours of full audiovisual, twin soundtrack, stereo video programming);
* flexibility of storage (the same disc may be used to store any combination of stills, moving images, text, or indeed any information that can be digitized – computer programs for example);
* easy and rapid access to specific items of information (as in the case of a gramophone record, the reading head may move instantly to any given track on the disc – in contrast to magnetic tape, which must be laboriously wound forward and back to locate specific sections);
* robustness and long life (the disc is pressed from strong plastic, surface scratches do not necessarily influence the quality of reproduction, there is

no contact between the disc surface and the reading head so wear is negligible, etc);

* cheapness when mass produced (pressed in large quantities, from a master, rather like a gramophone record) – however, the costs of the master are high, so the medium is not very attractive economically for short-run production;
* security of copyright (due to the high cost of the master and the special equipment used for pressing copies, the problem of copyright infringement is much smaller than is the case with videotape, audiotape and magnetic discs).

17.3 Modern systems of interactive video: the hardware

A modern interactive video system has four principal components:

1. a video player (despite the tendencies to 'think disc', we shall consider both videodisc and videocassette players as possible candidates – the cassette machines do also have some advantages in certain circumstances, as we shall see);
2. a computer (usually a microcomputer, as portability is desired and no very great computing power is required – nevertheless, any size of computer could be used);
3. an interface unit, or card, designed to link the player and computer;
4. a video monitor, or adapted television receiver (which will present video from the player, audio, computer-generated graphics or text and the computer programs during the programming or debugging of the system).

These are the basic hardware requirements, but of course there are many variations in system characteristics and complexity.

Four levels of interactive video instruction

Instructional systems incorporating videodisc are usually categorized in levels, according to their capabilities for interaction.

Level 1 systems are the simplest, comprised of a video player and a monitor. As the material is to proceed linearly, the player may be a videocassette player. Player has little memory capability and no processing power. Level 1 systems are essentially playback machines.

Level 2 systems consist of a videodisc player and a controller, either internal or external. Control data can be decoded from a videodisc with a limited command set to handle most of the requirements. Combines a simple program with branching. Best limited to simple dedicated applications.

Level 3 systems combine all the capabilities of the videoplayer and the microcomputer, providing increased memory and data processing capabilities. Sophisticated branching is possible.

Level 4 systems usually imply two videodisc players and advanced capabilities (multichannel audio, windowing, etc). These allow almost instantaneous feedback and more complex branching. Level 4 systems are constantly developing and evolving to include more advanced capabilities.

Figures 17.1, 17.2 and 17.3 illustrate three levels of sophistication, as envisaged by Richard Currier (1983). Figure 17.1 shows a typical 'Level 1' system – little more than a simple video playback system. There may be a special keypad to facilitate random access, but the accessing is done by the learner, possible with the help of instructions in a printed study guide.

Figure 17.2 shows a Level 2 system with an internal dedicated microprocessor, which allows it to branch automatically to certain segments. This branching may be activated in a variety of ways – by means of a keypad, a bar-code reader that picks up coded instructions from the printed study guide, a light pen that selects from a menu that appears on the screen (or alternatively a touch-sensitive screen activated directly by the learner's finger). Normally, a system of this nature would be equipped with but one or two of these options.

A 'Level 1' system

Level 1

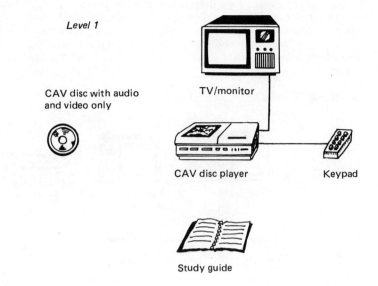

CAV disc with audio and video only

TV/monitor

CAV disc player

Keypad

Study guide

Figure 17.1 *A typical Level 1 interactive videodisc system*

A 'Level 2' system

Level 2

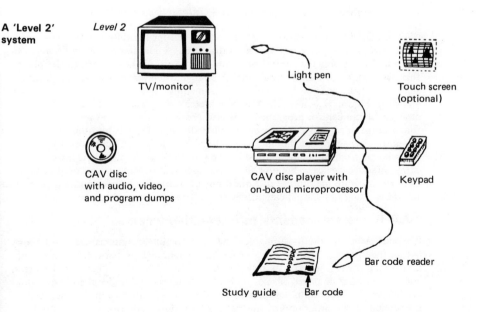

TV/monitor

Light pen

Touch screen (optional)

CAV disc with audio, video, and program dumps

CAV disc player with on-board microprocessor

Keypad

Study guide   Bar code

Bar code reader

Figure 17.2 *A Level 2 interactive videodisc system with some of the different available means of interaction*

Figure 17.3 shows a fully fledged interactive videodisc system. It has all the four principal components mentioned at the beginning of this section – player, micro, interface and one or more monitors. Depending on the number of monitors, the complexity of the control program and the other control options available, this system would classify as a Level 3 or Level 4 interactive video system.

Level 3 or 4

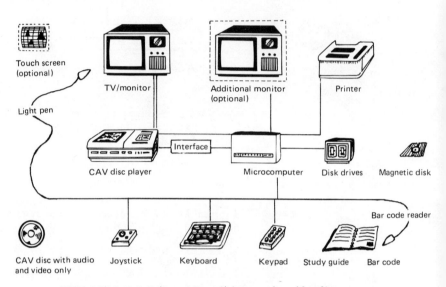

Figure 17.3 *A full (Level 3 or 4) interactive videodisc system*

Note the difference between the discs specified for the different versions illustrated. The Level 2 system uses a disc that, in addition to the audio and video material, also stores the control program for the disc. When loaded in the player, this program loads into internal storage built into the player and via the built-in microprocessor, controls the videodisc's branching. The more sophisticated system shown in Figure 17.3, uses the microcomputer's disc drives and magnetic disc storage for the control programs. It is also possible, however, to set up a system that stores the control program on the videodisc and loads it into the general-purpose micro, through the interface. The videodisc is then controlled by instructions from the computer, which flow in the reverse direction, through the same interface device. This type of system has been developed and used at the Ontario Institute for Studies in Education, in Toronto, Canada, and has proved to offer advantages of simplification, system cost reduction, and security.

17.4 Some antecedents to interactive video

It would seem that the most obvious characteristic of the systems described above is that they provide greater or lesser facilities for branching between segments of information, which may be in a variety of formats – moving images, still images, audio recordings or pages of text. This branching may be under the control of a predetermined programme of questions and answers, rather as in programmed instruction or the programmed tutorial modes of CAI. Alternatively, it may be under the complete control of the user, as in a random-access database, or 'audiovisual encyclopaedia'. A third alternative is that the microcomputer system keeps complete records of choices made and results achieved by a student, which may be compared with the records of others, analysing the error patterns and trends. This allows the system to 'learn' about the learner (as a human tutor would) and thus offer some modicum of 'intelligent' advice to the learner on the most appropriate choices at a given moment. We thus see that modern interactive video technology applies any one of the three control models that we defined for individualized systems of instruction in general (see Chapter 1):

* Prescriptive/autocratic control
* Learner-directed/democratic control
* Adaptively interactive/cybernetic control

As we have already seen all these systems of control in other forms of individualized systems, both print-based and computer-based, we may ask whether the only innovational aspect of modern interactive video is that it uses VIDEO as the principal medium of delivery. However, if we take the broad general meaning of interactive video, and thus interpret video as 'visual and audiovisual communication', we may question whether any aspect, apart from the disc, is really innovatory. Is not resource-based learning, which uses a high proportion of audiovisual instructional packages, and is controlled by an algorithm of criteria and test items to be mastered in any order, a close cousin of interactive video? Is the Postlethwait Audio-tutorial System not an early forerunner of today's systems? Is not the Zuni Indian school another example?

If automated control over branching is an aspect to consider, then the early branching teaching machines of the 1960s, such as the US Industries' AUTOTUTOR, were as interactive as some of the modern systems described above (but were of course limited to the presentation of printed and graphic information). However, the Lamson Tutor, developed at the same period by the British Royal Air Force, offered a form of interactive audiovisual instruction. This machine was essentially a linear teaching machine, presenting a sequence of programmed 'frames' printed on continuous 'fanfold' (computer printer) paper. The paper was driven past a viewing window by mechanical cogs and electric motors. However, the student could respond to multiple-choice questions by means of a series of buttons. These buttons could be programmed to do a variety of things:

* Skip forward a certain number of frames to another sequence.
* Switch on or off a number of electrically operated devices (there were seven outlets on the back of the machine, where one could plug in slide projectors, videotape players – black-and-white, reel-to-reel in those days – simulators, super-eight loop film projectors, even laboratory equipment and experiments).

The programming of the functions was done by punching holes in certain columns at the margin of the paper tape, using a hammer and metal punch – primitive technology, but effective and easy to learn. The Lamson Tutor was refined and marketed by a commercial company (Lamson Products), but despite its versatility it never really caught on.

A decade later, in the early 1970s, a new contender appeared on the market. This was the decade of the super-eight film in education (videotape was still very expensive, unreliable and generally only available in black-and-white). The eight-millimetre concept loop films were in vogue and indeed were most effective self-instructional media in many types of training situation. A firm called Educational Systems Ltd, in the UK (specialists in self-instructional programming), developed a machine initially named the ESL-PIP. The product looked so promising, however, that the patents were taken up by Philips of Eindhoven, for mass production. The PIP machine had much in common with some current interactive video systems. Basically, the PIP was a super-eight film projector, in which the frames of the film were moved on in response to pulses that were recorded on one track of a Philips-type audiocassette. The other track of the cassette was available for normal audio recording. This system enabled the presentation of the film at any rate of frame movement that was desirable. Still shots would occupy only one frame of the film, however long they were to last on the screen. Slow motion sequences could be programmed at any desired speed. Normal movie sequences could be presented in a much more economical manner, as the normal movie film speed is dictated by sound synchronization rather than movement fluidity requirements. When well programmed, a PIP film could present moving images, stills, pages of text with synchronized audio, and could branch forward, by rapid wind on, in response to commmands coded on to the audiotape. It was never a commercial success and, after a few abortive years, Philips ceased manufacture.

We are now in the 1980s and 'interactive video' has appeared on the market. What does it do? It shows moving images, stills, pages of text and presents audio messages, with the capability of branching between segments of information. What are its chances in the marketplace? To answer that, we might ask why the earlier forerunners described above did not make it.

Let's consider technological sophistication first. The Lamson Tutor was a primitive beast, to be sure. Even in comparison with the filmstrip-based branching teaching machines of its day, it appeared a clumsy, home-made contraption. However, it did work and was reasonably reliable. Also, it was the only machine on the market that offered automated interactive audiovisual possibilities at an accessible price. The PIP machine, on the other hand, was a beauty of design and functionality. It was a one-piece, small, modern-shaped back projector, taking a cassette-loaded film and a cassette-loaded audiotape that would automatically synchronize. In comparison to the typical synchronized slide-tape system of today, the PIP was simplicity itself to set up and use. In this respect, it compares favourably with most interactive video systems.

How about the task of preparing software for the machines? Once written, Lamson programs required careful typing and some manual hole-punching which was laborious, but not difficult to master. The PIP, on the other hand, required specialist equipment to put the program together. Users could not do it for themselves. This situation compares with today's interactive video scene – the technical production of a videocassette-based interactive video programme is about as difficult as programming the Lamson Tutor, while the technical preparation of a videodisc master and multiple copies must, like the PIP films, be generally entrusted to a specialist organization.

What about the instructional programming? Both the Lamson and the PIP machines were 'skip-linear' in operation, which means that they could only branch forward to new material, while modern interactive video systems (discand tape-based), can branch forward and back. This is not as big a technical drawback as may at first appear, provided that one is producing material to be used under *prescriptive control* conditions. In such a case, the author pre-programs all the branches that may be taken by the student, in the form of a control algorithm. If the author wants to send the student back to a previous segment, he just branches him forward to a repeat of the same segment. If the segments are short (as they should be in most instructional material), and 'branching back' is not all that frequent, the penalty of extra film footage is not too severe and the effect, as far as the learner is concerned, is quite natural. Incidentally, this little 'trick of the trade' may also be used in videocassette-based interactive video, to avoid the excessively long rewind times encountered when backtracking to an early part of the tape. It often pays to repeat the video clip again at a point adjacent to the one from which the learner will branch.

As regards the basic strategies and tactics that may be employed in the instructional design, there are some marked differences between the modern interactive video systems and their predecessors. The Lamson and PIP machines were born firmly in the programmed instructional tradition. They were designed to react to predetermined responses in fixed and predetermined ways. They are only capable of handling programmes designed for prescriptive control. Modern, computer-controlled interactive video, on the other hand, opens up the possibilities of totally learner-directed control (as in the case of searching a database or 'audiovisual encyclopaedia') and also of cybernetic control (as when a computer is used to create a database about the learners and offer an 'intelligent' learner-guidance service). Perhaps it is in the adoption of one or other of these two forms of control, as opposed to prescriptive, pre-programmed, control algorithms, that the really innovative aspects of modern interactive video systems may be found.

17.5 Some research results

Some support for this view may be found in the, as yet sparse, research evidence on the use of interactive video in education and training. One such study, described by Laurillard (1984), illustrates the importance that the type of control strategy used has on the effectiveness of interactive video as a medium of instruction. Laurillard describes experiments performed with students of the British Open University, to investigate three key questions: what should be the balance of student control versus program control with respect to –

Student control versus program control

(a) the sequence of presentation of content in the video part of the package (Laurillard refers to this as the 'receptive' mode of learning);
(b) the choice of how many practice exercises to do and at what level of difficulty (this is called the 'active' mode of learning);
(c) the strategy for alternating between these two modes of 'receptive' and 'active' learning.

For the purposes of the study, an existing, well tried and successful Open University course segment was chosen and the existing video lesson was converted into interactive video form. This was done by isolating concepts covered by the original lesson as separate self-contained video sequences and then developing, for each of these sequences, a CAL program to provide practice exercises with tutorial guidance where necessary. The teaching built into these CAL units did not go beyond that contained in the original texts which formed part of the OU course, used as a basis for the experiment. This structure enabled the researcher to verify that, in general, the interactive version of the lesson was at least as effective as the previously used lesson package. However, the principal findings are related to the three questions mentioned above.

Methods and results of student control of sequence

As regards the sequence of presentation of the video component, student choice was built-in in two ways. At the end of each segment of the video presentation, students could opt to do an exercise or to continue with another video segment. Similarly, they could choose more exercises or more video at the end of each CAL exercise. The choices were presented as menus of all the video segments and CAL exercises available, facilitating free choice of sequence to each student. Some groups received advice on the most 'appropriate' (in the author's view) choice and other groups did not receive such advice. The second way of allowing student choice of sequence was within a given video segment – students could interrupt the presentation and choose to repeat it, review the last 30 seconds, skip forward 30 seconds, or just leave it and move to another learning activity. In summary, the results were as follows:

* between segments, students expressed the desire for some advice on appropriate choices of study activity (those who did not receive such advice complained· of feeling 'lost');
* however, when such advice was given in respect of the next 'appropriate' video segment to view, it was followed in only 48 per cent of cases;
* the option to stop/rewind/skip parts of a given segment of video was hardly used at all by the students (they would tend to view-on 'passively' until the video stopped automatically).

Student control of practice exercises

As regards the freedom to choose practice exercises and do as many as were felt necessary, the students tended to exhibit yet more variety of individual choice. Once more, author's advice was given, in some cases, as to the most 'appropriate' moment for attempting a given set of exercises. Also, the number of exercises attempted in a given set, was under complete student control. The results were as follows:

* when advice on 'appropriate' moments for taking a given exercise set was given, it was followed in only 28 per cent of cases;
* students varied considerably in the number of exercises attempted in a given set (from one to 12 exercises) and these variations were highly complex, the same student attempting many exercises of one set and only one or two of another (this being clearly related to individual difficulties with individual concepts of the lesson).

Student control of overall learning strategy

As regards the overall strategy of alternating between the 'receptive' and 'active' modes of study, students showed a similar degree of individual preferences and styles. All in all, based on these experimental results and on interviews with the participating students, Laurillard concluded that

'students can make full use of most aspects of control and moreover make use of them in such a variety of ways that it becomes clear that program control must seriously constrain the individual preferences of students. To justify the use of program control, the designers must demonstrate that they 'know best' what the student needs at each stage, ie that program control gives improved learning outcomes. Given the continual failure of educational research to ever find an evaluation instrument sensitive enough to produce significant differences of this kind, it must be preferable to give students the benefit of the doubt. We should acknowledge that the unpredictability and variation in their learning behaviour could be derived from perfectly legitimate and effective learning strategies, and that these should be considered in the design.'

Laurillard was, of course, working with a fairly sophisticated group of adult learners and with certain specific types of objectives. It may be that with other types of students, on other types of learning tasks, the conclusions would not be as overwhelmingly in favour of the use of student-directed control strategies. However, it is clear that there are whole areas of education and training where the above-mentioned findings could be expected to hold good and it is perhaps in these areas that the special advantages of interactive video might be best exploited. The students in this experiment learned as well as or better than previous groups using non-interactive materials, *irrespective of the learning strategies adopted.*

Other relevant research on learning strategies

Other research on individual learning strategies, for example the well known experiments of Pask and Scott (1972) on 'serialist' and 'wholist' learners, confirms the real importance of matching presentation sequence and structure to the learner's own preferred learning style and suggests that often (though not always) the learner is the best judge of which learning strategy would be most appropriate for a given learning task. Such viewpoints have been incorporated in many schemes of resource-based learning over the last two decades (Postlethwait's Audio-tutorial System is one example – see Postlethwait et al, 1972). However, these systems tended to necessitate the use of complex, multi-media, learning carrels, equipped with audiotape recorders, slide projectors, video players and suchlike, all under student manipulation and therefore subject to loading problems, delays and frequent breakdowns. The modern, interactive video system, based on an optical laser disc, may overcome most of these technical difficulties. Furthermore, storage and reproduction costs are reduced drastically when one set of hardware and one storage medium may substitute all the complexity of a 1970s study carrel and non-print media library. Student-directed resource-based learning becomes an economically and administratively viable methodology through the adoption of interactive video. However, not all students in all situations are capable of proceeding in a totally self-directed manner. Laurillard's experiment showed that advice on video segment viewing was followed in just over half the cases that it was given. This is significant, as the students would only have the segment title on the 'menu' as guide to what to expect in the segment. Contrast this with the lower rate of advice-taking as regards practice exercises, due to a higher awareness of whether a particular segment had posed problems of comprehension or had appeared relevant enough to justify devoting time to its practice. Whether they followed it or not, students wished to have some form of advice. In Laurillard's experiment, this advice was based exclusively on the author's analysis of the subject matter and its internal structure. It did not take into consideration the knowledge-states, interests or preferred learning styles of individual students. A Level 3 system may, however, be programmed to track the learning style and choice pattern of each individual student and, through comparison with his learning success and perhaps with the paths and success rates of other students, offer advice on 'appropriate' options based not only on the subject matter's structure but also on the individual student's learning record. We are now in the domain of *cybernetic control strategies*.

The use of cybernetic control strategies

There are many possible ways of implementing such strategies. One is through the maintenance of a cumulative record of a given student's learning progress and comparison with previous fruitful and fruitless choices that had been made by previous students with similar learning patterns up to this point. Such an approach was used in PROJECT PLAN in the late 1960s to prescribe learning tasks in the normal classroom situation by means of a distant computer-managed instruction system (Flanagan, 1968). Another approach is used by Pask in the CASTE system of computer-assisted learning. CASTE presents students with a map or network of all the topics included in a course and all valid paths for going from one topic to another (an 'entailment mesh'). Students may exercise their options as regards what they study and in what order they proceed from one topic to another. The system may however be implemented under two different control strategies – Pask refers to them as 'loose CASTE' and 'strict CASTE'. In 'loose CASTE' the student is quite free to direct his own study, but in 'strict CASTE', once he opts to study a given topic, the system will instigate a series of questions and tasks planned to make the student 'prove' (to both the system and to himself) his readiness to study the chosen topic. This may lead the student to opt for an earlier prerequisite topic which, 'he now knows, he does not dominate adequately enough to proceed to his original choice'. Early versions of the CASTE system, dating back to the 1970s, presented the learning material, under the system's control, from a bank of several random-access slide projectors. A modern implementation would no doubt use an interactive videodisc delivery system.

From an instructional design viewpoint, therefore, we see that the videodisc, under computer control, makes possible many of the ID strategies we have dreamed of but failed to implement for cost or complexity reasons.

17.6 Some existing approaches to ID in interactive video

'When a new technology is introduced, there is a tendency to use it in the same manner as the technology it has replaced' (Tennyson and L'Allier, 1980). These authors made this remark in relation to computer-assisted instruction and we have seen, in the previous part of this book, that most CAI methodology up to quite recently bore out the truth of this statement. One of the same authors (L'Allier, 1983) has felt it necessary to repeat the same observation in relation to current approaches to the utilization of interactive video. We shall, in this section, give a brief overview of the early applications of the interactive video technology, concentrating on those applications that may lend support to L'Allier's views. In the following section, we shall examine some of the more recent trends, which indicate that practitioners in this new field are indeed beginning to learn to use interactive video in novel ways that do exploit its special characteristics.

Reinventing programmed instruction

Early enthusiasts for any innovation, tend to apply it in all and every situation. This is not altogether to be decried. It may form part of a necessary 'learning curve', part of the research and development required to establish, in practice, the real benefits and limitations of a particular technology. This may, however, go too far. When institutions, funding agencies and publishers get caught up in the net of over-enthusiasm, a self-generated impetus may be given to the technology, leading to rapid growth for a time, until the sponsors and promoters of the 'enthusiasts' finally come to realize that the promised results or benefits are not being delivered at the costs predicted. Then a rapid decline sets in. There are many examples one may quote. As editor of the *International Yearbook of Educational and Instructional*

Teachnology (promoted by the Association for Programmed Learning and Educational Technology of Great Britain) from 1967 to 1975, I was able to trace the rise and fall of programmed instruction during that period, as measured by the number of new titles of programmed texts published in the English language, worldwide. The graph shown in Figure 17.4 maps the very rapid growth in publication, followed by an equally rapid decline. We note, also, a slight recovery, or rebirth of activity, after 1969, indicative of the 'real' market for programmed texts of quality, written on appropriate topics (Romiszowski, 1974).

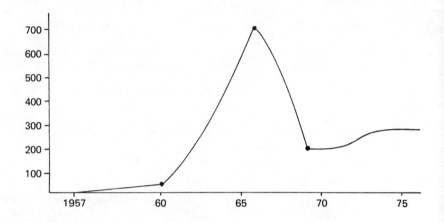

Figure 17.4 *Number of programmed texts 'in print' (1957-1974)*
(Romiszowski, 1974)

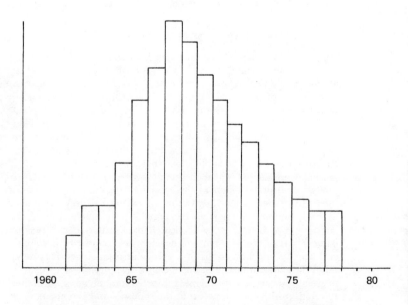

Figure 17.5 *Number of effective ETV broadcasting stations*
in Latin America (Tiffin, 1980)

The phoenix syndrome

A similar trend was identified in the use of educational/instructional television, during the same period in the USA. The rapid rise, sudden death and slow rebirth of a technology, 'from its own ashes' as it were, has been called the 'phoenix syndrome' – the enthusiasts must so abuse the technique that it inevitably dies, to be reborn, painfully, through the work of later more systematic not-so-blindly enthusiastic specialists, who identify and exploit the technique's true worth.

The phoenix-like rebirth is not necessarily guaranteed, however. The early over-enthusiasts may do such an effective job of assassinating the technique that it takes years to recover, by which time it may well be obsolete. Figure 17.5 shows data on the rise and fall of educational television in Latin America (Tiffin, 1980). To date, there is no sign of the heralded rebirth and current economic difficulties in the continent make it most unlikely that the investment necessary to rise from the ashes and learn from earlier mistakes will occur in the near future. In the meantime, the problems that effective ETV systems could address continue and grow, the investment so far made in existing systems of ETV continues to be unproductive and new developments suggest that by the time the technology can pull itself out of its present burnt-out state, it will be largely obsolete, as compared to new, upcoming, ideas such as local area networks, interactive 'live' teleconferencing networks, slow scan video intercommunication networks and the like.

Can we predict and avoid the phoenix syndrome? Can we take a critical look at what practitioners are doing with a newly emergent technology (in this case, interactive video) and avert the premature incineration of all the early, enthusiastic effort devoted to its first stages of development? We can but try. Let us look at the ways in which interactive video has typically been used in its first half-dozen years of enthusiastic promotion. Let us see whether we can notice any signs of impending suicide. Let us decide what may be done to avert any danger which threatens.

The four 'most popular' interactive video modes

In an article published in November of 1983 Samuel Howe (1983) mentions four 'current ways of using interactive video in the classroom'.

1. As a FILE. This is no more than the rapid access of specific 'clips' from a file of video segments. These would then be incorporated as 'audiovisual aids' in a normal classroom teaching situation, or alternatively, students may individually access the segments, following some study guide (as in 'conventional' resource-based learning systems) or without a systematic study guide (as in 'conventional' use of a library). The access may be computer-controlled, but really a booklet listing the titles of the segments and their location is all that is necessary, together with a Level 1 video set-up, to make the system work.

2. As an INSERT. This involves the insertion of text or questions into an existing video presentation. Howe suggests: the insertion of textual frames explaining or describing an upcoming segment of the video; a few key questions at the end of the segment, to test overall comprehension; extra information to 'enhance' the video message or adapt it to the needs of a given target population; the inclusion of a glossary of terms which may be used by viewers who do not understand some of the terminology employed in the video; and so on. Some other authors (eg Lawrence, 1981) call this approach the 'repurposing' of an existing videotape. It has the advantages of cheapness, as existing materials are utilized, ease of production and the possibility of non-professional, teacher-made productions. Generally, such repurposed materials would gain little by transfer to videodisc, as they are essentially a linear sequence of segments (the original tape was planned for linear, continuous presentation) interspersed by some reading assignments and exercises.

3. As PROGRAMMED LEARNING. This is essentially the preparation of branching sequences, as used in Crowder-type branching texts or in the programmed-tutorial CAI mode, except that some, or even all, of the information is presented in video form. These 'video information frames' are interspersed by computer-generated 'practice' and 'test' frames. In terms of instructional design strategies, we are back in Part 2 of this book and probably in the 'Rul-eg' rather than the mathetics or other more sophisticated methodology. In terms of video production technology, we are in the 'information communication' game, rather than in true 'instructional video design'.

4. As SIMULATION. This is essentially the CAI simulation mode, with enhancement of video segments to present the case material and also the debriefing commentaries. This has been especially well applied in simulations of human interaction situations. Howe quotes the example of first aid training in which video segments show the accident and victims, CAI segments teach and test how to deal with the situations illustrated and video is also used for demonstration of proper techniques, etc. This is an example of a simulation exercise based largely on expositive instructional strategies. There are however many examples of experiential simulation exercises presented in videodisc. In medical diagnosis and treatment (a favourite field of application at present) there are many videodiscs which allow the viewer to make decisions which then affect the outcome of the treatment being given. The viewer may 'cure or kill' the patient (and in many different ways) depending on the pattern, and even the speed, of the decisions taken. The viewer learns general principles of medical diagnosis and decision-making through the experience of a variety of alternative cases, each with alternative scenarios. These are true experiential learning situations.

So what is new in these four modes?

If Howe is correct in his analysis of the current 'state of the art' as regards the chief trends in the practical use of interactive video, then we may observe that L'Allier's comments quoted at the beginning of this section are, in the main, correct. The FILE mode is no more than the automation of access to a collection of non-print media. For convenience, they are stored on one videotape or disc and when the size of the collection warrants it, are better accessed by means of a

A file is a file is a file

computer than the alternative booklet index or card file. The rapid access times and relatively cheaper per-slide storage costs offered by videodiscs are very important administrative considerations in large non-print media collections. They may offer significant benefits in terms of large-scale duplication and dissemination, but hardly make the headlines from an instructional design or development viewpoint. The simple cataloguing and organizing of existing media does not, of itself, lead to improved instructional design.

Insert what in what?

The INSERT mode, or 'repurposing' of existing video material, may be a useful method of getting more 'mileage' out of earlier investments in video production. If the earlier materials are instructionally sound, but a bit out of step with current needs, this may be a most cost-effective approach to ID. If the existing materials are not 'instructional', but merely 'informational', then the repurposing required may be quite extensive. It is possibly better to think of this case as the generation of new instructional designs (usually CAL designs in the present context) which utilize excerpts of existing video as part of the informational content. The instructional materials developed on the computer are not then 'inserts' in an existing video programme, but segments of the existing video programme become 'inserts' in the newly designed instructional programme. This latter approach is, of course, a completely legitimate ID procedure. Why reshoot footage when existing material serves the bill? But it is hardly an example of the true meaning that Howe ascribes to the 'insert' mode. This approach is now seen to have its limitations. To what extent can one just 'face-lift' an existing video programme to suit new objectives and new target audiences, without descending to the level of design detail which is characteristic of new media development?

Do we need the videodisc to do inserts?

Before we leave the INSERT mode, let us just observe that as long as we are limited to the making of relatively small commentaries, introductions and explanations, test questions at the end of significant segments of an essentially linear sequence of video segments, providing glossaries, etc, we may often be able to do all the 'repurposing' we need, without the use of any modern interactive video hardware. All this might conceivably be as effective if printed in an accompanying study guide – and let the student start, stop and, very occasionally, rewind the videotape. This has, indeed, often been done in the past with existing videotapes, or even films. Where is the novelty, from the ID point of view? And where is there any other benefit? As Vazquez-Abad, Winer and Mitchell (1982) observe, 'no computer is cost-effective when it is used as a mere "page turner", or even a "teaching machine" in the sense of a provider of programmed instruction'... which takes us on to the third mode of use mentioned by Howe.

Beware the fate of PI
The earlier comments on the 'rise and fall' of programmed instruction should not be taken as an indication that we feel (as many apparently do) that the programmed instruction movement is long dead and forgotten. On the contrary, the whole flavour of earlier sections of this book should show clearly that we believe the movement had, and indeed continues to have, a profound influence on current instructional design methodologies. If anything, we bemoan the fact that too little attention is currently paid to all that was learned in those early, pioneering days of instructional technology. We are, however, alarmed at the rather indiscriminate way that some of the early methodologies are currently being revived and applied in the production of CAL courseware. We have already shown, in Part 3 of this book, how the majority of CAL courseware authoring modes (especially those most employed in commercially distributed packages) are closely based on programmed instruction methodology of the 1960s. We have indeed also indicated that some of the more promising and powerful methodologies of those days (eg mathetics, information mapping and structural communication) have so far found very little application in the design of CAL materials, although some of them have very definite advantages to offer in this context. An analysis of CAL packages currently on the market, in qualitative terms, shows that a significant proportion suffer from the same ills that contributed to the rapid demise of programmed texts in the late 1960s. A quantitative analysis, if performed, may well reveal that 1987 or thereabouts will be the turning point in the growth graph for 'generic' CAL packages publication, just as 1967 was the turning point for the 'generic' programmed text. We hope this will not be the case, but fear it might be, if the principal growth area continues to be in the use of computers to do what can equally well be done by texts. Incidentally, the buying market rejected such texts some 20 years ago. The Lamson Tutor and the PIP machine both died in attempts to link audiovisual to such programmed instruction.

It is doubly worrying, therefore, to see that one of the principal growth areas for the new interactive videodisc technology is seen to be in programmed-instruction-type designs with the added 'bells and whistles' of video. This does not bode well for the long-term health and survival of interactive video.

Some light on the horizon – SIMULA-TION
Fortunately, the fourth mode quoted by Howe – SIMULATION – holds out hope for a better future. Without a doubt, the use of CAL in the simulation mode has, so far, made the greatest general impact as an instructional innovation. There are, of course, some other specific developments of great local importance (eg the use of LOGO in primary schools) or of as yet little practical application (eg the use of EXPERT SYSTEMS as instructional tools) which also hold great promise for the future. But it is simulation which has, here and now, demonstrated the power of the computer as an instructional tool in a whole range of subject areas and with all manner of student groups. However, computer-based simulations have been somewhat limited to subjects which can effectively be presented on computer output devices (principally on monitor screens). Physical and other scientific phenomena may often be adequately portrayed by means of formulae, schematics, tables, graphs or even animated computer graphics. But there are many learning situations which defy presentation on standard computer-terminal equipment. One approach, useful in the psychomotor skill and equipment operation domain, is the construction of computer-controlled simulators (flight simulators, driving simulators, etc). However, this tends to be an exceptionally costly pastime, only justifiable in a small number of special training cases.

The incorporation of video segments, large collections of still shots filmed in high resolution, audio and all the other media options offered by the videodisc, into computer-based simulations, opens up a vast new field of possibilities. Many machine-operation simulations which would not be cost-effective if attempted by the construction of special simulation equipment, may be attempted through video-based simulations. Many diagnostic skills that require colour or shape perception to a high degree of perfection may now be developed (eg the spate of interactive video-based medical diagnosis simulation exercises). But perhaps most significantly, simulations which involve the interpretation of the behaviour of other human beings (which up to now would require role-play or other real-life interaction sessions) are now open to the development of self-study, individual, computer-based simulation exercises. The simulation field in general and the development of interactive social simulations, may yet save interactive video from the phoenix syndrome.

17.7 What are we learning about the effective use of IV?

Although the interactive video movement is only some years old, new ideas and innovative design guidelines *are* beginning to emerge. Those specialists who have put pen to paper as regards the novel ID aspects of interactive video, are beginning to show strong agreement on a few cardinal principles, although they differ strongly on one or two other points.

13 principles for IV projects

In one of the earliest comprehensive manuals on the design of videodisc/microcomputer courseware, DeBloois (1982) derives a list of 13 design principles. These were derived from the systems analysis of an ongoing interactive videodisc design project. In summary, these principles are:

1. An interactive videodisc system is not merely a merging of video and computer – it is an entirely new medium with a host of special characteristics. (DeBloois lists some two dozen of these special characteristics, advantages and disadvantages – we have mentioned most of these in earlier sections of this chapter.)
2. To develop interactive videodiscs, adequate resources are required. (DeBloois refers to both material and human resources. We shall see that other specialists differ in what is to be considered 'adequate'.)
3. The objectives of interactive videodisc systems must reflect the instructional needs of a wide variety of potential learners. (DeBloois talks of preparing a 'matrix' of materials at various levels of difficulty and complexity, with a variety of possible objectives – cost of the medium may be one reason for this, but another is to profit from the vast branching potential of the medium. Would you agree that this should be a general guiding principle for all videodisc productions?)
4. The design effort must be interdisciplinary in nature. (Agreed – no one person is likely to be a specialist in all the media options that may be stored on a videodisc, in all the possible instructional strategies, in graphic art, in computer programming, computer graphics, etc.)
5. The treatment of the subject matter must allow for frequent, meaningful, and 'upbeat' attractive opportunities for learner interaction. (More on this later.)
6. Feedback to learners should take both serious and humorous forms and should be a constant design feature. (Other specialists disagree on the 'humorous' aspect.)
7. Learner motivation is the *sine qua non* of interactive videodisc design. (Should it not be so for any well planned instructional design, or does DeBloois mean that the medium is firstly motivational and secondly instructional in its principal purpose?)
8. Instructional cues presented to the learners must be of the highest quality. (What DeBloois means by 'cues' other authors refer to variously as 'programme quality', 'production values', etc. Other authors disagree somewhat on the importance of this aspect in all interactive video productions.)

9. Training equipment and supporting software must be dependable. (Goes without saying, almost, but often overlooked in practice.)
10. Project planning and management is critical. (Ditto.)
11. Evaluation is essential for assessing quality, effectiveness, efficiency and worth. (Ditto.)
12. The use of hardware systems must be as transparent as possible. (What DeBloois calls 'transparent', others refer to as 'user friendly'.)
13. Formal and informal dissemination channels for information about the technology must be developed. (In order to keep designers up to date.)

The reader may observe that we have here a mixture of 'macro' level principles which apply to the planning and execution of projects, with some 'micro' level principles which relate directly to the ID decisions that must be taken. This 'micro' group interests us principally in the present context. It is composed of principles 3, 5, 6, 7 and 8. We shall look at these a little more closely, to see what other specialists, working with interactive video in other contexts, have to say as regards these five principles. We shall also see if they have any further items to add to this list.

Interactive video and instructional objectives

As regards the objectives of an interactive videodisc and, particularly, the variety of objectives that one specific disc may wish to achieve, one notes a lot of differences of opinion. A study of some of the interactive videodiscs which are currently available, reveals that they vary from the single, uniform set of objectives (akin to programmed instruction in its most classic form) to no specific objectives at all (the case of videodiscs which aim to act as a resource, or audiovisual database, to be used in any manner of ways by any number of different user groups). There are some productions which fall into the 'middle ground' identified by DeBloois in his fifth principle – productions which do have discernible instructional objectives, but which adapt to different student needs by the inclusion or omission of certain segments of the disc. These inclusions or omissions are generally selected by the student himself, from a menu of topic options. However, within any one of these topic options, there are clear and specific instructional objectives, if the disc has indeed any predetermined instructional goals. The major part of current commercial videodisc production (the 'generic' discs) does not however have specific instructional goals. The discs are intended as informational resources firstly and the interactive questioning is of very much secondary importance. The same is not true of videodiscs produced for specific clients to meet specific training needs. These, in our experience, tend to be as careful in the specification of objectives and of target population characteristics as most other forms of instructional materials designed according to the principles of the systems approach.

Interactive video and individualized instruction

What tends to be far more individualized than the objectives of a videodisc presentation, is the variety of paths through the material. Some producers who quantify this aspect, talk of several million alternative pathways through a given disc. Less often, but also quite frequently, one finds alternative presentations, leading to the same objectives, but at different levels of difficulty, with different types or different numbers of illustrative examples and so on. All in all, we do not see that current examples of *instructional* videodiscs have gone much beyond the principles of individualization that educators have striven towards for many decades (see Chapter 1). However, the interactive video medium offers new opportunities to 'do it right'.

Interactive video and interaction

As regards the opportunities for interaction, most authors agree that this is one of the most important design principles to be borne in mind. However, the *type* of interaction to aim for is emphasized by many. Hon (1983) makes the point that in order to achieve 'good' interaction between human and machine, we should study what makes for 'good' interaction between one human and another. He suggests that one studies the interaction styles of successful speakers, entertainers, salespersons and teachers. The instructional designer should strive to create the same types of successful interaction processes between student and mediated teacher. Gayeski and Williams (1984) also stress the use of human interaction processes as models for the design of man-machine interaction, but go further, in stating that the most successful and powerful interactive video programmes, in their experience, were those in which the 'user can vicariously interact with another person on the screen ... the video consists simply of medium shots and close-ups of a person in the programme reacting to the user's selections'.

Interactive video and feedback

The feedback element to the student is, of course, part and parcel of the interaction process. DeBloois suggests again that this should be ever present and should take both serious and humorous forms. Others warn against the indiscriminate use of humour, the first name of the user and a chatty style in an attempt to build rapport. No one 'enjoys talking to a friendly idiot'. Gayeski and Williams stress the need to build in a modicum of intelligence, or 'seeming intelligence' into the interactions. They abhor the package which rejects an answer that may be plausible but was not anticipated by the author. Several ways of avoiding this are available. One particularly useful way is through 'recursive programming' – asking the user to explain the unanticipated answer, put it in other words, agree or disagree that it is a synonym for the anticipated answer and so on – in short to attempt to recreate the sort of conversational interchange that would take place between two humans when one does not quite understand what the other meant to say. The final stage of this is, of course, to 'remember' the unanticipated answer given that was shown to be a plausible alternative way of responding and accept it if another future student were to respond in the same way. Thus, to an extent, the programme learns from the students as the instructional process proceeds. Gayeski and Williams suggest that it is not difficult, from the programming viewpoint, to build-in this type of recursive self-improving ability, especially if the final decision on whether a particular new unanticipated answer should be permanently built into the system, is monitored by the author or teacher in charge.

Another viewpoint on interaction and feedback is expressed by Hoekma (1983). He criticized the over-chatty and friendly styles so commonly used in CAI interactions suggesting that plain language and economy of words are the safest approach: '...feedback like "GREAT! YOU DID IT!" seems very friendly on the face of it – until you discover that the same feedback is given every time the user completes even the most rudimentary task!' Hoekma suggests that one should keep in mind the model of a 'good servant', a butler for example, when planning the style of CAI or IV interactions: '...capable and unobtrusive; considerate and respectful; presenting choices when necessary with a minimum of fuss and bother; responding immediately and appropriately; and anticipating the master's desires when possible'.

Frequency and con-stancy of interaction and feedback

As regards the frequency of interaction and constancy of feedback, some other authors have more precise views and suggestions. A distinction is made between interactions planned into the programme by the author and interruptions of the planned flow by the user. Hoekma calls these 'Normal Path Interactions' and 'User Interrupt Interactions'. The 'normal path' interactions should be as frequent as the topic demands and should not be 'idiot' questions put in just to use the interactive capabilities of the medium. There should be a clear plan of interactions and this should be made clear to the user. Often in practice, users are found staring at a still frame, not aware that the programme is requesting some form of response or choice to be made. As regards the 'user interrupt' interactions, these should be of a standard and simple-to-implement nature. The ideal is to have a specially labelled HELP button, available to the user at all times. However, in this case, to make the help received at all times useful and seemingly intelligent, requires a considerable programming and design effort.

Interactive video and motivation

As regards learner motivation, it is important to point out that, as with many other self-instructional media presentations, and especially video, the learner is effectively in control and can switch off or walk away if a given presentation loses appeal. In the interactive video medium, where the learner is encouraged to browse, skip forward, be selective in what he studies, the temptation to 'abort' any segment that does not immediately grab his attention is very strong indeed. For this reason, special attention must be paid to the aspect of learner motivation. Of course, the interactivity of the medium, if used appropriately, helps to maintain interest and motivation to learn – many writers tell of cases where visitors intending to spend a few minutes browsing through a videodisc ended up spending several hours, deeply engrossed in a subject matter that had no real intrinsic value for them.

How does one go about building in this level of attractiveness? There are several ways. David Hon stresses the affective aspects of interactivity. We already mentioned his suggestion of using human interaction as a model. Some of the specific principles he suggests might build affective rapport between man and machine are (Hon, 1983):

* RESPONSE TIMING – the quick reaction which shows interest and attention on the part of the listener, the quick retort whether humorous or not, etc.
* SPONTANEITY – a level of conversation/interaction that is both friendly and considerate of the views and feelings of the other person involved.
* LACK OF DISTRACTORS – an environment heavy with extraneous 'noise' makes it difficult to maintain interest on one task – this is relevant, for the planning of the learning environment, the hardware/software systems that deliver the programme and for the design of messages and feedback.

Hoekma (1983) has some other principles to add to this list. He argues that the user should feel 'in control' and 'comfortable' with the medium, and that the medium must keep up a constant level of interest. To achieve these aims, he suggests the following design principles:

* RAPID PACE – as motion sequences shot for videodisc must continuously earn the viewer's interest, and as slow and boring sequences are likely to be skipped, a 'natural' style for interactive video is 'rapidly paced and visually rich'.
* VISUAL IMAGERY – an approach that does not take full advantage of the realism of photography or the clarity of intelligently designed graphics, wastes a major strength of the interactive video medium.
* FRAME-ORIENTED MESSAGES – still frames should present one clear and self-contained message. Rather than attempting to fill the screen with information, it is better to present partly-filled screens, ensuring that the user 'punctuates' the message, by the necessary screen-change key punching, rather than interrupting a thought midway in order to 'turn the page' as it were.
* SELF-CONTAINED COMPONENTS – if a user can determine the sequence in which a number of different segments can be viewed, it follows that each segment must make sense to the viewer, regardless of which other segment had been viewed beforehand.
* SELF-EVIDENT STRUCTURE – menus, titles, local indexes, and frame-numbering conventions should be liberally used, in order to help the user to 'find his way about' the videodisc – getting lost and confused does not promote effective learning nor high motivation.

The question of high production values

Hoekma also espouses one further principle – that of HIGH QUALITY production. This tallies with DeBloois' eighth principle. Many writers stress this principle, but some disagree as to its importance. Hoekma's reasons for stressing the importance of 'production values' include:

* a lot of poor quality which may pass in a conventional video presentation, due to its rapid and 'one-time' screening, will be noticed by videodisc viewers as they repeat (and maybe even study frame-by-frame) a given sequence;

* today's video audiences bring with them some very high expectations as regards lighting, sets, camerawork, graphics and special effects, gained from their experience of commercial television;
* as the package producer has no control over the quality of the monitors and players used by the viewers, one should aim high in order to obtain clear communication even on imperfect equipment;
* as videodisc design and development is invariably a long and relatively expensive procedure, the extra time and cost spent in ensuring high quality levels in visual and recording presentations is but a small part of the total investment and may be crucial in avoiding negative impressions in viewers (which could mean that 'the many subleties built into the deeper structure of the program never have a chance to see the light of day').

Most professional video producers tend to support such views. For example, William Comcovich, in an interview with Sandra Devlin (Devlin, 1984), states 'I have never had a client complain when I deliver top quality ... Quality delivers the message more convincingly... When viewers see a sloppily produced programme, they start questioning the content... and that can be fatal to sales and marketing.' Gayeski and Williams (1984) in another article in the same journal, argue the opposite case: 'Because the viewer's attention is held by the quality of the interactivity rather than by the programme's production values, special effects are not necessary: in fact, they are often undesirable. We've even produced interactive video programmes without any editing at all by merely having the system access the desirable brief cuts of video on the raw tape. We had the opportunity to present two interactive videotapes on banking skills ... one tape was meant for mass distribution and uses professional actors, special effects and a perfectly lit set – a very high quality production approach ... it cost about 75 thousand dollars to shoot and edit ... the other tape was produced by in-house trainers at the bank ... actual bank employees served as first-time actors ... the entire production cost under 10 thousand dollars... Both the bank employees and other audiences within and outside the banking profession seemed to prefer the 'home-grown' programme because they could identify with the people and the situations more easily'.

Conclusion: the dog-food syndrome

Both views are of course probably correct in their proper context. Gayeski and Williams are describing 'hard' training applications, where other job-related factors tend to motivate viewers to study and persevere. Comcovich has to 'sell' his product to the organization where the trainees work. It's the old 'dog-food syndrome' – the buyer and the ultimate consumer are not one and the same and often have different criteria for judging what is 'right'.

We are now at an early stage in the history of interactive video. Enthusiasts and prophets abound and are doing a good job in recruiting converts in the education and training fraternity – the 'buyer' or 'client' systems. But what will count in the long run are the reactions and the learning results at the 'ultimate consumer' level – the trainees and students who will use our productions. How will they react?

The answer to the question will depend on the way in which we, the instructional designers and developers, make use of the medium and the opportunities it offers.

We have, in this chapter, outlined some of the real potential of interactive video. We have indicated where it is little different from earlier media systems which failed, and where it offers new opportunities for interactive study under student-directed democratic control or even under 'intelligent' cybernetic control. We have also tried to outline the growing body of design principles which are springing from the work of successful early practitioners. It is now up to us to apply the growing body of experience and theoretical principles to develop materials which satisfy both the 'client' and the 'consumer'. By bearing in mind the 'dog-food syndrome', we may save interactive video technology from the 'phoenix syndrome'.

PART 5.
The Evaluation of
Instructional Materials

Overview

This last section of the book deals with several aspects of evaluation.

Chapter 18 analyses the concepts of evaluation as they relate specifically to the development of instructional materials. A distinction is drawn between *validation*, when experts give their opinions on content, accuracy, relevance, pedagogical treatment and so on, and *evaluation*, which is specifically concerned with measuring the results achieved by the materials when used by the learners. A distinction is also drawn between two forms of materials evaluation: the *developmental* testing of each segment of the material as it is written, and the *field*-testing of units, or total course packages in simulated or real-life conditions. This conceptual structure is used in the later chapters as an organizing framework.

Chapter 19 is a 'how-to-do-it' chapter, outlining techniques and procedures for the planning and execution of both field-testing and developmental testing of self-instructional materials. The final sections deal with the interpretation of the results of evaluation sessions and the formative actions that may be taken to revise and improve the materials.

Chapter 20 is the final chapter of the book and, as such, takes a more global amd macro-level view of the whole process of instructional materials design and development. The first three sections are an extension of earlier chapters, dealing with some of the more general aspects of materials evaluation which should not be forgotten when dealing with the detail of developmental and field testing. These aspects are:

- The affective reactions of the learners to the type of material being envisaged, the author's general style of writing or communication, the media being used, the study style demanded from the learner, and so on.
- The costs involved in the development of instructional materials, which require close control, lest they skyrocket and get out of control.
- A variety of unexpected (but often potentially predictable) 'surprise effects' that may occasionally be encountered in projects using auto-instructional materials.

The latter sections of Chapter 20 take a yet more global viewpoint, examining the present and future trends in the instructional materials design/development process as a whole. Several important trends are identified; among them are:

- The tendency to get carried away by new technologies, often misapplying them, in the early stages.
- The tendency to forget important lessons learnt by previous generations of instructional developers.
- The tendency of the computer to be increasingly used, not only as a delivery system, but as a design tool (expert system).
- The resultant potential changes that may occur in the structure of educational/training systems and in the roles of teachers.
- The resultant trends that are apparent in the role of the instructional designer.
- The way in which instructional materials and systems design will evolve and what this implies for the skill of the future instructional materials developer.

18. The Validation and Evaluation of Instructional Materials

18.1 Developmental and field-testing

We have already mentioned some aspects of the evaluation of instructional materials, in Chapter 13 of our companion volume *Producing Instructional Systems*. In Exhibit 13.2, we presented a questionnaire for the pedagogical evaluation of the instructional materials to be included in a lesson. As we shall be referring to this questionnaire several times, we reproduce it again here, as Exhibit 18.1 (at the end of this chapter). This questionnaire is intended to be used by an instructional designer, or an instructor, at the moment of selecting existing materials to be used in a given lesson. There are, however, two other moments at which we might evaluate the materials:

The two stages of materials evaluation

* During the DEVELOPMENT of the materials and
* During their actual USE in teaching.

We shall call these two stages of materials evaluation, respectively:

* DEVELOPMENTAL TESTING and
* FIELD-TESTING.

Both these stages involve the verification of the effectiveness, efficiency and suitability of the materials, in relation to the instructional objectives to be achieved (hence 'TESTING'), but they differ both in the moment at which they are carried out (during DEVELOPMENT or during use in the FIELD) and in the methods employed.

18.2 Validation and evaluation

In this chapter, we shall discuss both developmental testing and field-testing. Sometimes we meet the term 'validation' used to describe these activities. It became very popular, in the early days of programmed instruction, to refer to all aspects of materials testing as 'validation'. Some authors on training have gone even further, to make a discrimination between the verification of the results of instruction (validation) and the verification of the long-term benefits of training in terms of job performance or organizational benefits (evaluation) – see Boog (1980) quoting Kirkpatrick (1977) and Whitelaw (1972).

Value and validity

However, we do not think that this distinction is particularly useful or accurate. The methods of testing in the long and short term may be different, but surely the basic idea is the same – to measure the VALUE of the training either to the individual trainee or to the organization. Thus we believe it valid to use the term 'evaluation' in both cases.

The last sentence illustrates a different and important use of the word validation. When we say that 'we believe it *valid*', we mean that we believe it to be *technically correct*. An important aspect of materials development is to check that the content and the treatment given to this content are technically correct. Perhaps we should reserve the term 'validation' for this part of the materials checking procedure.

We shall make the distinction, in our work, between:

1. *Materials validation*, which includes the checking of the content for correctness, by an expert in the subject matter, and the checking of the

treatment given to the content by an instructional designer or experienced teacher, in order to verify that the language is well chosen for the intended population, the examples and explanations are relevant to their interests and prior learning, the visual and graphic presentations really communicate what they should and the materials really do relate to the learning objectives that gave rise to the materials development project;

2. *Materials evaluation*, which includes both the developmental-testing and field-testing stages that we described earlier.

18.3 A systems view of materials evaluation

We included the 'pre-evaluation' of existing instructional materials as part of Level 3 of the instructional design process. It is an essential aspect of lesson planning. Exhibit 18.1 shows quite clearly the types of questions that should be in the instructional designer's mind at this stage. Note that the questions in this exhibit would, in the main, be classified into our concept of 'validation', in that they attempt to compare certain aspects of the materials with some theoretical 'model' of how such materials should really be structured. For example, question 11 asks whether 'good use is made of spacing, blocks, underlining, colour, print style or size and other aids to communication'. We are expected to make a *technical judgement*, as experts, of the use that the author has made of these aids to communication and not to *evaluate* the extent to which the materials actually communicate what is intended. This would require a test of the materials on a sample of students. Such a test actually will be made, the first time that the lesson, using the materials, is given. This is what we refer to as a FIELD-TEST: the evaluation of the materials in the real-life situation in which they are intended to be used. Thus, we consider that the *field-testing* of materials is part of *Level 3 design*.

Level 4   We descend to *Level 4 design*, when there are no existing suitable materials and the decision is taken to *develop a specific package*. In such a case, the validation, which would normally be performed as part of materials selection, will also be performed during the development stage. The materials will undergo vetting by technical experts, to check the content and by pedagogical and/or media experts to check the treatment given to the content.

In addition to this validation, performed by the members of the design team, the development stage includes a detailed evaluation of the probable instructional results of the material. In reality, this is mainly performed by *sample members of the intended students* (the 'target population'). Since the objectives of instruction are specified in terms of student behaviour and performance and since evaluation, in our usage of the term, means the verification of whether the intended objectives have been achieved (and if not, why not), it is essential to observe and measure student behaviour and/or performance during any evaluation of our materials. This implies that the development of any instructional materials should include their testing-out, on a sample of typical students. This is what we refer to as *developmental-testing*.

We shall see that developmental-testing and field-testing are very different in terms of the procedures to be followed. In the following sections we shall describe and attempt to justify these procedures. Before we go on, however, it may be useful to establish perspective, by creating a schema that illustrates how all the various aspects of validation and evaluation that we have mentioned, 'hang together'. Figure 18.1 attempts to present such a schema.

A systems diagram of the process   Figure 18.1 illustrates several important aspects of evaluation. Firstly, it distinguishes clearly between the lesson-planning, Level 3 and the materials development, Level 4 of the instructional design process. Secondly, it shows quite clearly the different roles of validation and evaluation. Thirdly, it shows field-testing as part of the lesson delivery process, but suggests that sometimes it may be necessary to simulate this lesson delivery, as part of the materials development

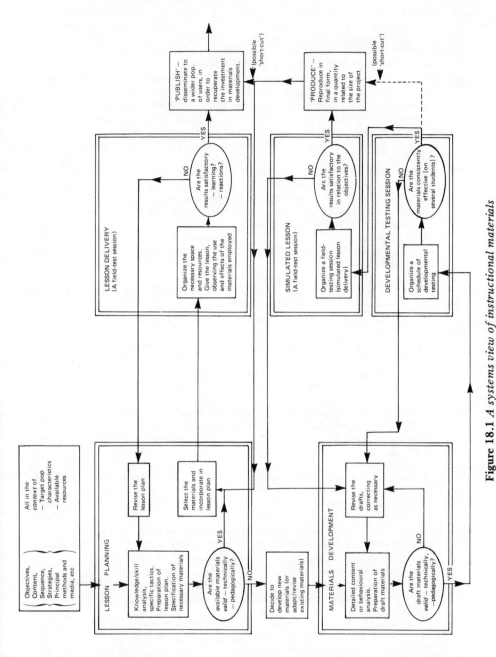

Figure 18.1 *A systems view of instructional materials evaluation and validation*

process. This last point may need further explanation. When a teacher is developing instructional materials for his/her own use, the normal approach would be to do some developmental testing on a small number of students, chosen from the target population and then to carry out field-testing during the actual use of the draft materials in the classroom. The results would be used to revise the materials for the next time that the lesson is given.

In the case of the development of materials to be used in many different localities, with many different groups of students, it is as well to test out the prototype version of the materials more thoroughly before investing in the production of many copies. In this case, the production team may choose to set up a special experimental field-testing exercise, possibly not as part of a regular course, but as a simulation of the real situation in which the materials are to be used.

Some 'short cuts'   It is normally not necessary to field-test the materials both under real and simulated conditions. When a teacher is preparing materials for his own use, he would usually field-test them during his own lessons, under the real operating conditions which he encounters in the classroom. Any revisions in the lesson plans, or in the materials used, would then be implemented for the *next* time that the lesson is to be given. The group of students used in testing would not necessarily suffer irredeemable harm, as the teacher would be paying exceptional attention to formative evaluation of all aspects of the lesson and correcting any weaknesses 'on-line', as they occur. If the materials used do not succeed in reaching all the objectives expected, the teacher would resort to other means of instruction. The materials would later be revised, in order to 'do a better job' next time round. In effect, our teacher has taken the lower 'short-cut', shown at the bottom-right of Figure 18.1.

A project set up to produce instructional materials for general publication, may operate under other restrictions. The design team may not be active teachers of the content being 'packaged' or may not have ready access to real classrooms of students who match the characteristics of the proposed 'target population'. It may simply be the 'wrong time of the year' to encounter real students in need of the materials being designed. In such cases, the design team should seek to set up a 'simulated' lesson specifically for the purposes of field-testing the materials prior to publication. Our designers are, in effect, taking the upper 'short-cut' in Figure 18.1.

'Short-cuts' to avoid   Thorough field-testing before publication, is very often *not* performed. There are several reasons for this. Publishers may not wish to devote funds to complex evaluation exercises that may not, in fact, yield as positive results as the marketing department would wish to publish. The producers of audiovisual and video materials, among others, consider that it is economically impossible to perform a special field-test and then remake the materials, due to the expense of the media concerned. They consider that the materials should be made 'right first time' and so lay great store on the technical validation of the packages and less on the practical evaluation of results. The writers of programmed self-instructional materials, on the other hand, place very strong emphasis on the field-testing of their materials under realistic conditions of use, before final revision and publication.

It is quite true that economic and other factors may make it impractical to revise materials after the full scale implementation (the delivery of the lessons). This, in our view, makes it all the more important to identify and correct any weaknesses in the materials during the actual development process. Hence, the developmental testing step is considered crucial for all types of instructional materials. However, the type of information furnished by a field-test is different and equally important. We believe that one should never take BOTH the 'short-cuts' shown in Figure 18.1.

18.4 Validation: technical and pedagogical

Two types of validation

As mentioned earlier, we reserve the term 'validation' for the inspection and approval of the materials by an expert – a subject-matter expert to judge whether the materials are technically valid and correct and an instructional design expert to judge whether the treatment given to the subject matter is pedagogically valid and creative.

Two levels of validation

Figure 18.1 showed that validation may occur at two different levels – at Level 3, as part of the lesson planning process, in order to SELECT suitable existing materials, and at Level 4, in order to CORRECT the initial drafts of materials, before evaluation on sample students. Very often, the materials may undergo both these validations. The first occurs during initial development, when the pedagogical validation is performed by the author and other teaching specialists that he may invite, and the technical validation is performed by the subject-matter experts on the design team.

The second takes place after the materials have been made widely available and are being considered for inclusion in a given course. In this case, the teacher who is planning the lessons acts as both pedagogical expert and subject-matter expert.

The approach to adopt is, however, very similar at both these levels. As regards technical validation, the only technique that can be applied is to read, view or otherwise inspect the material and analyse the content for *correctness, completeness and usefulness*. As regards pedagogical validation, the experts also inspect the materials and, in the light of their prior experience as teachers of the proposed target population, or as designers of similar materials, analyse the treatment in relation to *strategies, techniques and tactics* adopted by the author.

Use of checklists

It is helpful to use some form of checklist or questionnaire, to guide the validation process. In the case of pedagogical validation, the list presented in Exhibit 18.1 may be used. The questions in this list are the basic minimum that should be asked when validating any type of instructional materials. There may, however, be a need to go deeper, during materials development, when every decision taken by the designer should be validated. Every sentence of written materials should be checked for vocabulary and structure, in relation to the existing levels of linguistic knowledge and skill of the target population. Every example or analogy to be used should be similarly checked. Every type of material has structural and design characteristics that require to be checked. In the case of programmed instruction materials, for example, a range of supplementary questions should be asked regarding such design characteristics as frame size, visual layout, strength and frequency of prompts, type of response elicited from the students, the feedback supplied and so on. In the case of audiovisual materials, a similar list of good design characteristics may be drawn up, with somewhat different questions for video productions and for slide-sound productions.

It is best if each instructional designer draws up personal lists of design characteristics that should be validated, thus reflecting his/her personal style and approach to the design of specific types of materials. In a large team project, it may be necessary to draw up a common list of factors. However, the exercise of drawing up this list is, in itself, an excellent way of integrating the team members and furnishing an opportunity for exchange of viewpoints and experience. The lists are not, after all, algorithms to be rigidly applied, but on the contrary, are items to be taken into account, heuristically, during materials development. They represent the conceptual schemata that guide the designer (or team).

Working with a subject-matter expert

In a small-scale materials development project, the author may of course also act as the subject-matter expert. In larger projects, it is quite common to work in a team, composed of subject-matter experts and instructional designers. In such cases, problems sometimes occur due to a lack of mutual understanding between these two quite different areas of expertise. Subject-matter experts often feel that they are also experts in the teaching of the subject, even if they have never actually taught it. Designers often underestimate the contribution that the subject-matter experts can make. Both specialists use their own jargon and many often

Exhibit 18.1

*A questionnaire for the pedagogical evaluation
of the instructional materials to be included in a given lesson*

Questions: answer only the questions that apply to a given case	Your evaluation			
	Yes	Partly	No	Cannot say
General aspects 1. Do all the materials specified in the lesson plan actually exist, and are they available for inspection?				
2. Have the materials passed the technical validation stage, and are they considered suitable in content?				
3. Bearing in mind the strategies and tactics specified in the plan, is the type of material appropriate?				
4. Does the material appear to be adequate to achieve the objectives set for it in the lesson plan?				
5. Are the materials really necessary? Do they contribute to the lesson in a manner that justifies the time/cost?				
6. Are the media employed appropriate for the objectives to be achieved and the content?				
Questions for written text 7. Are the style and language used appropriate for the target students?				
8. Does the material really say what it sets out to say? Are the sentences logically structured, well-organized, etc?				
9. Are sufficiently well-chosen examples included to define all new concepts clearly and adequately?				
10. Are there sufficient well-designed illustrations and graphics to support or summarize the text adequately?				
11. Is good use made of spacing, blocks, underlining, colour, typography and other aids to communication?				
12. Are worked examples, self-tests, practice exercises, introductions, summaries, glossaries, references and indexes used adequately and to good effect?				
Questions for audio recordings 13. Do the scripts obey the criteria of style, language and structure defined above for the target population?				

Exhibit 18.1 *(continued)*

Questions: answer only the questions that apply to a given case	Your evaluation			
	Yes	Partly	No	Cannot say
14. Are the recordings of a reasonable length, generally not exceeding five minutes?				
15. When appropriate, is the first person singular used to address the learner? Is informality maintained?				
16. Are special effects, such as background music, used sparingly, to ensure that they do not interfere with the main message?				
17. Are the voices of the speakers clear and pleasant in tone, varied to emphasize important points, etc?				
18. Is the quality of recording adequate for clear comprehension in normal conditions of use?				
Questions for visuals 19. Are the notation, conventions and graphic symbols used in visuals known to the learners?				
20. Is the intended message well conveyed by the visual? Does it use the visual medium to good effect?				
21. Are the size and style of lettering used appropriate to the size of projection screen/room?				
22. Is the layout of the visuals clear and uncluttered, illustrating the key points of the message?				
23. Is the quality of the artwork and the use of colour and shading of adequate quality for communication?				
Questions for audiovisuals 24. In addition to meeting the above criteria, do the audio and visual elements complement each other?				
25. Are the rate of change of visuals and the quantity of commentary well planned and varied?				
26. Are the structure of the shots, their sequence, special effects, etc designed for pedagogical, rather than artistic, effectiveness?				

even use the same words but in quite different contexts and with quite different meanings. In order to get the most out of the validation process, it is necessary to smooth out these differences. The designer must learn the content to be communicated, at least sufficiently to be able to converse on equal terms with the subject-matter expert. The subject-matter expert must learn enough about the instruction design approach being used to be able to appreciate the designer's contribution to the project and be able to supply useful help, when requested. Once more the joint drawing up of a list of 'duties' may be a step towards greater mutual understanding.

19. Developmental and Field-Testing: Organization, Execution and Interpretation

19.1 The organization of developmental testing

The principal objective of developmental testing is to correct any defects or weaknesses in the materials being developed at the earliest possible opportunity, before the materials are printed and published, or filmed, or drawn in final art form, etc, and also before the same defect/weakness is repeated several times over.

When? The earliest opportunity is just as soon as there is enough material written in its first draft to make a meaningful presentation to a student. Let us imagine that the material is divided in 'chapters', each dealing with one specific topic or leading to the achievement of one specific objective. These 'chapters' may, in reality, take the form of practical exercises, or a series of sequences in a proposed dramatic presentation on videotape, or indeed a chapter or module of self-instructional programmed text. As soon as the first chapter is drafted out, it should be tested out on one or more students. This testing will lead to two types of revision:

1. A revision of the structure and content of the chapter being tested, in the light of the comments or difficulties of the students, and
2. A revision of the author's view of the target population, in terms of the prerequisite knowledge and skills actually possessed, the learning skills and habits and the interests and sources of motivation that will influence their acceptance and understanding of the materials.

How to plan? The implementation of developmental testing is best planned, right at the beginning, in the form of a schedule, as shown in Figure 19.1.

We see that a series of testing sessions are programmed at specified intervals of time. At the first of these sessions, the author tests out the first version of the first chapter. This is then revised, in the light of the data collected, and a further chapter is drafted. At the second session, the same student (or small group) that was used to test the first draft of Chapter 1, now studies and comments on the first draft of Chapter 2. The new version of Chapter 1 is tested again, but using a different student or small group. With luck, the second draft of Chapter 1 will now teach much better and may indeed be fit to publish and use. There may appear some new problems, however, that were not encountered in the first testing. It is a good policy to take each chapter through as many stages of testing as is necessary to eliminate all weaknesses and obtain consistent learning results with consecutive groups of students. The minimum recommended number of testings is three, even if no problems are encountered on the first testing. The feedback that the author gathers on Chapter 1 will also help to make the second and subsequent chapters better adapted to the target population right from the start, so the later chapters may need less repeats than the earlier ones. However, one should remember the need for continuity – in order to study a later chapter, the student usually needs to have knowledge of earlier parts of the materials. It is best, therefore, to plan a developmental testing schedule for a minimum of, say, four or five repeats of each chapter, thus using the same four or five groups of students throughout the project.

How many students? This raises the question of how many students to use. Often, in materials development projects, it is difficult to obtain adequate numbers of typical members

TESTING SESSIONS (REALISTICALLY PROGRAMMED IN TIME) →

CHAPTERS (OR MODULES OR UNITS) OF MATERIAL ↓

	1	2	3	4	5	6	7	8	9
Ch.1	Draft (1) Student (a)	Draft (2) Student (b)	Draft (3) Student (c)	Draft (4) Student (d)					
Ch.2		Draft (1) Student (a)	Draft (2) Student (b)	Draft (3) Student (c)	Draft (4) Student (d)				
Ch.3			Draft (1) Student (a)	Draft (2) Student (b)	Draft (3) Student (c)	Draft (4) Student (d)			
Ch.4				Draft (1) Student (a)	Draft (2) Student (b)	Draft (3) Student (c)	Draft (4) Student (d)		
Ch.5					Draft (1) Student (a)	Draft (2) Student (b)	Draft (3) Student (c)	Draft (4) Student (d)	
Ch.6						Draft (1) Student (a)	Draft (2) Student (b)	Draft (3) Student (c)	→ etc
Ch.7							Draft (1) Student (a)	Draft (2) Student (b)	→ etc
Ch.8								Draft (1) Student (a)	→ etc
Ch.9									

Figure 19.1 *A schedule for developmental testing of instructional materials, using four students, to test up to four drafts/revisions*

of the target population, at the right stage of development in terms of prior learning, just when needed for developmental testing. Luckily there is no need for large numbers at this stage of testing. Our intention is to go deeply into any problems encountered, by means of close observation of the students while they are studying, or by personal discussion of any learning difficulty encountered or error made. Large groups hinder, rather than help such a 'clinical' approach to evaluation. We could say that the minimum number of students for each test session of a chapter is one and the maximum in most cases is three. However, some types of student (notably adults) are more likely to be honest about their difficulties when on their own, so often the maximum number is also one student.

A 'clinical' approach

The last paragraph has indicated that our approach during a developmental testing session is 'clinical', in the sense that we wish to obtain the maximum of in-depth information on all aspects of the students' interaction with the materials. We need to observe the students while they are studying. Any blank expressions, long delays in responding to questions, or other non-verbal signs of difficulty should be noted and investigated. Any incomplete information or inadequate examples should be completed and corrected by the tester at once, during the testing session, so that the learner approaches each new teaching point with a full understanding of the previous ones. This is of importance, in order to avoid the gradual build-up of difficulty due to incomplete mastery of earlier material, making it difficult for the evaluator to pinpoint the real reasons for a given learning problem. Thus, the evaluator really rewrites the material *during the testing session*, in looking for better ways of presentation or explanation that overcome the learning difficulty encountered. This aspect of developmental testing has several implications.

How to carry out develop-mental testing?

(a) It is essential that the developmental testing is performed by the author of the material and not by some specialist evaluator on the design team.

(b) It is essential to use some form of recording the discussions, comments and suggested revisions that occur during the testing sessions. When the major part of the material is verbal and the main part of the feedback obtained is through discussion, it is useful to use a tape recorder to record all the comments and suggestions made during the testing session for further analysis and use during the writing of revised drafts. When the materials are visual, the recording system may need to be capable of storing alternative 'live' demonstrations made by the evaluator during the session. When the subject matter deals with the development of physical or interpersonal skills, it may be necessary to record aspects of the learner's responses. In both these cases, the use of film or video may be of great assistance to the evaluator. Often, however, such techniques are not available or are impractical. An alternative approach is to develop a methodology of detailed note taking. This may break the rhythm of study and evaluative discussion, if overdone. The best approach is to have a team of evaluators, one making notes while another enters into discussion with the student. A certain amount of practice is required in order to run a testing session which is sufficiently deep and thorough, without leading the student to take fright and 'clam up'.

(c) It is essential therefore to choose the students with due attention to their suitability as informants, as well as their suitability in terms of prior learning, etc. The students used in developmental testing should be free of any form of inhibition, undue introversion (though it is useful to be introspective and self-critical) or excessive tendency to accept as correct anything that is presented to them by 'superiors'. On the other hand, they should not be 'troublemakers' or 'nitpickers' that have an excessive tendency to find fault with everything presented to them.

(d) It is essential to run the developmental test sessions in a cordial atmosphere that will not create barriers to communication between students and evaluators. It is a good idea to consider the test students as 'members of the design team' and promote an atmosphere of joint cooperation in the search for quality in the final product.

Project control

Before leaving the subject of developmental testing, one should mention a very important by-product of the testing schedule. The test sessions are planned at predetermined intervals and a given amount of revision and new materials development must occur between any two sessions. Therefore, the testing schedule is also the production schedule and may be used for project control. It is therefore important that the dates of the testing sessions be programmed realistically at intervals throughout the project, so that all the revisions of earlier 'chapters' and the writing of a first draft of a further chapter may be concluded within the interval. It is, of course, possible to revise the dates of subsequent testing sessions if, by chance, the project gets delayed. However, any such re-programming should follow a full analysis of the causes of the delays.

The developmental-testing schedule should, thus, be initially developed from a project timetable, possibly based on a PERT network analysis. Once developed, the schedule acts as the principal control document for a 'management-by-objectives' approach to control of the project's productivity.

19.2 The organization of field-testing

Figure 18.1 (in the last chapter) showed that field-testing of instructional materials may take place either in the real classroom conditions or under simulated conditions. We have already mentioned some of the factors that might favour a small one-school project's decision to carry out field-testing as part of the normal teaching, in real classroom conditions and, conversely, some of the constraints that may force a large-scale project to opt for a simulation of the real conditions under which the materials are to be used, once produced and published. We also mentioned some of the factors which often lead materials producers to 'skip' field-testing altogether and our own opinion that this is a dangerous practice, which often leads to weaknesses in the materials that could easily be avoided. In this section, we shall examine just how a field-test session should be set up in order to yield the maximum of useful information whilst avoiding the excessive costs or time penalties so feared by publishers and media producers.

Let us start by contrasting the type of information which the instructional designer hopes to gain from developmental and field-testing.

The different functions of field and developmental testing

In the developmental testing situation, we adopt a 'clinical' approach, in attempting a deep analysis of the instructional effectiveness of our materials, using a relatively small number of students as 'informants'. We get into conversation with the students, seeking and accepting their suggestions, involving them as 'partners' in the instructional design process. We observe carefully all the time, seeking to identify any non-verbal signs of difficulty or boredom and, if identified, investigate the causes by questioning the students and suggesting possible modifications for their approval. At the end of a developmental testing session, there is no point in setting a post-test, as we would not be quite sure whether the students responded correctly because of the discussions and extra comments supplied by the designer during the session. The function of the developmental testing session is to get to know the target population better, improve one's ability of communicating effectively with the target population and 'catching the big problems' in the materials at the earliest possible opportunity. It is as much an opportunity for the instructional designer to develop his/her techniques and skills, as for the improvement of the instructional materials.

The field-testing session, on the other hand, seeks to supply information on how the materials function in the reality for which they have been planned. We now no longer collect in-depth data on student reactions, difficulties and opinions, but are more interested in measuring exactly what the students have learned as a result of using the materials that we are designing. We therefore need larger groups of students, in order to yield useful results and so we apply tests, both during and at the end of the learning sessions, in order to ensure that the results are due to the use of our materials and not to any other extraneous events.

A typical field-testing session should therefore commence by carefully selecting the group of students. They must be ready to commence the study of the content to be presented and they must be relatively 'green' in relation to the specific objectives to be achieved through the study of our materials. It may be necessary to use two forms of entry test:

* A *prerequisite test*, to check that all essential prior learning has indeed taken place.
* A *pre-test* on the objectives to be mastered in the test session, to check that the material to be presented is not yet known.

Only students who score highly (near 100 per cent) on the prerequisite test and achieve very low scores on the pre-test, should be selected for field-testing.

At the end of the testing session, one would apply a post-test, which should be equivalent in content and difficulty to the pre-test, both being based, rigorously, on the specific objectives of the section of materials being tested. The pre-test and post-test may sometimes be identical, but it is as well to avoid this when possible. In the case of materials that teach factual or simple conceptual content, it may be impossible to vary the tests, as there may be only one valid question to ask in relation to a given objective. In the case of most conceptual learning, however, and especially in problem solving, it is possible to prepare 'alternative-equivalent' questions to be used in the pre-test and the post-test.

In addition to these two tests, it is a good idea to prepare further 'alternative-equivalent' items, to be used for the testing of each objective during the process of instruction. The reason for this is that it is often quite difficult to distinguish between poor explanation and insufficient practice as possible causes of poor performance on the post-test. Has a student failed to answer a given question correctly at the end of the session because he failed to understand the materials, or did he simply forget, due to lack of use and repetition, by the time he reached the end of the session? If he responded correctly to a question at the time of initial explanation, but failed an equivalent question at the end, the chances are that we do not need to revise our initial explanations or demonstrations, but need to revise, practise or otherwise reinforce the initial learning more frequently, at later points in the materials. If, on the other hand, a student could not respond correctly both during and at the end of the study session, then it is probable that the materials were not clear enough in their explanations or demonstrations. It is somewhat more difficult to interpret the unlikely, but possible, case of the student who responds incorrectly during the session, but correctly at the end. However, such chance results get evened out, when one takes the frequency of errors, by a group of students, as the basis for decisions.

We may picture the field-testing process in the form of a block diagram, as shown in Figure 19.2. We shall refer to this figure again, later on, when we consider some of the measures and types of tests to be used. For the time being, we stress only that the testing process should be designed in such a way as to yield the maximum of useful information to the instructional designer, enabling him/her to identify any weaknesses in the materials and their probable causes.

The need to gather all useful information may lead to the inclusion of somewhat more testing activity in the lesson than would normally be the case. However, this should be considered as the only admissible departure from the lesson's normal structure. After all, any of the factors that are normally present in the lesson and in the environment in which the lesson is given, may contribute to the effectiveness, or otherwise, of the materials being tested. A noisy environment, or a classroom in which the intensity of illumination cannot be controlled, may adversely affect the learning from an audiovisual presentation. If, however, such conditions are known to be the norm, we should test our materials under these conditions. For example, it is now quite common to use videotape for on-the-job training on factory production lines. The new trainee watches a very short video

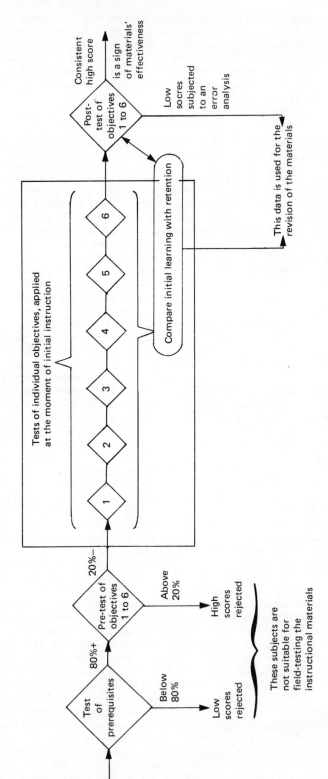

The figure illustrates a scheme for field-testing a unit of materials that has six main instructional objectives. In principle, this scheme could be used for any type of instructional materials, utilizing any media.

Figure 19.2 *The tests used in the field-testing process*

demonstration of one operation in a given task and then attempts to practise it in reality, replaying the demonstration as often as proves necessary. This approach, a hybrid of the TWI methodology and programmed instruction, with the added communication benefits of video, has proved to be extremely effective. However, the factory floor is not designed to promote learning. It is noisy, there are innumerable distractions, the illumination levels are excessively high for viewing video, etc. As these factors cannot be controlled or avoided, it is necessary to simulate them in the field-testing session, in order to evaluate their real effects on the effectiveness of the materials. This is what we meant, when, in Figure 18.1, we specified 'real' or 'simulated' conditions for the field-test.

Some problems and difficulties

At this point, we begin to sympathize with the materials producers who maintain that field-testing of some types of materials, notably film and video, is not possible in practice. In order to simulate all aspects of the final instructional session, one must in effect have the materials ready and produced in a form equivalent to the final product. With print-based materials, it is quite acceptable to use typed and photocopied first drafts in the place of the final books, so that a full field-test may be carried out and modifications made, before the expensive stages of typesetting, printing and binding are undertaken.

In the case of audiovisual materials, however, the situation is more difficult. Is a cheap, black-and-white version, acted by amateurs, really of much use as a test version of a production that will finally be shot in colour, on professional quality equipment and with professional actors? The media experts generally think not and further argue that the cost of production of a simple test version is still very high, in relation to its value.

– And how to overcome them

However, we beg to differ. Whenever we are really talking of instruction, rather than of some general informational programme, we feel that we cannot escape from the obligation to field-test and evaluate the instructional effectiveness of our materials. Some way must be found to overcome the economic/time constraints, which make it difficult to carry out a true field-test. One approach is to look for ways of less complete simulation. We mentioned, in Chapters 10 and 11, that an enormous amount of very useful research into CAI was performed, in the early days, without the use of a computer, by the simulation of the interactions that would normally occur between a student and a computer terminal. A similar approach may be used in field-testing some forms of audiovisual materials. It is quite common, for example, to field-test a slide-tape presentation by using the slides in a lesson and speaking the planned commentary 'live'. In this way, the learners experience a situation similar to the final product, enabling the designer to alter the slide sequence or the commentary before embarking on the more permanent and costly steps of the production process.

A similar approach can usually be taken to the field-testing of a video instructional programme. Such a programme is (or should be, if well designed) replete with demonstration shots of machinery, equipment, interpersonal conflict situations, or whatever. In the final product, these shots are edited together in a given sequence, linked by 'talking head' shots of the presenter, or simply by 'voice-over' commentaries. It is possible to use the separate shots as part of a 'live' presentation, supplying the commentary from a prepared script, keeping to planned timing, etc. In this way, the effectiveness of the individual shots, their sequence in the overall programme and the commentaries can be evaluated in near-simulated conditions, before the expense of editing and recording of soundtracks is embarked upon.

19.3 Treatment and interpretation of the results

Developmental testing and field-testing yield two quite different types of information. In the case of developmental testing, our 'clinical' approach of observation and in-depth interviewing of a very small number of students, furnishes a great deal of qualitative information such as lists of difficulties, suggestions for improvement, etc. In the case of field-testing, however, the principal information gathered is in the form of test results and, possibly, opinion questionnaires – quantitative data such as test scores, error rates and percentages in favour of certain aspects of the package being tested.

Qualitative analysis

The treatment of the information collected is therefore also quite different. The qualitative insights collected during developmental testing do not require any mathematical/statistical treatment in order to be interpreted. The instructional designer performs an intellectual analysis of the events which occurred during the session and the comments made by the students, coming to some heuristically-based evaluation of the significance of each event or comment. If several students make the same errors or comments, then the designer is inclined to treat the matter more seriously and make appropriate modifications to his draft materials. However, just one student might make a comment which, on analysis, proves to be of great value and leads the designer to make radical and far-reaching changes in his approach. We have seen cases where one isolated observation led to the discovery of a powerful analogy that could be used as a mediating device for a whole body of subject matter, or where one isolated comment has led to a complete change of strategy for a whole module.

Quantitative analysis

In the case of field-testing, our approach is quite different. The raw test data collected must be analysed objectively, in order to make sense. Some form of mathematical/statistical treatment of the results may be required. The interpretation of the results is based on tendencies encountered in the group of students used for testing, not in individual results. We mentioned earlier the not very common, but possible, case of a student who showed weakness and high error rates during the process of instruction, but resolved the final test items satisfactorily. If one or two students in a group register such results, we can dismiss the occurrence as accidental. If, however, the majority of the group were to show this tendency, then we would wish to investigate more deeply in order to discover what aspects of our materials may need to be modified.

How much statistics?

This investigation may lead us to reorganize the results and to make calculations. However, the depth of statistical treatment we require is not very great. We should remember that we are NOT engaged on a project of research, attempting to prove or disprove some hypothesis related to the design of instruction. Our task is one of 'product development' and we operate by means of a cyclic, or 'iterative', process of testing, modifying, retesting and so on. We do not need to apply tests of statistical significance to our results. On the contrary, if our results show differences or tendencies that are so small as to require some form of advanced statistical analysis, we can be almost certain that the said difference/tendency is of no practical interest. We are seeking to make effective, or rather, cost-effective changes in our materials. Any change that leads to only one or two percentage points of improvement (and may therefore require advanced statistics in order to be identified at all) is unlikely to be worth the cost of the time required to reformulate the materials. The only test of significance that we need to apply is 'if a significantly large part of the test group make a particular type of error. Then something should be done to improve the materials'.

Mastery-learning and its implications for evaluation

This raises the question of 'what do we mean by a significantly large part of the group' and the associated question of 'how large a group to use'. To answer these questions, it is well to recall the principles upon which most forms of individualized instruction are based. Instruction, we know, is a pre-planned process which aims to achieve specific, predetermined, learning objectives. The effectiveness of the instructional process is, thus, measured in terms of objectives achieved, or in other words, mastery. The concept of mastery-learning, as proposed by Bloom (1968) and other writers on evaluation, suggests that, given the necessary learning time and other resources, all (or nearly all) students who have the necessary prerequisites, may be led to master all (or nearly all) of the intended learning objectives. In the conventional (non-individualized) classroom, the content/objectives and the learning time/methods are fixed. This leads, naturally, to variable achievement levels. In individualized instruction, the content and objectives may remain fixed (though sometimes they vary too), but the learning time and the learning methods are allowed to vary in relation to individual student characteristics, so that all students reach high levels of achievement.

Figure 19.3 presents a graphical contrast between the conventional and individualized situations.

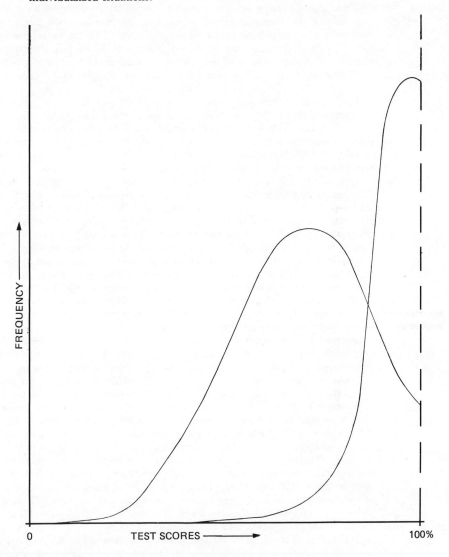

Figure 19.3 *A contrast between typical test score distributions expected under 'conventional learning' and 'mastery-learning' conditions*

A full discussion of the mastery-learning concept and its implications for educational measurement and evaluation may be found in our earlier volume, *Designing Instructional Systems*. In the present context, we need only emphasize that mastery-learning implies that all of the intended objectives should be achieved, not just a sample of them. This, in turn, implies that all the objectives should be individually tested during formative evaluation. We already stressed this aspect of test design for formative evaluation, in Chapter 13 of *Producing Instructional Systems*. Our description, earlier in the present chapter, of the organization of field-testing sessions, once more stressed the need to use tests that separately measure the mastery of each of the objectives.

Double percentages

Thus, the answer to our first question – what do we mean by a significantly large part of the student group? – should take into consideration that nearly all the students should master nearly all the objectives. This can be expressed in the form of a *double percentage*, say 90-90, which is a way of defining the expected standard of achievement – 90 per cent of the students in the target group should score at least 90 per cent on a properly structured post-test that covers all the objectives.

Such 'double percentage' criteria became much used in the early days of the programmed instruction movement, as one way of expressing the 'efficiency' of a set of materials. The designer would revise and re-test all sections of the materials until a predetermined criterion was consistently achieved. It was quite common to aim for a 90-90 criterion, but often some other, less ambitious 'double percentage' was selected. In reality, the criterion should be related to the importance of total mastery of the objectives, to the students or to society or employer. A 90-90 criterion may be considered too low for certain aspects of airline pilot or medical training. We may wish to adopt 100 per cent as the criterion for job competence. However, we may admit that our selection procedures are imperfect and our time and resources for training are not unlimited. Thus a realistic criterion may be 70-100, or, 'at least 70 per cent of students should reach total mastery of all the objectives' (these 70 per cent are the ones who will qualify). In some other course, say language skills, where total mastery is not an absolute necessity for satisfactory performance, but where a basic minimum vocabulary must be learned, a criterion of 100-70 may be more appropriate. Each case merits individual analysis.

What size of test group?

Our second question – what size of group is required to perform a field-test? – is now easier to answer. We require a group large enough to ensure that the double percentage, or any other criteria which we decide to use, are reliable. In a very small group we cannot be sure that the results obtained are typical of the whole target population. We may have to apply some test of significance, in order to verify to what extent we can extrapolate the results obtained by the group to the poulation in general. Using a larger group, however, we may be more confident. The larger the group, or groups, used in field-testing the greater the level of confidence that we can place on the results obtained. Authors differ in their suggestions of the minimum group size that should be employed, some suggesting 50 to 100 and others maintaining that anything over 20 is quite adequate. The number is not all that critical when we remember that our evaluation-revision process is iterative, allowing us to correct any false conclusions in subsequent field-tests. Given a limited number of sample students, it is as well to remember that we may need to re-test materials after revision. Thus, if we have 100 or so students available, we should first select a group of about 25 or 30 by means of our prerequisite tests and pre-tests.

If this first field-test uncovers any errors or weaknesses in the materials, we can revise as necessary and still have students for two more field-testing, sessions. If, on the other·hand, the first field-test indicates that the materials are satisfactory, we can replicate the exercise with another group of students, to check the reliability of the initial findings, and still have a group in reserve in case this second field-test uncovers the need for some revisions.

Other criteria

The 'double percentage' is not the only form of criterion that may be used to judge the quality of instructional materials. Various authors suggest the use of a measure of the gain in learning that the student group achieves. The gain is expressed as the difference between post-test and pre-test scores:

$$\text{GAIN} = \text{PRE-TEST SCORE} - \text{POST-TEST SCORE}$$

This may be calculated for individual students, but usually, one calculates the MEAN GAIN for the group as a whole.

Another, somewhat more sophisticated, criterion is *McGuigan's Ratio*. This takes into account that a gain of, say, 50 per cent, from zero to half marks, is not really the same thing as a gain of 50 per cent from half marks to full marks. In the first case the students knew nothing to start with and learned half of what was intended. In the second case, the students already knew half of what the materials set out to teach, but learned all that they did not know at the beginning. McGuigan's Ratio attempts to take all this into account, by calculating the ratio of WHAT THE STUDENTS ACTUALLY LEARNED to WHAT THEY COULD HAVE LEARNED, or,

$$\frac{\text{POST-TEST SCORE} - \text{PRE-TEST SCORE}}{100 \text{ PER CENT} - \text{PRE-TEST SCORE}}$$

Once more, we can calculate McGuigan's Ratio for individual student scores, but it makes more sense to work in terms of the mean scores of the group.

In practice, these two gain measures and other possible variations are not all that useful or needed, if the students selected for the field-testing session satisfy the criteria, defined earlier, of near-zero mastery of the intended objectives. In that case, the raw gain score tends to be equal to post-test score and McGuigan's Ratio tends to equal the post-test score expressed as a percentage.

Another reason against the use of such ratios and 'one-figure' representations of the field-test results is shown in the following case study.

19.4 A case study in field-test data interpretation

A publisher was considering the publication of a set of modularized, self-instructional programmed materials, which had been prepared by a teacher, initially for use in his own school. The teacher had amassed a considerable amount of data on the use of these materials with his own students and presented excellent field-test data to the publisher as proof that the materials were worth publishing. The publisher, however, felt the need to repeat the field-test with a wider range of target groups, implementing any further modifications that proved necessary, in order to guarantee effectiveness across a wide range of target populations. It was felt that the author's data, though probably quite genuine, referred to the reality of only one school, in a particular social context and to groups of students already familiar with the teaching and communication style of the author. It is perhaps a pity that more publishers do not take as seriously their responsibility for the quality of the instructional materials they publish.

An independent consultant was contracted, who organized the use of the draft materials in four schools in different localities, with different social structure and using different approaches to the conventional teaching of the subject matter in question. This subject matter was binary arithmetic and the target population was defined as secondary school children, in general, who had reached the required level of competence in a series of prerequisite mathematical skills (as defined by the prerequisite test which had been developed by the author and validated by other teachers).

Structure of the materials
The material was divided into three modules, each requiring between two and three hours of study. Each module had its own post-test. The modules were arranged sequentially so that the post-test of the first module was an adequate prerequisite test for the second module, and so on.

Organization of the field-test
Figure 19.4 shows the overall structure and content of the materials, as field-tested by the consultant. The three modules were studied during three consecutive weeks, one post-test being taken at the end of each week. The time spent on study varied in the range of two to three hours, per module, depending on the individual learning rates of the students. The students could organize their own

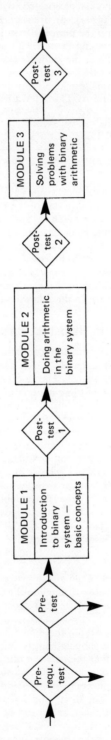

Figure 19.4 *The structure and content of the materials*

division of time, but could only study the materials during the normal mathematics classes (45 minutes per day), under the supervision of their usual mathematics teachers and with the presence of the consultant or his assistant. Every care was taken to ensure that the materials were studied exactly as intended by the author. The teachers and consultant were available to answer questions concerning the way of study or other doubts, but did not give full-scale tutorial assistance as the materials were intended to be capable of leading all the students to mastery of all the specified objectives.

A first analysis of the data
After the three weeks of study, all the data collected was analysed. The first analysis performed was a frequency distribution of scores on each of the tests. It was found that the four groups had performed in very similar ways, so the results were merged into one table (Figure 19.5).

(A) Frequency distribution of scores on three post-tests, relating to three consecutive modules of a self-instructional course. (The content of the course is binary arithmetic, for secondary school children.)

TEST 1			TEST 2			TEST 3	
score	frequency		score	frequency		score	frequency
20	40		20	23		15	15
19	15		19	13		14	10
18	7		18	8		13	12
17	3		17	11		12	10
16	8		16	6		11	5
15	1		15	3		10	11
14	2		14	8		9	10
13	2		13	4		8	4
12	1		12	–		7	2
11	1		11	2		6	2
10	1		10	3		5	1
9	3		9	1		4	2
8	–		8	1		3	–
7	–		7	1		2	1
6	–		6	1		1	–
5	1		5	–		0	–
4	–		4	–			
3	–		3	–			
2	–		2	–			
1	–		1	–			
0	–		0	–			

[*Observation:* The maximum score on tests 1 and 2 is 20. The maximum score on test 3 is 15. The number of students in the test group was 85.]

(B) Table comparing the three tests in relation to an 80% criterion.

	TEST 1	TEST 2	TEST 3
80% (or more) was scored by the following percentage of the student group	85%	73%	55%

Figure 19.5

In part A of this figure, we see the frequency of scores obtained on each of the tests. A close study of these distributions reveals that Test 1 yielded a narrow-spread distribution, near to full marks, rather like the theoretical curve for 'mastery-learning', presented in Figure 19.3. The scores on Test 2 are somewhat more spread out and those obtained on Test 3 even more so, tending towards the shape of the 'conventional' curve of Figure 19.3.

This led the consultant to observe, in his interim report to the publisher, that module number 1 appears fit to publish, whilst the other two modules, especially the third, should be analysed more closely and revised.

Part B of Figure 19.5 shows a further calculation of the percentage of students who achieved 80 per cent or more on each of the post-tests. If we apply an '80-80' criterion as our measure of quality (this was suggested by the consultant) we see that, once again, module number 1 is up to the desired criterion, while the other two modules, especially the third, do not make the grade.

In order to identify the causes of weakness in the modules, the consultant performed a further analysis – see Figure 19.6.

Module number	Items responded incorrectly by three or more students (5% or more of group)	Frequency of error
1	131	6
	37	5
	33	4
	8, 11, 22, 52, 106	3
2	21, 33	7
	131, 157, 166, 167	6
	9, 29, 155, 184	5
	28, 30, 52, 104, 128, 163, 174, 188	4
	36, 55, 67, 69, 77, 78, 79, 80, 92, 95, 99, 133, 136, 142, 150, 151, 159, 168, 178	3
3	29, 30	10
	175	9
	21	8
	25, 159	7
	11, 55, 115, 133, 153, 188	6
	9, 32, 38, 40, 52, 64, 70, 104, 176, 187	5
	15, 21, 50, 78, 79, 106, 144, 157, 163, 167, 182	4
	4, 8, 18, 28, 30, 31, 39, 43, 46, 69, 77, 83, 84, 85, 114, 115, 124, 136, 142, 179	3

Figure 19.6 *Distribution of errors on the self-assessment questions in the three modules of self-instructional materials on binary arithmetic. (Note that the item reference numbers are listed at the left and the frequency of error at the right.)*

A second analysis

As the materials being field-tested were in the form of small-frame programmed instruction, on the 'Skinner' model, the consultant decided to use a criterion of programme quality suggested by Professor Skinner himself – the error rate on responses to individual frames of the programme. This, somewhat antiquated and today largely discredited, criterion is based on the principle of 'error-free learning'. It was felt that learning is enhanced when the learner is prevented from making wrong responses. The principle holds good in certain circumstances, but there are also many instances when one may learn most effectively from one's mistakes. Be that as it may, however, the early exponents of programmed instruction believed in writing materials so that most students would get most of the responses to the frames in the body of the programme 'right first time'. A common criterion was that 95 per cent of the students should respond correctly to every question in the programme. Frames that did not reach this criterion were rewritten. Our consultant, then, adopted this 95 per cent criterion and set about identifying all the frames in the three modules that had been answered incorrectly by more than 5 per cent of the students. As the total number of students in the four groups was 85, he looked for the frames which had been answered incorrectly by three or more students.

Remembering that there were 85 students, each studying three modules, requiring more than 100 individual responses in each module, we can understand why it took the consultant several days of quite boring work to prepare the data presented in Figure 19.7. Unfortunately, once prepared, this analysis did not prove to be as useful as expected. Several of the frames identified as suspect were found to be embedded in the early parts of sequences that supplied feedback and more practice, so that the student appeared, by his/her later responses, to have overcome any early misunderstanding. Other suspect frames were found to be inessential to the mastery of the principal objectives that were tested in the post-tests. The author, in an attempt to enrich the materials and lead the students a little beyond the specific instructional objectives, had included at times some questions like 'what do you think are the implications of...?' or 'name a further use that you can see for...'. Naturally, these questions 'stumped' more than a handful of the students. In addition, while performing the frame analysis, the consultant had noticed that many students who had scored poorly on the post-tests had registered low error rates on the frames in the body of the programme. Were the frames 'over-prompted' so that it was possible to respond correctly without really understanding the principles involved? Or had some students simply written in the correct answers after looking them up, without attempting to work them out for themselves? Or were the students responding honestly and with understanding to the questions in the frames, but due to limited practice and insufficient repetition, forgetting some key concepts and procedures before reaching the end of the module?

A third analysis

At this point, the consultant realized that although he had obtained an overall feeling for what had and had not been learned by the students, his analyses so far had not presented this crucial information in a clear and objective manner. It was obviously most important to know exactly which objectives had and which had not been mastered by the majority of the students. This would perhaps identify those sections of the materials that were most in need of revision. If this had been done earlier, a lot of time spent in tabulating errors on individual frames would have been saved.

The last analysis performed is presented in Figure 19.7 – a distribution of error rates on individual test items. Each item tests only one objective, so we have the data we need.

Even a cursory glance at Figure 19.7 reveals some surprises. In particular, the error rates on some of the items in Test 1 are seen to be quite high. It now appears that module number 1 may not be all that fit to print. In the previous analysis, the extremely low error rates on the early items in the test served to disguise the poor mastery of the objectives tested by the later items. An analysis of the module, or of the test items themselves, immediately shows that the part which was answered correctly by nearly everybody, is composed of verbal learning – basic concepts such as what does the term 'binary' mean, what is the basis of our common 'denary' (or some would prefer 'decimal') system, etc. The second part of module number 1 deals with the application of these concepts to the transformation of denary numbers into binary numbers and vice versa. It appears that these procedures were not mastered adequately by up to a quarter or so of the students in our groups.

Did these students do badly on the second module, which deals with the four arithmetic operations, using binary numbers, because that module is too difficult or unclear, or did they err due to an incomplete grasp of the content of the first module?

Defects of the field-test design

This, and many similar questions, cannot be answered from the data collected. It would appear that another field-test should be organized after revising all three modules. We now know which objectives have not been mastered, but our data does not help us much to ascertain why they were not mastered. A further

TEST 1		TEST 2		TEST 3	
Item No.	Frequency of error	Item No.	Frequency of error	Item No.	Frequency of error
1	4	1	9	1	29
2	5	2	6	2	36
3	—	3	11	3	8
4	2	4	4	4	20
5	2	5	36	5	37
6	2	6	32	6	31
7	9	7	29	7	38
8	5	8	2		
9	7	9	16		
10	18	10	—		
11	25	11	7		
12	4	12	11		
13	17	13	12		
14	19	14	15		
15	25	15	11		
16	2	16	15		
		17	12		

Figure 19.7 *Frequency distribution of errors on each test item
in the three post-tests used for field testing
the self-instructional materials on binary arithmetic*

question that springs to mind is concerned with the content of the third module. This deals with using binary arithmetic to solve problems. Do the errors committed spring from an incomplete mastery of the concepts and procedures of binary arithmetic, which were dealt with in the first two modules, or from poor levels of mathematical problem-solving skills, which are not formally taught in any of the modules? Probably from a combination of both factors. Were problem solving skills of the type required in module 3 tested for in the prerequisite test? No! Should they have been? Yes! Could it be that the much better results presented to the publisher by the author of these materials, were due to superior levels of problem-solving skills in the author's students? Quite possibly, but our field-test was not set up to answer this!

General principles of the design of field-testing

We have probably gone far enough with our analysis to indicate the degree of care that should be taken in the planning of a field-testing session and of the data collection and analysis procedure. Our case study has illustrated a number of points to be borne in mind by the instructional designer, when faced with the task of evaluating instructional materials of any form. We have used programmed instruction materials as our example, because the testing process can be carried out rigorously in all its theoretical aspects, something that is not always possible in the case of other materials. However, the lessons we have learned apply in general to the field-testing of any type of material presented in any medium. We list just some of the observations that may be made, on the basis of our case study.

1. Raw test scores, gain scores, double percentage criteria and the like are sometimes useful to describe in a succinct manner the quality expected or achieved by a given set of materials. However, they do not help us to identify specific areas of weakness and still less why these weaknesses are present. Their use is summative, whereas the main purpose of field-testing any materials is to carry out a formative evaluation.

2. Errors made during learning are not necessarily related in an obvious manner to final learning results. One can waste a lot of time and effort by opting to analyse the instructional PROCESS that is being promoted by the materials, before checking out whether the PRODUCTS in terms of

objectives mastered, indicate that specific parts of the process are indeed faulty.

3. The most fruitful way to commence a formative evaluation is to measure the products of learning and identify the extent to which each specific objective is being mastered. The 'mastery learning' concept, in practice, means just this – all the specified objectives should be mastered. Any objectives that in general fail to be fully mastered, act as pointers to those sections of the materials that require a deeper analysis.

4. This deeper analysis should be carried out with due regard to the specific content of study, the type(s) of learning that are being promoted, the characteristics and general skills possessed by the target population and so on. Beware of global mechanistic techniques such as frame error counts, number of slides presented per minute, ratio of talk to demonstration, etc. They are all useful in specific circumstances but may lead to wasted time and wrong conclusions if misused.

5. Do not put all your eggs in one basket, in the sense of using up all your stock of available 'guinea pigs' in one huge field-test. In the case analysed, it would obviously have been much more sensible to first test module 1 only, with one or two of the available groups. This would have revealed the weaknesses in the second part. Only after the elimination of these and re-testing on a further group, can one be sure that the errors made in the second module did not occur for reasons associated with the first module. Some form of schedule, similar to that used to control the developmental testing stage, should be planned for field-testing, otherwise we run out of guinea pigs before the materials run out of bugs.

6. Finally, remember that all the instructional design skills which are used in the planning and development of instructional systems in general and instructional materials in particular, should be brought to bear on the task of evaluation. One should attempt to put oneself in the shoes of the designer of the materials being evaluated. The planning of the testing sessions should be preceded by a detailed pedagogical analysis of the materials, in order to understand fully the author's intentions, treatment, style of writing, and any other aspects that may influence WHAT should be measured and HOW to measure it. Naturally, all this is much easier when the author is the evaluator, or is at least involved in the process. It is of course essential that, in formative evaluation, the person to be responsible for carrying out any revisions to the materials, should be involved in all aspects of the evaluation.

The publisher's role in materials evaluation

One final comment we might make on the case study presented, is that it is, in essence, a true story. The data presented was actually collected and the initial field-test was set up and executed as described. Due to the errors identified, the whole exercise was repeated again with other students, on the basis of a more careful plan: an excellent example of learning through one's mistakes, no doubt. However, the main point we wish to make is that the publisher involved was prepared to make the necessary investment in evaluation. This, as we mentioned earlier, is all too rare in the business world. There is no similar tradition associated with the more common, informational, materials that make up the bulk of published material, both printed and audiovisual (including the bulk of educational publishing). Such materials may only require a technical and pedagogical validation. Instructional materials, however, are a special case, requiring evaluation in field-test conditions, against mastery criteria derived from the objectives, before the decision to publish is taken.

20. Evaluation: Final Words

20.1 Reactions

To close our discussion of materials evaluation, we consider several other aspects that the instructional designer should evaluate whilst performing the developmental and field-tests.

The evaluation of reactions

Firstly, let us consider the question of student reactions to the materials. We have already discussed the evaluation of reactions in general, in Chapter 13 of *Producing Instructional Systems*, when we presented some questionnaires that can be used for the formal gathering of data on student opinions. Such questionnaires are usually applied at the end of a course or unit, to evaluate the overall reactions. It is possible, however, to prepare a similar form of questionnaire which refers specifically to the materials being field-tested and apply it as part of the field-testing procedure. Some care must be exercised in interpreting the data from such questionnaires, when the type of materials being tested are unfamiliar to the students. Novel ways of learning may create a 'Hawthorne Effect' that leads students to praise excessively the new materials simply because they are new and different. After some familiarity, this initial enthusiasm may well disappear. Cases also occur in which the novelty inspires anxiety and negative attitudes, which also tend to diminish with time.

There is also the problem of incompatability between the students' normal, or preferred, learning styles and those demanded by the materials. Some students prefer to learn by listening and discussing in groups and may have little experience or skill in learning alone from a book. Such differences are often cultural, promoted by the way that the local educational system operates.

One ignores such factors at one's peril. It is not always that easy to collect this sort of information by means of a standard questionnaire. Rather, it is necessary to observe the students, during both developmental and field-testing.

20.2 Costs

Evaluation of development costs

Another aspect that should be evaluated during materials development is the cost. It is probable that some initial estimate of development and production costs was made in order to decide whether or not to develop the new materials. During the process of development, the instructional designer should keep records of time spent and other costs incurred at every step of the project. This should enable the identification of excessive costs before it is too late to reformulate the project or modify the budget. Many projects leave this aspect uncontrolled, with the result that the materials are not ready on time, or use up all the available resources in the first modules.

The cost of analysis

One aspect that greatly influences the costs of materials development is the *level of analysis* applied by the designer. The more detailed, deeper forms of analysis, such as those used for the development of fully programmed instruction, are very time consuming and therefore costly. We have known projects that got so carried away in the analysis stage that time and money ran out before any production was attempted. One project spent three full years analysing the tasks and skills involved in all the jobs normally executed in a large industry (telecommunications). The original plan had budgeted three years for the whole project – analysis, design, development and implementation. Worse still, as the analyses referred to tasks in an industrial sector experiencing considerable changes (due to new developments in electronics, microprocessors, data processing and transmission), the bulk of the analyses so laboriously prepared in the earlier part of the project were quite out of date before they ever came to be used.

The instructional designer may sometimes need to 'cut his suit according to the cloth', modifying his initial plans in the light of formative evaluation of the rate of production and the rate of consumption of available resources during the project. An effective 'MBO' system of control is required.

20.3 Surprises: some case studies

The evaluation of unexpected effects

The failure to monitor the progress of production, or to identify incompatability between the materials and target population, may lead to some nasty surprises. Other sources of shock include the failure to recognize that the materials are unacceptable to the teachers who will be expected to use them in their teaching, or that they come into conflict with some aspect of the context in which they shall be used, or that, although the production resources exist, there is a lack of resources necessary to disseminate the materials and put them to widespread use. Many of these aspects are really Level 1 and 2 considerations which should have been carefully analysed before the decision to develop the new materials was taken. However, we all make mistakes, and so the author of the materials should repeat these evaluations during the development stages, in order to avoid any nasty 'surprises' later in the project. Among the 'surprises' that we have known to occur, we may mention the following examples:

The case of the over-verbal maths materials

A set of self-instructional printed materials, developed for use in primary school mathematics, did not prove to be as effective in large-scale use as early field-tests suggested. Further study showed that the learners were improving their reading skills more than their arithmetic skills, due to the over-verbal treatment given to the topic. A full rewrite was required, giving more attention to the existing literacy levels in the target population.

The case of lack of study skills

A set of programmed materials were developed and field-tested in the capital of a developing nation. The materials were field-tested in several city schools and were shown to be very effective. Later dissemination to rural schools proved to be a total failure. These schools used very few textbooks and therefore the children, although already literate, had not developed the necessary skills for learning from books. When the same material was read aloud by teacher and children in unison, and the responses chanted by the group as a whole, learning was as effective as it had been under individualized self-study conditions in the city schools.

The case of surprise attitude changes

A nationally broadcast set of radio programmes, entitled 'French for Tourists', was field-tested by initial use in recorded form with groups of specially invited adults who planned to visit France for their vacations. Learning was found to be excellent. A deeper study, however, revealed that many of the group had developed anxieties and inhibitions in relation to speaking French and, as a result, did not go to France at all.

The case of poor dissemination

An ambitious project to develop distance education materials for primary and secondary grades, to be used as the basis of the formal education system of a developing country, was launched. The materials – TV, radio and supporting printed materials – were developed over a period of some years, to cover all the curriculum of the 8 to 14 years age group. All the materials underwent very rigorous developmental and field-testing. Once implemented, it was found that poor radio and TV reception in many areas, coupled to the poor state of receivers and maintenance facilities, resulted in 70 per cent of the programmes being missed by most schools during the first year of operation. This situation was, in part, predictable and should have led the authors to assign a less essential role to the broadcast element in the system.

The case of poor financial planning

Another project developed a large quantity of self-instructional texts, to be used as remedial material in the schools of a nation that had a chronic shortage of teachers and, therefore, many children entered the secondary schools with large 'gaps' in their mastery of primary school curriculum content. The self-instructional modules were developed by a team of specially trained writers, over a three-year period – a total of over 20 man-years of investment in the development stage of the project. The resultant material – over 40 programmed texts, running to many thousands of frames – was extensively field-tested in three large schools, about 300 students being used in successive developmental and field-testing groups. The final material was excellent. Unfortunately, the 300 'guinea pigs' were the only students ever to benefit from it, as the Ministry of Education who had sponsored the production never found the funds necessary to print and disseminate the materials on a large scale. The sheer number of pages in the finished texts ran to more than the total of pages in all other official school books in use at the time. This too could probably have been predicted from a context analysis, in time to alter the authors' methodology and produce slimmer, perhaps somewhat less effective, but more viable materials.

The case of the antagonistic instructors

In the early years of the Industrial Training Boards in the UK, much investment was made in the development of new training materials. An early application of programmed instruction for the training of engineering apprentices met with great success, and resulted in the publication of a series of booklets in linear programme format. They presented the theory topics normally required as part of the (principally practical) training of engineering apprentices. Although these materials were thoroughly field-tested, they suffered from many of the defects of early programmed texts (see Chapter 5) and would not today be considered as very good examples of the art. They were, however, very popular in their day, and the publishers sold many thousands of sets. A little later, another set of materials for engineering apprentices was prepared by a professional instructional design company, with the backing of one of the training boards. These materials were in every way superior to the earlier booklets. Although still basically linear programmed texts, they used much more sophisticated writing techniques and even today would still be considered as excellent examples of instructional design.

However, these materials were a total commercial failure. Of the several thousand copies printed, only a few dozen were sold. Analysis of the causes of this failure revealed that the very aspects of the materials that made them 'stand out' pedagogically, had doomed them from the start. The authors had succeeded in integrating the necessary theoretical content with the practical training, by writing the programmes so that they were used 'on the job', guiding the learner through a series of practical machining exercises and introducing the necessary theory at the moments when it had a real practical value. The *planned* result – totally individual, self-paced, mastery-based learning, totally eliminating lectures and guaranteeing learning success in reduced time (proved by thorough field-test results published proudly in the preface). The *unplanned* result – instructors receiving an inspection copy felt their jobs seriously threatened and so prepared negative appraisal reports, finding all manner of imaginary fault with the content, in order to ensure that their companies did not adopt the materials.

The case of the unprepared instructors

The earlier, not so well designed, but nevertheless reasonably effective, programmed texts, were not seen as a threat, as they only took over the theory teaching, thus releasing instructors to spend more time on the practical part of training, which most instructors considered to be more important, more dependent on the instructor's amassed experience and more 'fun'. However, to get the best out of these booklets, it was necessary to integrate them carefully with the rest of the training, making their study related to the practical content, so that the students studied the appropriate booklets at the most appropriate moment. It was also necessary to evaluate learning, to ensure that students were not simply copying in the correct answers, without really studying the material. Many other aspects of self-study had to be understood and taken into account by the instructors.

We had the opportunity to perform a research study on the way that these booklets were being used in practice. We found that in the large majority of cases, best use was not being made of them, largely because the instructors were not accustomed to working with programmed materials. What was required, and had not been produced by the initial project, was an instructor's manual which explained the principles of the method, a summary of the content of each text (so it would not be necessary for the instructors to read through all the texts in order to find out what they teach and how) and a fully worked out control system that could be directly applied by the instructors to implement and manage the individualized learning system correctly and, incidentally, also enable training managers or supervisors to maintain overall control of the way the course was being executed and the learning results being achieved. The need for such preparation of instructors and the other human resources involved could, of course, have been predicted during the development stage of the project, and the exact content of such a manual should have been researched during the field-testing of the materials.

20.4 Evaluating the materials design process

A 'systems approach' to evaluation

In this section of the book we have attempted to show that materials evaluation, as any other formative evaluation, is a continuous process of monitoring one's decisions and actions, predicting the probable consequences and using these observations and predictions to improve one's performance. Some specific techniques do exist, and some specific moments can be identified as the best for performing a formal developmental or field-test. However, this does not release one from the obligation to keep a constant watch on one's progress. We have seen how the evaluation process plays a critical role in the effective application of a 'systems approach' to instructional design and development. He have also seen how it is important to maintain a 'systems view' of one's project and its interaction with its operational context. Only then are we able effectively to evaluate all factors that may influence the project's future success. The 'CIPP' model, presented in Chapter 13 of *Producing Instructional Systems*, is just as useful an organizing framework for materials evaluation as it is for the overall evaluation of education and training projects.

A 'systemic' view of ID

Hopefully, the organization of this book (and of its predecessors – *Designing Instructional Systems* and *Producing Instructional Systems*) around a four-level model of the instructional design and development process, has helped to promote a 'systemic' (rather than merely systematic) approach. We have presented several techniques and some instruments to aid the instructional designer, but these are not intended to be taken as recipes to be slavishly applied. Rather, they are mere examples of how one *might* wish to proceed. The reader should use these examples as a springboard from which to develop a personal set of techniques and tools.

The variety in the system

The number of possible techniques, tools and approaches, both to instruction itself and to instructional design, are probably unlimited. Both instruction and its design are, after all, *heuristic* processes, where many routes may lead to various equally 'correct' solutions. One should, of course, work from well-tested basic principles, but the application of these principles may be as varied as human creativity allows. We can gain an insight into the rich variety of solutions open to us if we remember the contrast we drew, in the preface to this book, between instructional design and chess.

We remember that the abandoned attempt to design a 'complete' training design algorithm (Dodd, LeHunte and Shepard, 1974) isolated several hundred specific rules that should be applied when designing simple psychomotor skill training and the simplest levels of concept learning. Probably several times that many rules would be indentified if ever a fully detailed analysis of the teaching-learning process were to be performed. How many different solutions could be provided by the creative application of these rules?

The game of chess is, after all, defined by only six basic rules that govern the movement of the pieces around an 8x8 matrix of 64 cells. This simple system can lead to the construction of the most varied strategies and tactics of play. No one, not even a 'grand master', would claim to have learnt all there is to learn about the game. Anyone, even a highly skilled player, continues to learn and to improve his play, given an appropriately stimulating and challenging problem to analyse and solve.

Instructional design, based as it is on a much larger body of basic rules, which govern the behaviour of a system (ie the learner) that is very much more complex than a two-dimensional matrix, must surely be many times more complex, more varied and more creative than chess. No one should claim to have learnt all there is to learn about teaching. Anyone, given appropriately stimulating and challenging problem situations, will improve his skills of instructional design.

The 'PIP' Thomas Gilbert (1978) has coined the concept of the 'PIP' (Potential for Improving Performance), which he defines as the ratio of 'exemplary performance' to 'typical performance'. This concept has very many attractive features, not least the implication that aiming to get people to perform 'typically' (the basis of all norm-referenced systems of assessment and teaching) is wasteful of human talent and of the opportunities that offer themselves to the human resources specialist. What car manufacturer would set about designing a new model that performs only 'typically'? Yet this is exactly the approach of most educational systems. The 'PIP' concept gives a measure of the *opportunity* that exists for improvement. Thus we should evaluate our success as teachers, trainers, or human resources developers, not in terms of what fraction of our students reach a predetermined standard of 'typical' performance, but rather in terms of how much we succeed in closing the gap between this typical performance and the best possible, or imaginable standard of performance.

The 'PIP' for materials development We mention the PIP at this point, in the last chapter of the book, for two reasons. Firstly, we shall use an example, quoted by Gilbert, that is of great relevance to our earlier argument. In many skill areas the PIP is quite small. Typical typing performance is usually defined as of the order of 40 words per minute, or 200 characters per minute. Some exemplary 'world record' typists are, however, known to type at 100 words per minute (500 c/m) and the level of 80 wpm (400 c/m) is achieved by a small but significant proportion of top secretaries. The PIP is therefore of the order of 2 to 2.5, indicating that there exists the potential to more than double the productivity of typists in general. Gilbert quotes a large number of PIPs that he has encountered in various projects. Some of these are:

– Filling a cosmetic order (accuracy)	1.3
– Microfilm reading	1.75
– Metal fabricating plant management	3
– Print shop management	6
– Encyclopaedia sales	12
– TRAINING MATERIAL DEVELOPMENT	25
– Teaching some mathematics topics	30

– And the reasons for this 'PIP' Need we say more? It would seem that the potential for improvement of typical instructional designers is greater than in many other, much more 'valued' professions. No doubt this situation exists in part exactly because education and training are not that valued. Society does not pay its educators in relation to the importance that it *says* it places on education and on human resources development.

The ID learning process Another factor that, no doubt, contributes to the extremely high PIP that Gilbert noted in relation to instructional materials design, is the sheer complexity of the instructional design and development process. We tried to illustrate this earlier by means of the comparison with chess. This factor of complexity implies that the learning of instructional design, like the learning of chess, is a lifelong process, which requires the development of ever more powerful heuristics for decision making. The development of such heuristics only takes place efficiently under appropriate learning conditions. These can be found described in earlier books of this series. In summary, they are:

- A thorough grasp of basic concepts and principles and the development of powerful cognitive schemata that interrelate these concepts and principles into complex decision-making strategies.
- Ample opportunity to learn experientially, by applying one's cognitive schemata and strategies to ever more varied problem situations.
- Ample opportunity to analyse solutions, both good and bad, both one's own and those proposed by others.
- Ample opportunity to evaluate these proposed solutions, both in practice and intellectually, through discussion.
- Ample interchange of ideas and experiences with other instructional designers who are at about the same, or higher, level of competence.

Such ideal self-development conditions are not always easy to arrange in practice. And many who are called upon to design instruction are not in a position to devote so much time to their own self-development. They must 'publish or perish', due to the economic constraints of making a living and producing materials within predetermined budgets/time limits. Thus, like the 'weekend' chess player, they 'muddle through', sometimes 'winning' and sometimes 'losing', when presented with typical, previously encountered, types of instructional design problems. But when confronted with a radically new problem, as for example the advent of interactive video, they seldom 'win' until they have made sufficient errors to learn 'the hard way'. Unfortunately, by then (as was illustrated earlier with the examples of programmed instruction, educational television and, possibly, computer-based learning), their errors may have 'killed the goose that lays the golden eggs', in that the promising innovation is so discredited that we do not get another chance to exploit its potential. Whether our dead goose will ever rise phoenix-like from its ashes is debatable.

It is fortunate that, just as most weekend chess games are not played against grand masters, the instructional problems most of us encounter are rarely out of the ordinary. However, just as most of the chess games played by our weekend player are either lost or, at best, won in an uneconomically inefficient number of moves, so the general state of instructional design and development continues at a low level of cost-effectiveness. The PIP remains high. The potential for improved performance is there, but it is not being realized. What might we do to reduce the PIP?

As we cannot entrust all the ID tasks to the handful of outstanding designers, we must work at ways of improving the performance of our typical designers. It is impractical to expect that all of these would have the opportunity, or indeed the willingness, to invest the effort required to improve their skills through the complex learning processes outlined above. Indeed, there is no guarantee that a satisfactory reduction in the PIP would result anyway. We must therefore seek other approaches.

20.5 The future of instructional design and development: how the ID process may change in the future

In *Designing Instructional Systems*, we presented a schema for 'front-end' analysis of performance problems. This schema is reproduced in Chapter 3 of the present book. A glance at this schema (see Map 3.4) will reveal a whole range of alternative solutions to performance problems. Some of these are appropriate when the cause of the problem is a genuine lack of knowledge or skill. Others are more appropriate when the cause is poor motivation, poorly planned job conditions, inappropriate consequences, and so on. A detailed analysis of the performance of specific instructional designers, working on specific projects, might reveal all manner of complex, multiple causes of poor performance. Here, however, we shall concentrate on just one alternative to the *development* of instructional design skills to the requisite level of proficiency. This alternative is the job-aid.

Job-aids   A job-aid is some document or reference material, or instrument, which reduces the performer's reliance on previous learning when performing the job. This definition distinguishes between tools used to carry out the job and 'tools' used to decide what to do and how to do it. It is the latter we refer to as job-aids or, perhaps more appropriately, performance-aids. Since their development (in the early 1960s) as a formal part of the 'armoury' of the performance technologist, job-aids have been increasingly applied in all manner of industrial and commercial contexts. One very well-known type of job-aid is the decision tree, or 'algorithm', used to map out all the alternative courses of action and the circumstances in which each should be taken. These may be presented in the form of decision tables, flow charts, or any other form that is easy to use as reference material. Some authors (eg Holden, 1983) estimate that anything up to about 80 per cent of industrial performance problems normally 'treated' by means of training, could be better and more economically resolved by means of appropriate job-aids. At the least, one would reduce the extent of the training required. Could we effectively use the job-aid approach to reduce the PIP in the instructional design/development job area?

Job-aids for the ID process
Many attempts have been made to do this. Specific tasks of the overall ID process have been analysed, and checklists, flow charts or decision tables developed. For example, several detailed algorithms have been developed to assist the task of media selection (including one by the present author – Romiszowski, 1968; 1974). Checklists abound for the selection or evaluation of off-the-shelf instructional materials, computer-based courseware and the like. Most books that deal with materials development or production (eg Brown and Lewis, 1973; DeBloois, 1982; Horn, 1974; Kemp, 1985; Heinich, Molenda and Russell, 1982; Gerlach and Ely, 1980) offer a variety of job-aids to guide both procedural and decision-making tasks.

Some limitations
The procedural job-aids work quite well, in general. The decision-making aids do not. This is because most of the decision-making processes in instructional design are complex, probabilistic, multi-faceted, heuristic processes, which defy reduction to a relatively simple, immutable, algorithm. This is not to say that the flow charts for instructional decision making that are encountered in the literature are of no value at all. They are often very useful summaries and memory aids to the chief factors that should be taken into consideration. They do *not*, however, guarantee (as true algorithms should) a correct decision in all cases. They must be used with 'judgement' and applied 'in the light of experience'. One cannot escape the ultimately heuristic nature of the decision-making process involved in ID.

Flow charts and decision tables have some undesirable aspects when used to describe highly heuristic decision making. They impose the decision-making style and sequence of the job-aid's author on the user. In procedural tasks this is not undesirable, as procedures do have a 'best' sequence of execution, or at least one that is known to be efficient. This is not necessarily the case with heuristic decision making. Efficiency often depends in these cases on a flexible, holistic approach, not bound by a standard sequence of analysis. Flow charts and decision tables (more so the former) present a fixed sequence of analysis, which tends to be adopted by the job-aid's user as the only sequence. Still more dangerous is the use of an algorithmic decision-tree form of flow chart in cases where several of the alternative decisions may be valid at one and the same time for different aspects of the problem under analysis. The case of performance problem analysis, mentioned above, is a good example – a poor performer may not know how to do a job, but even if he did, the way that the job is organized might prevent him from performing adequately. In such cases, the use of a flow chart may be extremely misleading as, by its nature, a flow chart is to be followed to one, and only one, of the alternative outcomes.

Conceptual schemata as job-aids
One alternative job-aid, with which we have been experimenting for some years, and which we have used profusely in this series of books, is the conceptual schematic approach. The circular diagram (Map 3.4) which describes our model for performance problem analysis is one such visual representation of a set of interrelated factors which should be taken into account in a holist manner when

making decisions. Such conceptual visualizations have proved to be very useful as learning and comprehension aids. They also act as effective memory-aids, helping the user to take into account all the relevant factors listed. They have been found to overcome some of the dangers and weaknesses of flow charts that we described above. However, they do not overcome the need to develop skill in taking appropriate decisions. They help one to remember how to approach the decision-making task, but the user's judgement and experience still come into play in making the correct (or incorrect) decision. Thus we have not overcome the need for our instructional designer to develop skill in applying judgement, and to gain the experience necessary to be able to do this. But what if our designer is unable or unwilling to devote the time and effort required to develop these skills and gain this level of experience?

Until recently, there was no practical answer to this question, except for, 'replace the designer with someone more skilled and experienced', or, 'give the designer constant access to another, more highly skilled and experienced, designer'. The first of these replies is not very helpful if the labour market is almost completely devoid of such 'more highly skilled' experts. The second is a little more helpful, as it suggests organizational solutions that might make the best use of our scarce 'expert' designers. The growth of such units as local 'teachers' centres', or 'learning development units' in colleges and universities, of 'institutes of educational technology', and of many forms of 'training systems consultancy' organizations (both in-house and external) is a direct result of attempts to make the best of a scarce resource. However, with recent developments in knowledge engineering and so-called 'artificial intelligence', another reply to our question might be, 'build a computer-based expert system for the design of instruction'.

Expert systems as aids to heuristic decision making

An 'expert system' is a computer-based 'distillation' of the experience of an outstanding (human) expert or, more commonly, the pooled experience of a group of outstanding experts. It is basically composed of two systems:

1. The knowledge-base – a representation of all the rules, facts, levels of probability of certain outcomes, and so on, that are known to the expert/group of experts.
2. The inference engine – a representation of the way in which the experts in question actually do take highly complex, heuristic decisions in the domain under analysis.

Building an expert system is not an easy task. Even in a limited area of enquiry as, for example, the design of steel frameworks for bridges, or the packing of all the correct components in a given custom-designed computer application (both are examples of existing practical expert systems), the number of separate rules used by the 'experts' may run into several hundred. The task of identifying the rules (by interviewing the experts) and then transforming them into a computer-based expert system is extremely lengthy and costly. Nevertheless, the effort may pay off, as often the resultant computer-based expert system, when used as an adviser by a typical (not outstanding) performer, results in decisions that are more consistently correct, or qualitatively better, than those of the outstanding human experts whose wisdom was distilled to create the expert system.

Expert systems for instructional design

Expert system design and development has a history of some 30 years, but only just recently has computing power become sufficiently plentiful and cheap, and software for the more rapid development of expert systems become readily available, thus turning a 'dream' into a practical possibility. The Royal Navy's project, mentioned earlier (Dodd, LeHunte and Shepard, 1974), was an early attempt at identifying the knowledge-base that would be necessary to design an 'expert industrial training officer system'. The attempt was abandoned because the number of rules identified, even in the partial analysis carried out, defied the hardware and software capabilities available at the time. This is no longer so. The technology is available, albeit still relatively crude and very expensive. Whether it will 'pay off' is a moot point.

In terms of cost/benefit, the pay-off must come from the commercial or social value of the improved instructional designs which would result from its use, as compared to the costs of developing it. It is difficult to make this judgement as the value of instruction may be measured in so many ways, dependent on the content of the instruction, its use in society, the number of potential students, and so on. However, leaving aside the cost aspect (and hence the commercial/social value of the improvements in instructional designs), we may reflect a little on the real extent of the probable improvements which would be achieved.

The improvements would spring from the 'power' of the rules and inference systems built up out of expert knowledge and general instructional theory applicable to any content that can be classified into a certain taxonomy (similar to the categories of knowledge which compose our own schema of the structure of information). Once the knowledge is classified, certain specific instructional rules, or sets of rules, would apply. These would be tempered by what we know of the specific target population, the specific content, etc. However, it is probable that these 'specific' factors would introduce a host of special cases and exeptions to the general 'rules'.

To make the expert system highly effective and efficient, these special cases must be built-in as further rules to be taken into account. But how many special cases are there? Will there be new special cases for each new topic or task that we attempt to analyse and teach? Experience in the design of expert systems in much simpler task areas suggests that this may well be the case. If this is so, will the sheer number of special cases and exceptions make it unfeasible to construct a highly effective, general purpose 'expert instructional designer system'?

Some writers on expert system design argue that the answer to this question is already positive. Others disagree. However, on purely economic grounds, it is probable that some time will pass before the use of powerful, general purpose, expert systems for instructional design and development, becomes a commonplace reality. Ministries of education are unlikely to finance the very high development costs until the use of expert systems in other areas of decision making becomes so much of a routine that educational policy makers will be more or less forced to follow suit. In the meantime, researchers are exploring other, less costly approaches to the use of the computer as an assistant in the ID process.

The LDS system of lesson design

One approach to simplification is the one adopted by David Merrill and his collaborators on the development of a 'Lesson Design System (LDS)' (Merrill and Wood, 1984). This is a computer-based 'adviser' that is capable of prompting an instructional designer to take into consideration the tenets of a particular theory of ID, at appropriate moments in the ID process. LDS is an implementation of Merrill's Component Display Theory (Merrill, 1983). This theory deals with some forms of (mainly cognitive) learning tasks and is composed of a set of quite carefully defined ID principles that relate to these specific tasks. Essentially, the LDS system is composed of four editors:

1. A content outline editor, which assists the designer in mapping out the structure of the content to be taught.
2. A strategy editor, which assists in the selection of appropriate instructional activities and in establishing the sequence of these activities.
3. A display editor, which assists in the layout and organization of the instructional message, whether it be for a printed text or for on-line presentation by a computer.
4. An on-line lesson editor, which corresponds to the authoring languages currently used for generating CAL materials.

The LDS system is, thus, much more than an authoring language or system, such as PILOT or PASS. It is an 'advisor' in that it assists in the selection of analysis technique, instructional strategy, presentation tactics, and so on. However, it is not a true expert system in the sense of taking these decisions. It is a prompting system that enables an experienced instructional designer to be more efficient – 'LDS does not pretend to be an instructional design system which can be used by any subject matter expert without formal training in instructional design...the system assumes that the user has knowledge about the instructional design process and wishes to have assistance in making this process more cost-effective' (Merrill and Wood, 1984).

Thus, although LDS is not an 'expert system' in the true and full meaning of the term, it is a form of computer-based job-aid that assists (or at any rate, is designed to assist) the instructional designer in the taking of key decisions. It does this by prompting the designer into making specific choices and then carrying out specific procedures. For example, the content outline editor first prompts the designer into classifying the objective to be achieved into one of three categories recognized by Component Display Theory – concept, principle or procedure. Once classified, the content to be learnt is analysed by an appropriate method of analysis: hierarchical taxonomic 'tree' construction for a concept; hierarchical 'prerequisite' flow chart construction (á la Gagné) for a principle; algorithmic flow chart construction for a procedure. However, it is the designer/user who does the actual analysis. The system prompts him by presenting an appropriate pattern of analysis on the screen. As the designer builds up his content structure, missing or out-of-place topics are more easily identified as gaps or inconsistencies in the pattern of analysis shown on the screen.

This short example serves to highlight some of the advantages, and also the limitations, of this approach. With regard to the latter point, it is important to note that the system, in its attempt to be general as regards subject matter content, and yet easy and economical to implement, restricts itself to a fairly rigid, one-theory, approach to the teaching-learning process. This may seriously restrict its applicability.

20.6 The computer-based expert instructor: a case study and a look into the future

The one-subject expert instructor system

There may however be another approach, also based on the application of expert-system technology, but possibly more feasible in the short run. If, as we suspect, a truly general-purpose expert system for instructional design is as yet some way into the future and may, for economical reasons, never be developed, why not develop special purpose expert instructor systems by distilling the expertise of known outstanding instructors on a topic or subject area of outstanding economic or social importance? We limit the design problem to a specific topic of great economic or social importance and ensure cost-effectiveness if we succeed in developing outstanding levels of instructional effectiveness. If we are today able to distil the collective wisdom of engineers into an expert system as good as any one of them at solving complex bridge-design problems, or the wisdom of a group of doctors into a system better than any one of them at prescribing treatment for a particular group of diseases, why should we not succeed in similarly distilling the wisdom of outstanding teachers of, say, electronics?

The case study which follows did not originate as an attempt to do this, but the project has certain characteristics which would enable one to develop it along the lines suggested. The project originated as a semi-individualized approach to the retraining of electronics technicians for Brazilian industry.

The SENAI project

SENAI is the National Service for Industrial and Apprenticeship Training, an organization which is financed and managed by Brazilian industry and is responsible for pre-service and in-service training of apprentices, adults and skilled workers. Although SENAI is a national organization, it is decentralized in its mode of operation, and each state of the Federation runs its own schools and offers courses that are adapted to the needs of local industry. The state of Sao Paulo is by far the most industrialized and is a leader in the development of new technologies, so it is natural for the SENAI of Sao Paulo to take the lead in the development of new training courses and systems that meet the needs of industries which are emerging or undergoing technological transformation.

Over the last few years, the Sao Paulo SENAI has been preparing itself to meet the growing demand for high-level training in the fields of digital electronics, microprocessors, computer processing, robotics and numerical control (SENAI, 1984).

The modular structure of these training programmes was dictated by the characteristics of the target population and their disparate needs, identified in the early stages of analysis. Each programme is composed of several modules of about two weeks in duration. Trainees may enter the programme at any point that is suited to their present needs and skills.

Each module has both a theoretical and a practical component, emphasis in the latter being placed on simulation of the tasks actually performed in the industrial context. The theoretical component is thoroughly documented in the form of structured, or 'mapped' materials, suitable for both initial learning and later on-the-job reference. These materials are developed in accordance with the rules and suggestions of the Information Mapping (TM) system of structured writing, together with some additional rules developed specially for this project.

In addition to the use of these 'maps', trainees are encouraged to make full use of other materials available in the training centre library, especially articles in specialist journals – often the most up-to-date sources of information in a fast-developing field like microelectronics. They use this resource to follow up topics of individual interest that go beyond the objectives of the module, to gain further technical knowledge on topics dealt with in condensed form in the course, to review concepts that were not fully understood in classroom sessions, or simply to access reference data not required to be learned but nevertheless necessary for the execution of practical assignments.

Another characteristic of this system that is of especial value in the SENAI project, is the facility with which 'mapped' materials may be updated or modified. The requirement for flexible, non-linear use implies that each block of information must be written in such a way that it makes sense on its own, when removed from the context of the material that immediately precedes and follows it. This feature, coupled with the division of the material into relatively short maps that deal with only one topic (or aspect of a topic), facilitates the processes of modification and updating of the content.

Yet another characteristic that we are exploiting in the SENAI project, is the use of clear map titles as indicators of the key words that may be used to relate a given map to a given book, chapter or article, available in the library. The authors of the mapped materials take great pains to analyse the subject thoroughly and structure the maps so as to include all relevant key words in the map titles and subtitles. This knowledge structure serves as a guide to authors when writing the materials, as a guide to trainees when using the materials and as a classification tool for librarians when indexing the other related materials available (Arce and Romiszowski, 1985).

The instructional design It is important to note that there are in fact two independent, though interrelated structures:

1. The structure of the course modules, composed of a sequence of laboratory activities, or 'units', carried out under instructor supervision in a specially equipped digital laboratory (we may call this the *task*, or *skill* structure, which was, of course, derived from the occupational/task/skill analyses of current industrial requirements).

2. The structure of the instructional (and informational) materials, composed of a network/schema/entailment mesh of key facts and concepts of the subject matter, digital electronics/circuits, which we may best refer to as the *knowledge* structure, and which we derive from subject matter/topic analysis.

These two structures are interrelated by a series of evaluation tests, which verify that trainees have the prerequisite knowledge necessary to commence a given laboratory activity and indicate the individual study needs of those trainees who do not have the prerequisites. However, the study materials may be linked to a variety of different courses and modules, for a variety of different target populations, simply by changing the evaluation test items and sequence. The same set of 'maps' may be used as material for almost any future training course in the field of electronics.

At a somewhat more detailed level of design, the knowledge structure is subdivided into two major categories:

* TE – theory, or conceptual knowledge that in general needs to be learnt, understood and used as a basis for problem solving and other higher order learning.
* TA – applied technology (tecnologia aplicada, in Portuguese), or factual knowledge about specific components, operational characteristics, symbols and conventions, or any other information that in general does not need to be learnt, but may be referred to when required (these maps act, therefore, as job-aids).

These two categories may overlap somewhat at times, but serve as a useful general organizing framework for the authors to use. Both categories may, of course, be found in the supporting library-based materials.

At a yet more detailed level, the information contained in one map, concerning one specific topic, is itself modularized. This is a result of the specific authoring rules adopted, to organize/present the information in 'maps'. A map presents, in structured format, all the information deemed necessary on one specific topic. The title of the map clearly indicates the topic in question, thus facilitating the reader's access to relevant information. The information that is presented is closely analysed and classified into types (eg definition, example, application, common synonyms, relation to other topics, etc). Each different type of information is presented in a separate and clearly marked-off 'block' (paragraph) and is identified by a descriptive subtitle that, once more, facilitates the reader's selection of relevant information. The material may be read in a flexible, non-linear manner, adapted to the user's needs and learning style.

The evaluation of the practical activities is generally based on the instructor's observations of the trainees and the analysis of their final reports. These observations are, however, only one part of the total evaluation and control system. Entry into a particular laboratory activity is controlled by a pre-test which checks the trainee's mastery of prerequisites that should have been learnt in preceding modules or units. These test items are listed in a classified item bank in a relational database, together with cumulative data on the use made of specific items, error frequencies and so on.

Thus, the database enables the same test to act both as a prerequisite test to control entry into laboratory activities and a post-test that provides both summative and formative data to the instructor and other interested parties. For example, materials developers may access the database in order to identify areas that require revision, instructors may use it as a powerful aid to diagnosis and treatment of specific individual learning difficulties and training administrators and client organizations may request a range of summative reports on the progress and overall outcomes of the training programmes.

Finally, of particular importance to our present discussion is the fact that all test items, all instructional and informational maps, and all the library materials are interrelated, in the same data base system, as the key words which describe the content and structure of the topic are common to all three. Any learner, reading any journal article, may at any moment access supplementary instructional material to help with the understanding of a term encountered in the article, simply by keying in the technical term that is causing difficulty. Alternatively, a learner who is reading an instructional map and would like to proceed beyond the objectives addressed in it, need only key in the map title in order to receive a detailed list of references to the other materials in the library that deal with the topic in question.

The system operates on an Apple-type microcomputer, with a Z80 enhancement running the CP/M operating system. The relational database management system used is D-BASE-2. With regard to the mass storage needs of the system we are using a ten-megabyte hard disc. This, as indeed are all the hardware components of our system, is Brazilian-manufactured.

The system described, and currently in operation, can be characterized as economical, practically feasible and largely adequate for the requirements of the parent institution. It is seen to be economical in that it uses only readily available hardware and software, of specifications that may now be obtained 'over the counter', as it were, in a growing number of countries, even those that have quite young microcomputer industries.

The system is also 'appropriate' in relation to the problem situation which led to its design. The computer is doing what it is best at doing, and other media, notably paper text, are used to do what they do better. There would be nothing to gain from the on-line storage, in electronic format, of the maps. Many of them are highly illustrated, containing complicated logic circuits – not easy to present on a monitor screen with clarity – and useful, on paper, to scribble on, modify and take away after training. Many of the maps become essential job-aids after trainees return to their places of work.

However, the design methodology adopted is such as to permit a number of future developments, given appropriate hardware and software availability and, of course, institutional interest and commitment.

Firstly, the instructional materials, as well as the test items, are so written that they may readily be stored and displayed by means of some computer-based system of delivery, should this prove to be desirable (for a distance-education system, perhaps). We may imagine offering the maps as a form of 'electronic dictionary' on electronics, perhaps as a service on a local network, or through a future satellite-based viewdata system available to all.

Secondly, the special characteristic of this system, which enables the cross-referencing of instructional and other related library-based materials, may be of especial interest to teachers in many subject areas, not necessarily technical nor vocationally orientated (especially in cases where learner control over the objectives of instruction is desired).

Thirdly, one must emphasize the importance of the 'by-product' of a structural representation of the subject matter, in the form of a key-word hierarchy or network. Apart from its use in the current project as an authoring job-aid and a materials analysis/classification tool, it may serve as the basis for yet other system enhancements. One we have mentioned already is the use of the structure by the trainee as a structural guide and advance organizer of learning strategies. Another, of particular importance to CAI, is the value of the structure as a basis for the transformation of the present system into a form of 'expert system', capable not only of presenting information on request, but also of diagnosing the user's information needs and offering tutorial guidance.

This would not, of course, be an 'expert' in the solution of electronics problems, but rather an 'expert teacher' of digital electronics, capable of 'inferring' a given user's needs for specific information and directing him to appropriate presentations. At present, the system is not very 'smart' in the advice it gives to the user. A simple questioning procedure may act as an 'inference engine' which would guide the user with greater precision to exactly those parts of the 'knowledge base' that he most needs. Further interactive questioning would test whether the information supplied was sufficient to meet the user's knowledge needs.

20.7 Conclusion – ID with a new face

The SENAI case study illustrates a number of important points which form an apt closure for this book. We have, in the earlier chapters, been focusing on the historical development of instructional materials design techniques. In this last chapter we have been analysing the latest developments of 'new' educational technology, which, in the view of some writers, heralds the total transformation of the instructional process and, therefore, of the instructional design process as well.

Will computers take over not only the bulk of the *presentation* of instruction, but also the bulk of the *design* of instruction? Will the age of computer-based expert systems render the instructional design profession, as we know it today, obsolete? Will the teaching profession as a whole become obsolete?

The 'old' media are not dead

Taking these questions one at a time, let us consider the presentation of auto-instruction first. The SENAI project illustrates that in many situations, as yet, the printed text is still the best medium to use. The current wave of popularity for CAI has led many designers to put into electronic format everything that can be so stored. Huge investments have been made to improve the resolution of computer graphics, so that now the computer can generate an image that rivals the colour photograph or the slide in visual quality. But is the production and use of such graphics cost-effective, as compared to using photos or slides? In many cases the question may be answered positively, but in many others this is not so. A good compromise is now available – the videodisc. This stores visual images in a most economical manner. But for many real applications, there is no need to have access to the images under the control of a computer. Yet the enthusiasts are apt to get carried away and make all instructional videodiscs 'interactive', thus incurring extra costs and no extra benefits.

The case of the printed word is yet stronger – the computer screen is not a convenient manner to present text in large chunks. Paper offers several clear advantages in many situations. For every case in which the electronic *transmission* or *storage* of text is an advantage, there is a case where the electronic presentation of text has disadvantages. The wave of CAI enthusiasm may lead to such absurdities as the questions asked at the end of a session on interactive video production, which was presented at the 1985 ADCIS conference. The presenters had used printed workbooks as one of the media – these offered the advantages of leaving the learner a 'hard' copy of key points for later reference, as well as giving him a learning task to do whilst the video segments were being located (the particular project used videotape, which has rather long search times). But the bulk of the audience at the conference seemed to be preoccupied with methods of eliminating the paper texts. Someone asked, 'Didn't the learners object to having to study from paper text?'; another immediately commented that it would be easy to modify the system so that, using two screens, one could present the text (computer-generated) on one monitor, whilst the video segment on the other was being searched for. Our own view, however, is that it will be a long time yet before it really becomes more practical and pedagogically desirable to always use computer-based presentation of computer-generated text, graphics, animations, etc. Let us not forget, in our enthusiasm for the new media, to use the most *appropriate* presentation media for each specific case.

Demand for instructional designers will grow

As regards the automated design of instruction, we also see a whole series of limitations. There is no doubt that certain parts of the ID process may be automated, or at least assisted, by means of computer-based 'expert' or 'advisor' systems. Our discussion of the heuristic nature of the process, the enormous 'PIP' associated with it and the lengthy nature of the training required to develop high-level ID skills, all serve as arguments for the development of effective job-aids. It is one thing, however, to envisage aids that render the reasonably skilled designer more effective (as in the case of Merrill's approach in the LDS system), and quite another to postulate a fully automated general purpose instructional design system that would take over the ID role in all its aspects and for all manner of subject matter content. We do not see this as an imminent probability and do not, therefore, have any fears for the future employability of experienced instructional designers. On the contrary, the undoubtedly greater reliance on distance-education and on individualized self-instructional systems in the educational and training systems of the future, will in our view lead to a marked increase in the demand for the services of competent instructional materials designers.

The ID profession should become more professional

A further point worth making in relation to the 'health' of the ID profession is that, in our view, there is a growing professionalism and a growing realization that there is indeed a rich knowledge-base to be exploited. The development of expert and advice systems, albeit of partial application, will do much to avoid the past tendency of each generation of ID practitioners to start all over again, 'reinventing wheels' that have given a lot of mileage in the past. One of the main messages of this book has been to show how powerful techniques for ID, developed in the 1960s and 1970s, have been largely forgotten by the designers of the 1980s, although they contained so many good ideas which today could be used and yet further developed. We have tried to show, by way of an example, the many ways that the Information Mapping (TM) techniques may be incorporated into modern CAL methodology. The SENAI case study illustrated the power of these techniques for the preparation of a knowledge-base that may be used in a variety of different ways, to create both rigid, system-controlled, competency-based instruction and highly flexible, learner-controlled, interest-based instruction. Our discussion also hinted how one may extend the project to become a form of expert instructor on a specific subject matter area.

New ideas must build on the old
Other techniques that are due for revival in a similar way are the mathetics approach to instructional exercise design and the structural communication methodology developed by Tony Hodgson. This latter has at least as much untapped potential for the area of computer-based learning as the information mapping (TM) approach to message design – possibly more. New ideas and techniques are, of course, appearing on the scene. One of these, discussed here, is the 'expert system' approach, but there are others. Some of these bear a resemblance to earlier techniques, often by chance rather than by design or derivation. Others are truly original and creative innovations. But life is too short to ignore the lessons and skills learnt by earlier generations of instructional designers. John Dewey, in the 1920s, had already made the significant observation that in education we suffer, more than in most other professions, from the 'baby-and-bathwater syndrome' – every new generation throws out the previous ways of doing things, sometimes improving on earlier weaknesses, but inevitably losing strengths and the good points that were developed at a great expense of time and effort. If this book helps to save some of the 'babies' of the last 20 years or so, before they get irretrievably sucked down the bath plug of oblivion, then one more of our objectives will have been achieved.

Expert systems and the role of the teacher
Finally, our last question dealt with the future of the teaching profession as a whole, in the light of the possible development of computer-based 'expert teachers'. Here, we feel that progress will be slow for a number of reasons. But progress will, inevitably, be made. If the expert, one-subject, teaching system is indeed a technical possibility (and we have no doubt that it is), then it is only a matter of time before such systems are developed. Their widespread use would change the role of the conventional teacher in many significant ways. It is unlikely that the human teacher would be replaced, especially in the early, formative years of schooling that are as important for socialization, culturalization and motivation as for instruction. However, the instructional function of the teacher may be much reduced in areas where effective expert systems exist. In adult professional education, where we are dealing with socially developed, highly motivated, goal-orientated learners, the replacement of the human instructor is more probable. Home-based study as a preparation for short, intensive, small-group seminars or meetings, may be the most common model of 'permanent education' in the future. If a highly effective 'expert teacher' is available to help you with your specific problem, at the exact moment, day or night, when the problem is significant to you, why should you pay (more dearly, no doubt) to attend a large-group course that deals in generalities, may or may not address your specific problems or interests, is less effective and much less flexible?

The new roles of 'SME' and 'ID' professional
If the predictions voiced in the last paragraph are anywhere near the truth, then we have another reason to be confident in relation to the future of instructional design as a profession. After all, someone will have to develop the expert systems. This requires both experts in the learning problems associated with a specific topic and experts in the development of computer-based expert systems. Thus, an effective 'expert teacher' of, say, digital electronics, would be created by the cooperative efforts of:

- Experts in the content of digital electronics (there are many of these around, if we pay the going price).
- Experts in the teaching of digital electronics (these are the current 'best teachers', who are of course experts in the content, but are much more than this and are much rarer than the subject-matter experts).
- Experts in the design of expert teaching systems (these do not exist at present – but the profession is just being born).

We may predict, therefore, some marked changes in the jobs of both instructors and instructional designers. Some instructors will become progressively more 'animateurs' (there is no exact equivalent in the English language – but there will have to be in the future) and some will become, at least at times, key members of an instructional design/development team. The cross-disciplinary 'professional' instructional designer will, on the other hand, have to add some new skills to his tool box. The development of expositive self-instructional messages, structured informational messages, experiential exercises, or audiovisuals will not become dying art forms. But in addition to these competencies, the designer of the future will require many computer-related skills and, no less important, knowledge-engineering skills.

The evolution of the ID profession

The 'behavioural engineer' of the 1960s, in order to survive the 1970s, had to evolve in two directions: to 'performance engineering' at the 'macro' level of decision making, and to 'cognitive engineering' at the 'micro' level. The performance engineering trend is well illustrated by the writings of Tom Gilbert, Geary Rummler, and others. The cognitive engineering trend (though not often called by that name) is illustrated by all those who have striven to put the principles of cognitive psychology into systematic practice – here we would include the work of Hodgson and others on structural communication, Dienes and others, who built the ideas of Piaget into systematic procedures for instructional materials design, and even Horn, Scandura and many others, who have stressed message design principles that take into consideration what we know about how learners use structure in learning.

Now, in the 1980s, the need is for evolution in two other dimensions. The 'information engineering' dimension deals with how new electronic technologies are influencing the way we store, transmit and communicate information in our societies. The 'knowledge engineering' dimension deals with how what is known about a given topic is related to, and can be accessed from, what is known about other related topics. This last trend is significant in many ways and may well be the key to progress in educational and communication technologies in the next decade. It is the key to the construction of effective auto-instructional systems that may operate under learner-control, for whatever objectives the learner may have at the time of learning. It is also the key to the designing of materials and messages that may more effectively cross the interface between 'information', as it exists and is structured 'out there' in the world at large, and 'knowledge', as it is formed 'in here' in the mind of the individual learner.

Bibliography

Agar, A (1962) Instruction of industrial workers by tape recorder (original in Swedish). *Affarsekonomi*, 10. See also (anon) Swedish training system breaks the language barrier, *Industrial Training International*, September 1966.

Allen, D and Ryan, K (1969) *Microteaching*. Addison Wesley, Reading, Massachussetts.

Arce, J F and Romiszowski, A J (1985) Using a Relational Data-base as a Means of Integrating Instructional and Library Materials in a Computer Managed Course. Proceedings of the 20th ADCIS International Conference. Association for the Development of Computer-based Instruction Systems (ADCIS), Philadelphia, April.

Atkinson, R C and Schiffirn, R M (1968) Human memory: a proposed system and its control processes. In Spence, K W and Spence, J T (eds) *Advances in the Psychology of Learning and Motivation Research and Theory*, Volume 2. Academic Press, New York.

Ausubel, D (1963) *The Psychology of Meaningful Verbal Learning: An Introduction to School Learning*. Grune and Stratton, New York.

Ausubel, D (1967) *Learning Theory and Classroom Practice*. OISE Bulletin No.1. Ontario Institute for Studies in Education, Toronto.

Ausubel, D (1968) *Educational Psychology: A Cognitive View*. Holt, Rinehart and Winston, New York.

Ausubel, D and Robinson, F G (1969) *School Learning: An Introduction to Educational Psychology*. Holt, Rinehart and Winston, New York.

Baggaley, J (1980) *Psychology of the TV Image*. Gower Publishing Company, Farnborough, Hampshire.

Baggaley, J and Duck, S (1976) *Dynamics of Television*. Saxon House, Farnborough, Hampshire.

Bates, T and Robinson, J (1977) *Evaluating Educational Television and Radio*. Open University Press, Milton Keynes, Buckinghamshire.

Bigge, M L (1982) *Learning Theories for Teachers* (4th edition). Harper and Row, New York.

Biran, L A (1974) The limitations of programmed instruction. In Romiszowski, A J *The Selection and Use of Instructional Media*. Kogan Page, London.

Bitzer, D L (1976) The Plato System. Paper presented at international conference on distance education and individualization in higher education, Caracas, Venezuela. The Open University, Venezuela.

Bitzer, D L and Skaperdas, D (1972) *The Design of an Economically Viable Large-Scale Computer Based Education System*. CERL Report X-5. Computer-based Education Research Laboratory. University of Illinois, Urbana, Illinois.

Blake, C *et al* (1975) *A Systems Approach to Teaching and Learning Procedures*. UNESCO Press, Paris.

Bloom, B S (1968) Learning for mastery. *Evaluation Comment*, **1**, 2. University College of Los Angeles (UCLA), California.

Bloom, B S, Engelhart, M D, Hill, W H, Furst, E J and Krathwohl D R (1956) *Taxonomy of Educational Objectives, Handbook I: Cognitive Domain*. McKay, New York. Reprinted (1972) by Longman, New York and London.

Boog, G G (1980) Validacao e avaliacao de treinamento (Validation and evaluation of training). In Boog, G G (ed) *Manual de Treinamento e Desenvolvimento*. Brazilian Association for Training and Development (ABTD). McGraw Hill, Sao Paulo, Brazil.

Bower, G H and Hilgard, E R (1981) *Theories of Learning* (5th edition). Prentice Hall, New York.

Bretz, R (1984) Slow-scan television: its nature and uses. *Educational Technology*, **24**, 7, July 1984.

Briggs, L J (1970) *Handbook of Procedures for the Design of Instruction*. American Institutes for Research, Pittsburgh, Pennsylvania.

Brown, J W and Lewis, R B (1973) *AV Instructional Technology Manual for Independent Study* (5th edition). McGraw Hill, New York.

Brown, J W, Lewis, R B and Harcleroad, F F (1977) *AV Instruction: Technology, Media and Methods* (5th edition). McGraw Hill, New York.

Bruner, J S (1966) *Towards a Theory of Instruction*. Norton, New York.

Brunstrom, C (1974) Developments in training design. In Romiszowski, A J (ed) *APLET Yearbook of Educational and Instructional Technology 1974/75*. Kogan Page, London.

Clarke, J (1970) The learning of mathematics in the primary school through the use of concrete materials. *Aspects of Educational Technology IV*. Pitman, London.

Covington, M V and Crutchfield, R S (1965) Facilitation of creative problem-solving. *Programmed Instruction*, **4**, 4.

Crowder, N A (1963) Intrinsic programming: facts, fallacies and future. In Filep, R T (ed) *Prospectives in Programming*. Macmillan, New York.

Currier, R L (1983) Interactive videodisc learning systems. *High Technology*, November.

Davies, I K (1971) *The Management of Learning*. McGraw Hill, London.

Dean, C and Whitlock, Q (1983) *A Handbook of Computer-based Training*. Kogan Page, London.

DeBloois, M L (1982) *Videodisc/Microcomputer Courseware Design*. Educational Technology Publications, Englewood Cliffs, New Jersey.

De Haan, R F and Doll, R C (1964) Individualization and human potential. In Doll, R C (ed) *Individualizing Instruction*. Association for Supervision and Curriculum Development, Washington, DC.

Devlin, S (1984) A touch-screen disc: Devlin interviews the producer. *EITV Journal*, May.

Dick, W and Carey, L (1978) *The Systematic Design of Instruction*. Scott, Forseman and Company, Glenview, Illinois.

Dienes, Z P (1960) *Building Up Mathematics*. Hutchinson, London.

Dienes, Z P (1973) *The Six Stages in the Learning of Mathematics*. OCDL, Paris. English language edition by the National Foundation for Educational Research, London.

Dodd, B T (1967a) *Programmed Instruction for Industrial Training*. Heinemann, London.

Dodd, B T (1967b) A diagnostic branching system for remedial training in the manipulation of vulgar fractions. *Programmed Learning*, **4**, 1.

Dodd, B T, LeHunte, R J G and Shepard, C (1974) Decision making in instructional design. In Baggaley, J *et al* (eds) *Aspects of Educational Technology VIII*. Kogan Page, London.

Dormant, D (1980) *Rolemaps*. Educational Technology Publications, Englewood Cliffs, New Jersey.

Dwyer, F M (1972) *A Guide for Improving Visualized Instruction*. Learning Services, State College, Pennsylvania.

Dwyer, F M (1978) *Strategies for Improving Visual Learning*. Learning Services, State College, Pennsylvania.

Egan, K (1976) *Structural Communication*. Fearon Publishers, Belmont, California.

Ellington, H, Addinall, E and Percival, F (1981) *Games and Simulations in Science Education*. Kogan Page, London.

Ellis, P and Romiszowski, A J (1972) *Vectors* and *Matrices*. Two volumes of self-instructional materials for adult learners. Council of Europe, Committee for Out-of-School Education, Strasbourg.

Evans, J L, Homme, L E and Glaser, R (1962) The RULEG system for the construction of programmed verbal learning sequences. *Journal of Educational Research*, 55, June/July.

Flanagan, J C (1968) Project PLAN. In *Technology and Innovation in Education*. Praeger, New York.

Flavell, J H (1963) *The Developmental Psychology of Jean Piaget*. Van Nostrand, Princeton.

Fleming, M and Levie, W H (1978) *Instructional Message Design: Principles from the Behavioral Sciences*. Educational Technology Publications, Englewood Cliffs, New Jersey.

Gagné, R M (1965) *The Conditions of Learning*. Holt, Rinehart and Winston, New York.

Gagné, R M (1974) *Essentials of Learning for Instruction*. Holt, Rinehart and Winston, New York.

Gagné, R M and Briggs, L J (1974) *Principles of Instructional Design*. Holt, Rinehart and Winston, New York.

Gallagher, J M and Reid, J K (1981) *The Learning Theory of Piaget and Inhelder*. Wiley, New York.

Garrison, C I (1970) *1001 Media Ideas for Teachers* (2nd edition). McCutchan Publishing Corporation, Berkeley, California.

Gayeski, D M and Williams, D V (1984) Interactive video: accessible and intelligent. *EITV Journal*, June.

Gerlach, V S and Ely, D P (1980) *Teaching and Media: A Systematic Approach* (2nd edition). Prentice Hall, Englewood Cliffs, New Jersey.

Gibbons, M (1971) *Individualized Instruction: A Descriptive Analysis*. Teachers College Press, Columbia University, Washington, DC.

Gilbert, T F (1961) Mathetics: the technology of education. *Journal of Mathetics*, **1** and **2**.

Gilbert, T F (1967) PRAXEONOMY: A systematic approach to identifying training needs. *Management of Personnel Quarterly*, **6**, 3. University of Ann Arbor, Michigan. Reprinted (1972) in Davies, I K and Hartley, J (eds) *Contributions to an Educational Technology*. Butterworth, London.

Gilbert, T F (1969) Some issues in mathetics: saying what a subject matter is. *NSPI Journal*, **8**, 2.

Gilbert, T F (1978) *Human Competence: Engineering Worthy Performance*. McGraw Hill, New York.

Godfrey, D and Sterling, S (1982) *The Elements of CAL*. Press Porcepic, Toronto.

Guerrier, D and Richards, J (1969) *State of Emergency*. An interactive, 'programmed thriller'. Heinemann (hardback) and Penguin (paperback), London.

Guilford, J P (1967) *The Nature of Human Intelligence*. McGraw Hill, New York.

Harris, C W (ed) (1960) *Encyclopaedia of Educational Research*. Macmillan, New York.

Heines, J M (1984) *Screen Design Strategies for Computer-Assisted Instruction*. Digital Press, Digital Equipment Corporation, Bedford, Massachussetts.

Heinich, R, Molenda, M and Russell, J D (1982) *Instructional Media and the New Technologies of Instruction* (3rd edition 1985). John Wiley and Sons, New York.

Hodgson A M (1968) A communication technique for the future. *Ideas*, 7. Curriculum Laboratory, Goldsmiths' College, University of London, London.

Hodgson, A M (1971) An experiment in computer-guided correspondence seminars for management. *Aspects of Educational Technology V*. Pitman, London.

Hodgson, A M (1972) Structuring learning in social settings: some notes on work in progress. *Programmed Learning & Educational Technology*, **9**, 2.

Hodgson, A M (1974) Structural communication in practice. In Romiszowski, A J (ed) *APLET Yearbook of Educational and Instructional Technology 1974/75*. Kogan Page, London.

Hodgson, A M and Dill, W R (1970) Sequel to the misfired missive. *Harvard Business Review*, November/December.

Hoekma, J (1983) Interactive videodisc: a new architecture. *Performance and Instruction Journal*, November.

Holden, E (1983) *Training for Performance and Profit*. Cybersystems, Montreal.

Hon, D (1983) The promise of interactive video: an affective search. *Performance and Instruction Journal*, November.

Horn, R E (1973) *Introduction to Information Mapping*. Information Resources Inc, Lexington, Massachussetts.

Horn, R E (1974) *Course Notes for Information Mapping Workshop*. Information Resources Inc, Lexington, Massachussetts.

Horn, R E et al (1969) *Information Mapping for Learning and Reference*. Information Resources Inc, Lexington, Massachussetts.

Horn, R E et al (1971) *Introduction to Probability*. Information Resources Inc, Lexington, Massachussetts.

Horn, R E and Cleaves, A (1980) *The Guide to Simulations/Games for Education and Training* (4th edition). Sage, Beverly Hills, California.

Howe, S F (1983) Interactive video. *Media and Methods*, November.

Huntingdon, J F (1979) *Computer Assisted Instruction Using Basic*. Educational Technology Publications, Englewood Cliffs, New Jersey.

Industrial Training Research Unit (1975) *CRAMP: A Guide to Training Decisions*. ITRU, University College, London.

Jenkins, R (1983) Preface. In Megarry, J et al (eds) *The World Yearbook of Education 1982/83: Computers and Education*. Kogan Page, London.

Jonassen, D H (ed) (1982) *The Technology of Text*. Educational Technology Publications, Englewood Cliffs, New Jersey.

Keller, F S (1968) Goodbye teacher! *Journal of Applied Behavioral Analysis*, 1.

Keller, F S and Sherman, J G (1974) *Keller Plan Handbook*. Benjamin, California.

Kemp, J E (1980) *Planning and Producing Audiovisual Materials* (4th edition). Harper and Row, New York.

Kemp, J E (1985) *The Instructional Design Process*. Harper and Row, New York.

Kirkpatrick, D (1977) Evaluation of training programs: evidence vs. proof. *Training and Development Journal*, November.

Krathwohl, D R, Bloom, B S and Masia, B B (1964) *Taxonomy of Educational Objectives, Handbook II: Affective Domain*. McKay, New York. Reprinted (1972) by Longman, New York and London.

L'Allier, J J (1983) An opportunity to shape a new technology. *Performance and Instruction Journal*, November.

Landa, L N (1974) *Algorithmization in Learning and Instruction.* Educational Technology Publications, Englewood Cliffs, New Jersey.

Landa, L N (1976) *Instructional Regulation and Control: Cybernetics, Algorithmization and Heuristics in Education.* Educational Technology Publications, Englewood Cliffs, New Jersey.

Langdon, D G (1978) *The Audio Workbook.* (The Instructional Design Library, 5). Educational Technology Publications, Englewood Cliffs, New Jersey.

Laurillard, D M (1984) Interactive video and the control of learning. *Educational Technology*, June.

Lawrence, J S (1981) Videodisc: here at last. *Electronic Learning*, April.

Leedham, J and Unwin, D (1971) *Programmed Learning in the Schools* (3rd edition). Longman, London.

Leith, G O M (1969) *Second Thoughts on Programmed Learning.* National Council for Educational Technology, London.

McPhail, P (1974) Building a role-playing exercise: the after-the-party scene. In Stadklev, R (ed) *Handbook of Simulation Gaming in Social Education.* Institute of Higher Education Research Services, University of Alabama.

MacPherson, E D (1972) How much individualization? *The Mathematics Teacher*, May.

MaCrae, D L, Monty, M R and Worling, D G (1981) *Television Production: An Introduction.* Methuen, Toronto.

Mager, R F (1962) Preparing Instructional Objectives. (Reprinted with this title in 1975.) Fearon Publishers, Belmont, California.

Mager, R F (1964) *The Language of Computers.* A programmed text. Learning Systems Ltd, London.

Mager, R F and Beach, K H (1967) *Developing Vocational Instruction.* Fearon Publishers, Belmont, California.

Mager, R F and Pipe, P (1970) *Analysing Performance Problems.* Fearon Publishers, Belmont, California.

Maier, N R F, Solem, A R and Maier, A A (1975) *The Role-Play Technique: A Handbook for Management and Leadership Practice.* University Associates, San Diego, California.

Mechner, F (1965) Science education and behavioral technology. In Glaser, R (ed) *Teaching Machines and Programmed Learning II: Data and Directions.* Department of Audio-visual Instruction (DAVI), National Education Association of the United States, Washington.

Merrill, M D (ed) (1971) *Instructional Design: Readings.* Prentice Hall, Englewood Cliffs, New Jersey.

Merrill, M D (1983) Component display theory. In Reigeluth, C M (ed) *Instructional Design: Theories and Models.* Erlbaum, Hillsdale, New Jersey.

Merrill, M D and Wood, L E (1984) Computer guided instructional design. *Journal of Computer-Based Instruction*, **11**, 2, Spring.

Miller, G A (1967) *The Psychology of Communication.* Penguin Books, London.

Minor, E and Frye, H R (1970) *Techniques for Producing Visual Instructional Media.* McGraw Hill, New York.

Moore, O K (1962) *The Automated Responsive Environment.* Yale University Press, New York.

Neil, M W (1970) A systems approach to course planning at the Open University. In Romiszowski, A J (ed) *A Systems Approach to Education and Training.* Kogan Page, London.

Odiorne, G S (1970) *Training by Objectives: An Economic Approach to Management Training.* Macmillan, New York.

Papert, S (1980) *Mindstorms: Children, Computers and Powerful Ideas.* Basic Books, New York.

Pask, G (1960) The teaching machine as a control mechanism. Transactions of the Society of Instrument Technology, London, June.

Pask, G (1976a) Conversational techniques in the study and pratice of education. *British Journal of Educational Psychology*, **46**. Reprinted (1978) in Hartley, J and Davies, I K (eds) *Contributions to an Educational Technology 2.* Kogan Page, London.

Pask, G (1976b) *Conversation Theory: Applications in Education and Epistemology.* Elsevier Publishers, New York.

Pask, G (1984) Review of conversation theory and a protologic (or protolanguage), Lp. ERIC/ECTJ Annual Review Paper. *Educational Communications and Technology Journal (ECTJ)*, **32**, 1.

Pask, G and Curran, S (1982) *Microman.* Macmillan, New York.

Pask, G and Scott, B C E (1972) Learning strategies and individual competence. *International Journal of Man-Machine Studies*, **4**, 3.

Pask, G and Scott, B C E (1973) CASTE: a system for exhibiting learning strategies and regulating uncertainties. *International Journal of Man-Machine Studies*, **5**.

Pfeiffer, J W and Jones, J E (1979) *The 1979 Annual Handbook for Group Facilitators.* University Associates, La Jolla, California.

Phenix, P H (1964) *Realms of Meaning: A Philosophy of the Curriculum for General Education.* McGraw Hill, New York.

Piaget, J (1957) *Logic and Psychology.* Basic Books, New York.

Piaget, J (1965) *The Child's Conception of Number.* Norton, New York.

Polya, G (1945) *How to Solve It: A New Aspect of Mathematical Method.* Princeton University Press, Princeton, New Jersey.

Polya, G (1963) On learning, teaching and learning teaching. *American Mathematical Monthly,* **70**.

Postlethwait, S N, Novak, J and Murray, H T Jr (1972) *The Audio-Tutorial Approach to Learning.* Burgess, New York.

Pressey, S L (1926) A simple device for teaching, testing and research. *School and Society,* 23.

Rackham, N and Morgan, T (1977) *Behaviour Analysis in Training.* McGraw Hill, London.

Read, D A and Simon, S B (1975) *Humanistic Education Sourcebook.* Prentice Hall, New York.

Romiszowski, A J (1965) Living with programmed instruction. *Technical Education and Industrial Training,* August, London.

Romiszowski, A J (1966) Programmed learning in industry today. *Visual Education,* June, London.

Romiszowski, A J (1968) *The Selection and Use of Teaching Aids.* Kogan Page, London.

Romiszowski, A J (ed) (1970) *The Systems Approach to Education and Training.* Kogan Page, London.

Romiszowski, A J (1973) Interim Report on Research Into Alternative Programming Styles for the Presentation of the Correspondence Course Sections of the Proposed Multi-Media Course on Vectors and Matrices. Council of Europe, Committee for Cultural Cooperation, Strasbourg.

Romiszowski, A J (1974) *Selection and Use of Instructional Media.* (Currently in revision for republication in 1986.) Kogan Page, London.

Romiszowski, A J (1976) A Study of Individualized Systems for Mathematics Instruction at the Post-Secondary Levels. Unpublished PhD thesis. Available in the library, The University of Technology, Loughborough.

Romiszowski, A J (1980) A new approach to the analysis of knowledge and skills. *Aspects of Educational Technology X1V.* Kogan Page, London.

Romiszowski, A J (1981) *Designing Instructional Systems: Decision Making in Course Planning and Curriculum Design.* Kogan Page, London.

Romiszowski, A J (1984) *Producing Instructional Systems: Lesson Planning for Individualized and Group Learning Activities.* Kogan Page, London.

Romiszowski, A J and Atherton, B (1979) Creativity and control: neglected factors within self-instructional programme design. *Aspects of Educational Technology XIII.* Kogan Page, London.

Romiszowski, A J, Ellis, P and Howe, A (1974) Programmed learning: a study of change. In Baggaley and Jamieson (eds) *Aspects of Educational Technology VIII.* Kogan Page, London.

Rowntree, D (1974) *Educational Technology in Curriculum Development.* Harper and Row, London.

Rushby, N J (1979) *An Introduction to Educational Computing.* Croom Helm, London.

Schiffirn, R M and Schneider, W (1977) Controlled and automatic human information processing, part II: perceptual learning, automatic attending and a general theory. *Psychological Review,* 84.

Schramm, W (ed) (1972) *Quality in Instructional Television.* University Press of Hawaii, Honolulu.

Schramm, W (1973) *Big Media, Little Media.* Agency for International Development, Washington, DC.

SENAI (1984) *Comunicacao/SENAI,* 58, Year 10, July/August 1984. Servico Nacional de Aprendizagem Industrial – Sao Paulo (SENAI-SP), Avenida Paulista 750, Sao Paulo, Brazil.

Servais, W and Varga, T (1971) *Teaching School Mathematics: A UNESCO Sourcebook.* Penguin Books, London.

Seymour, W D (1954) *Industrial Training for Manual Operations.* Pitman, London.

Seymour, W D (1966) *Industrial Skills.* Pitman, London.

Seymour, W D (1968) *Skills Analysis Training.* Pitman, London.

Shane, H G (1962) The school and individual differences. In Henry, N B (ed) *Individualizing Instruction,* 61st Yearbook of the National Society for the Study of Education. University of Chicago Press, Chicago.

Shannon, D and Weaver, W (1949) *The Mathematical Theory of Communication*. University of Illinois Press, Urbana, Illinois.

Shaw, M E, Corsini, R J, Blake, R R and Mouton, J S (1980) *Role Playing: A Practical Manual for Group Facilitators*. University Associates, San Diego, California.

Simpson, E J (1967) Educational objectives in the psychomotor domain. In Kapfer, M B (ed) *Behavioral Objectives in Curriculum Development*. Educational Technology Publications, Englewood Cliffs, New Jersey.

Simpson, E J (1969) *Psychomotor Domain: A Tentative Classification*. University of Illinois, Urbana, Illinois.

Skinner, B F (1954) The science of learning and the art of teaching. *Harvard Educational Review*, 24, Spring.

Skinner, B F (1968) *The Technology of Teaching*. Appleton Century Crofts, New York.

Sleeman, D and Brown, J S (1982) *Intelligent Tutoring Systems*. Academic Press, New York.

Stadsklev, R (1974) *Handbook of Simulation Gaming in Social Education*. Institute of Higher Education Research and Services, University of Alabama.

Suppes, P *et al* (1968) *Computer Assisted Instruction: Stanford's 1965/66 Arithmetic Program*. Institute for Mathematical Studies in the Social Sciences, Stanford University, Stanford, California.

Suppes, P and Morningstar, M (1972) *Computer Assisted Instruction at Stanford 1966-68: Data, Models and Evolution of the Arithmetic Programs*. Academic Press, New York.

Taylor, L C (1971) *Resources for Learning*. Penguin Books, London.

Tennyson, R D and L'Allier, J J (1980) Breaking out of the Skinner box. *NSPI Journal*, **19**, 2.

Thiagarajan, S (1977) *Group Programs*. Educational Technology Publications, Englewood Cliffs, New Jersey.

Thomas, R M and Thomas, S M (1965) *Individual Differences in the Classroom*. McKay, New York.

Thorndike, E L (1927) The law of effect. *American Journal of Psychology*, 39.

Thornhill, P (1965) *The Waterloo Campaign*. A programmed text. Methuen, London.

Tiffin, J (1980) Educational television: a phoenix in Latin America? *Programmed Learning & Educational Technology*, **17**, 4.

Underwood, B J (1982) *Studies in Learning and Memory: Selected Papers*. Praeger, New York.

Underwood, B J and Schulz, R W (1960) *Meaningfulness and Verbal Learning*. Lippincott, Chicago.

van Ments, M (1983) *The Effective Use of Role-Play*. Kogan Page, London.

Vazquez-Abad, J, Winer, L R and Mitchell, P D (1982) Integrative CAL: multi-computer and multi-media. *Computers and Education*, Volume 6. Pergamon Press, Oxford.

Wallach, M A and Kogan, N (1965) *Modes in Thinking in Young Children*. Holt, New York.

Washburne, C and Marland, S P (1963) *Winnetka: The History and Significance of an Educational Experiment*. Prentice Hall, New York.

Whitelaw, M (1972) *The Evaluation of Management Training: A Review*. Institute of Personnel Management, London.

Wilkinson, C E (1971) *Educational Media and You*. GLC Educational Materials and Services, Toronto, Canada.

Wittich, W A and Schuller, C F (1973) *Instructional Technology: Its Nature and Use*. (5th edition). Harper and Row, New York.

Wohlking, W and Gill, P J (1980) *Role Playing*. Educational Technology Publications, Englewood Cliffs, New Jersey.

Wyant, T G (1973) Syllabus analysis. In Budgett, R and Leedham, J (eds) *Aspects of Educational Technology VI*. Pitman, London.

Wyant, T G (1974) Network analysis. In Howe, A and Romiszowski, A J (eds) *APLET Yearbook of Educational and Instructional Technology, 1974/75*. Kogan Page, London.

Zeitlin, N and Goldberg, A L (1970) *Structural Communication: An Interactive System for Teaching Understanding*. Educational Technology Publications, Englewood Cliffs, New Jersey.

Subject Index

Author Index